Space-Time Block Coding for Wireless Communications

Space-time coding is a technique that promises greatly improved performance in wireless communication systems by using multiple antennas at the transmitter and receiver. *Space-Time Block Coding for Wireless Communications* is an introduction to the theory of this technology.

The authors develop the subject using a unified framework and cover a variety of topics ranging from information theory to performance analysis and state-of-the-art space-time block coding methods for both flat and frequency-selective fading multiple-antenna channels. They concentrate on key principles rather than specific practical applications, and present the material in a concise and accessible manner. Their treatment reviews the fundamental aspects of multiple-input, multiple-output communication theory, and guides the reader through a number of topics at the forefront of current research and development.

The book includes homework exercises and is aimed at graduate students and researchers working on wireless communications, as well as practitioners in the wireless industry.

Space-Time Block Coding for Wireless Communications

Erik G. Larsson

and

Petre Stoica

CAMBRIDGE
UNIVERSITY PRESS

CAMBRIDGE UNIVERSITY PRESS
Cambridge, New York, Melbourne, Madrid, Cape Town, Singapore, São Paulo

Cambridge University Press
The Edinburgh Building, Cambridge CB2 8RU, UK

Published in the United States of America by Cambridge University Press, New York

www.cambridge.org
Information on this title: www.cambridge.org/9780521824569

First published 2003
Reprinted with corrections 2005
This digitally printed version 2008

A catalogue record for this publication is available from the British Library

ISBN 978-0-521-82456-9 hardback
ISBN 978-0-521-06533-7 paperback

CONTENTS

ABOUT THE AUTHORS

Erik G. Larsson received the M.Sc. degree in Applied Physics and Electrical Engineering from Linköping University, Linköping, Sweden, in 1997, and the Ph.D. degree in Electrical Engineering from Uppsala University, Sweden, in 2002.

During 1998-99, he was a research engineer at Ericsson Radio Systems AB in Stockholm, Sweden, where he was involved in algorithm design and standardization of location services for the pan-European GSM system. During 1999-2000 he was a research and teaching assistant in the Dept. of Systems and Control, Uppsala University, Uppsala, Sweden. Since September 2000, he has been with the Dept. of Electrical and Computer Engineering, University of Florida, Gainesville, FL, where he is currently an Assistant Professor.

His principal research interests include space-time coding for wireless communications, digital signal processing, estimation and detection, spectral analysis and radar applications. He has some 20 papers in IEEE and other international journals and holds four U.S. patents related to wireless communications. He also is an associate editor for IEEE Transactions on Vehicular Technology.

Petre Stoica is a Professor of System Modeling in the Department of Systems and Control at Uppsala University (UU) in Sweden. Previously he was a Professor of Signal Processing at the Bucharest Polytechnic Institute (BPI). He received a D.Sc. degree in Automatic Control from the BPI in 1979 and a Honorary Doctorate in Science from UU in 1993. He held longer visiting positions with the Eindhoven University of Technology, Chalmers University of Technology (where he held a Jubilee Visiting Professorship), Uppsala University, University of Florida, and Stanford University.

His main scientific interests are in the areas of system identification, time-series analysis and prediction, statistical signal and array processing, spectral analysis, wireless communications, and radar signal processing. He has published 7 books, 10 book chapters and some 500 papers in archival journals and conference records on these topics. The most recent book that he co-authored is "Introduction to Spectral Analysis" (Prentice-Hall, 1997). Recently he has also co-edited two books on Signal Processing Advances in Mobile and Wireless Communications (Prentice-Hall, 2001).

He received a number of awards including co-recipient of the 1989 ASSP Society Senior Award for a paper on statistical aspects of array signal processing and recipient of the 1996 Technical Achievement Award of the IEEE Signal Processing Society for fundamental contributions to statistical signal processing with applications in time series analysis, system identification and array processing. In 1998 he received a Senior Individual Grant Award from the Swedish Foundation for Strategic Research. He is also co-recipient of the 1998 EURASIP Best Paper Award for Signal Processing for a work on parameter estimation of exponential signals with time-varying amplitude with application to speed measurement, a 1999 IEEE Signal Processing Society Best Paper Award for a paper on parameter and rank estimation of reduced-rank regressions, a 2000 IEEE Third Millennium Medal as well as of the 2000 IEEE W.R.G. Baker Paper Prize Award for a work on maximum likelihood methods for radar. In 2002 he received the Individual Technical Achievement Award from EURASIP.

He is on the Editorial Boards of five journals in the field: Signal Processing; Journal of Forecasting; Circuits, Systems and Signal Processing; Multidimensional Systems and Signal Processing; and Digital Signal Processing: A Review Journal. He was a guest co-editor for several special issues on system identification, signal processing, spectral analysis and radar for some of the above journals and for the IEE Proceedings, as well as a member of the international program committees of many topical conferences. From 1981 to 1986 he was a Director of the International Time Series Analysis and Forecasting Society. He also is a member-at-large of the Board of Governors of the IEEE Signal Processing Society, a Honorary Member of the Romanian Academy, a Fellow of the Royal Statistical Society, and a Fellow of IEEE.

PREFACE

The book in hand is the result of merging an original plan of writing a monograph on orthogonal space-time block coding with the ambition of authoring a more textbook-like treatment of modern MIMO communication theory, and it may therefore have the attributes of both these categories of literature. A major part of the book has resulted from the development of graduate courses at the universities where the authors are active, whereas a smaller part is the outcome of the authors' own research.

Our text covers many aspects of MIMO communication theory, although not all topics are treated in equal depth. The discussion starts with a review of MIMO channel modeling along with error probability and information theoretical properties of such channels, from which we go on to analyze the concepts of receive and transmit diversity. We discuss linear space-time block coding for flat and frequency selective fading channels, along with its associated receiver structures, both in general and in particular with emphasis on orthogonal space-time block coding. We also treat several special topics, including space-time coding for informed transmitters and space-time coding in a multiuser environment. Furthermore we include material about the existence and design of amicable orthogonal designs, a topic which is relevant in the context of orthogonal space-time block coding.

The text focuses on principles rather than on specific practical applications, although we have tried to illustrate some results in the context of contemporary wireless standards. MIMO communication theory is a rather multifaceted topic, and it has been a true challenge to present the material in an as simple and concise form as possible, without compromising mathematical rigor by more than a negligible extent. To maximize the readability of the book, we have made consistent use of matrix-algebraic notation and placed lengthy proofs and calculations in appendices.

The best way to read this book may be in a sequential manner, although it can also be used as a pure reference text. We believe that the book is suitable as the main text for a graduate course on antenna diversity and space-time coding, as well as for self-studies and as a reference for researchers and engineers. Each chapter is concluded with a number of problems that can be appropriate exercises or homework for students if the book is used as a teaching instrument. The level of difficulty of these problems is quite diverse. We have tried to place the "easier" exercises first, but such an ordering is always subjective. If the material is used in a classroom setting, the instructor may wish to cover some topics in more detail than others. For example, depending on the length of the course, the material placed in the appendices and many of the proofs can be left for self-studies or as reference material. We have included a general appendix (Appendix A) with a short review of matrix algebra and probability theory, that can serve as a good starting point for readers who need a refresher in these fields. Finally, we should note that this text and the book *Introduction to Space-Time Wireless Communications* by Paulraj, Nabar and Gore (Cambridge University Press, 2003) complement each other very well in several ways. A topical graduate course can therefore be based on a combined use of these two books. Researchers interested in the topic of space-time wireless communications may also benefit most by using both books as references.

Note to instructors: Our book has a web site (`http://publishing.cambridge.org/resources/0521824567`) where we have posted additional material for instructors, including an extensive set of overhead slides that summarize the results in each chapter.

We thank the Swedish Foundation for Strategic Research for supporting most of our research in the area of this book. We also would like to thank Dr. Girish Ganesan for his contributions to Section 9.6 and to Appendix B, for several useful and stimulating discussions, and for proofreading parts of an early draft of the manuscript. We are grateful to Dr. Philip Meyler of Cambridge University Press for being the best publishing editor we have ever worked with. Furthermore, we are also grateful to Prof. Ezio Biglieri, Prof. Helmut Bölcskei, Prof. Robert Heath, Prof. Mikael Skoglund, Dr. Thomas Svantesson, Prof. Lee Swindlehurst, Prof. Lang Tong, and Prof. Douglas Williams for their good constructive criticism on a preliminary version of the text. The goal of having this text in print as soon as possible prevented us from carefully following all their comments and suggestions for revision. However, we plan to do so in a new printing, or possibly in a new edition of this book. Finally, we would like to acknowledge Sweungwon Cheung, Dhananjay Gore, Jonghyun Won and Wing-Hin Wong for their help with proofreading the text and eliminating typographical and other errors.

Erik G. Larsson
University of Florida

Petre Stoica
Uppsala University

NOTATION

a	the scalar a
$\boldsymbol{a}$	the vector $\boldsymbol{a}$
$\boldsymbol{A}$	the matrix $\boldsymbol{A}$
$[\boldsymbol{A}]_{k,l}$	the (k,l)th element of $\boldsymbol{A}$
$\mathcal{A}$	the set $\mathcal{A}$
$(\cdot)^*$	the complex conjugate
$(\cdot)^T$	the transpose
$(\cdot)^H$	the conjugate transpose
i	the imaginary unit ($i = \sqrt{-1}$)
$\mathrm{Re}\{\cdot\}, \mathrm{Im}\{\cdot\}$	the real and imaginary parts
$\bar{(\cdot)}$, $\tilde{(\cdot)}$	shorthand for $\mathrm{Re}\{\cdot\}$ and $\mathrm{Im}\{\cdot\}$
$\boldsymbol{s}'$	the vector obtained by stacking $\bar{\boldsymbol{s}}$ and $\tilde{\boldsymbol{s}}$ on top of each other: $\boldsymbol{s}' = \begin{bmatrix}\bar{\boldsymbol{s}}^T & \tilde{\boldsymbol{s}}^T\end{bmatrix}^T$
$\log(x)$	the natural logarithm of x
$\log_2(x)$	the base-2 logarithm of x
$\mathrm{Tr}\{\cdot\}$	the trace of a matrix
$\mathrm{rank}\{\cdot\}$	the rank of a matrix
$\lvert\cdot\rvert$	the determinant of a matrix, or cardinality of a set
$\lVert\cdot\rVert$	Euclidean norm for vectors; Frobenius norm for matrices
$\boldsymbol{A} \geq \boldsymbol{B}$	the matrix $\boldsymbol{A} - \boldsymbol{B}$ is positive semi-definite
$\boldsymbol{A}^{1/2}$	the Hermitian square root of $\boldsymbol{A}$
$\propto$	proportional to

δ_t	the Kronecker delta ($\delta_t = 1$ if $t = 0$ and $\delta_t = 0$ if $t \neq 0$)
$\otimes$	the Kronecker product
$\odot$	the Schur-Hadamard (elementwise) product
$\mathrm{vec}(\boldsymbol{X})$	the vector obtained by stacking the columns of $\boldsymbol{X}$ on top of each other
$\boldsymbol{I}_N$	the identity matrix of dimension $N \times N$; N omitted whenever no confusion can occur
$\mathbf{0}_{M \times N}$	the $M \times N$ matrix of zeros; N omitted whenever no confusion can occur
$\lambda_{\max}(\boldsymbol{X}), \lambda_{\min}(\boldsymbol{X})$	the largest respectively smallest eigenvalue of $\boldsymbol{X}$
$\Pi_{\boldsymbol{X}}$	$\boldsymbol{X}(\boldsymbol{X}^H\boldsymbol{X})^{-1}\boldsymbol{X}^H$ (the orthogonal projector onto the column space of $\boldsymbol{X}$)
$\Pi_{\boldsymbol{X}}^{\perp}$	$\boldsymbol{I} - \Pi_{\boldsymbol{X}}$ (the orthogonal projector onto the orthogonal complement of the column space of $\boldsymbol{X}$)
$z^{-1}\boldsymbol{x}(n)$	$\boldsymbol{x}(n-1)$ (the unit delay operator)
$z\boldsymbol{x}(n)$	$\boldsymbol{x}(n+1)$ (the unit advance operator)
$\boldsymbol{H}(z^{-1})\boldsymbol{x}(n)$	$\sum_{l=0}^{L} \boldsymbol{H}_l\boldsymbol{x}(n-l)$ (linear causal filtering)
$\boldsymbol{H}(z)\boldsymbol{x}(n)$	$\sum_{l=0}^{L} \boldsymbol{H}_l\boldsymbol{x}(n+l)$ (linear anti-causal filtering)
$\boldsymbol{H}^H(z^{-1})\boldsymbol{x}(n)$	$\sum_{l=0}^{L} \boldsymbol{H}_l^H\boldsymbol{x}(n-l)$
$\boldsymbol{H}^H(z)\boldsymbol{x}(n)$	$\sum_{l=0}^{L} \boldsymbol{H}_l^H\boldsymbol{x}(n+l)$
$p_{\boldsymbol{x}}(\boldsymbol{x})$	the probability density function (p.d.f.) of $\boldsymbol{x}$
$P(\mathcal{E})$	the probability of an event $\mathcal{E}$
$E_{\boldsymbol{x}}[\cdot]$	statistical expectation over $\boldsymbol{x}$; $\boldsymbol{x}$ omitted whenever no confusion can occur
$\boldsymbol{x} \sim N(\boldsymbol{\mu}, \boldsymbol{Q})$	$\boldsymbol{x}$ is real-valued Gaussian with mean $\boldsymbol{\mu}$ and covariance matrix $\boldsymbol{Q}$
$\boldsymbol{x} \sim N_C(\boldsymbol{\mu}, \boldsymbol{Q})$	$\boldsymbol{x}$ is circularly symmetric complex Gaussian with mean $\boldsymbol{\mu}$ and covariance matrix $\boldsymbol{Q}$
$Q(x)$	the Gaussian Q-function: if $x \sim N(0,1)$, then $P(x > t) = Q(t)$
$O(x)$	if $f(x) = O(x)$, then there is a constant c such that $\lvert f(x)\rvert \leq cx$ for large x

COMMONLY USED SYMBOLS

s	complex symbol
$\mathcal{S}$	the symbol constellation ($s \in \mathcal{S}$)
$\mathcal{S}'$	the constellation for $\bar{s}$, $\tilde{s}$ (for "separable constellations")
$\boldsymbol{s}$	vector of complex symbols
$\boldsymbol{X}$	transmitted space-time block code matrix
$\mathcal{X}$	the constellation of all code matrices ($\boldsymbol{X} \in \mathcal{X}$)
$\boldsymbol{x}$	$\mathrm{vec}(\boldsymbol{X})$
$\boldsymbol{H}$	channel matrix (for frequency flat channels)
$\boldsymbol{H}(z^{-1})$	channel impulse response
$\boldsymbol{h}$	$\mathrm{vec}(\boldsymbol{H})$
$\boldsymbol{Y}$	received data matrix
$\boldsymbol{y}$	$\mathrm{vec}(\boldsymbol{Y})$
$\boldsymbol{E}$	received noise matrix
$\boldsymbol{e}$	$\mathrm{vec}(\boldsymbol{E})$
σ^2	noise variance
$\boldsymbol{\Lambda}$	noise covariance matrix
$\boldsymbol{\Upsilon}$	channel covariance matrix
P_s	probability of a symbol error
P_b	probability of a bit error

n_r	number of receive antennas
n_t	number of transmit antennas
n_s	number of symbols per STBC matrix
N	number of time epochs per block
N_0	length of data block (for block transmission)
N_t	length of training block
N_b	number of concatenated STBC matrices (for block transmission)
N_{pre}	length of preamble
N_{post}	length of postamble
N_p	length of preamble and postamble if they are of equal length ($N_p = N_{\text{pre}} = N_{\text{post}}$)

ABBREVIATIONS

AWGN	additive white Gaussian noise
BER	bit-error rate
BPSK	binary phase shift keying
BS	base station
DD	delay diversity
FER	frame-error rate
FIR	finite impulse response
GLRT	generalized likelihood ratio test
GSM	the global system for mobile communications
i.i.d.	independent and identically distributed
ISI	intersymbol interference
IT	informed transmitter
MIMO	multiple-input, multiple-output
MISO	multiple-input, single-output
ML	maximum-likelihood
MLSD	ML sequence detector
MMSE	minimum mean-square error
MPSK	M-ary phase shift keying
MS	mobile station
MSE	mean-square error
MUI	multiuser interference
OFDM	orthogonal frequency division multiplexing
OSTBC	orthogonal space-time block code/codes/coding
p.d.f.	probability density function

PSK	phase shift keying
QAM	quadrature amplitude modulation
QPSK	quadrature phase shift keying
SER	symbol-error rate
SIMO	single-input, multiple-output
SISO	single-input, single-output
SNR	signal-to-noise ratio
s.t.	subject to
STBC	space-time block code/codes/coding
STC	space-time coding
STTC	space-time trellis code/codes/coding
TR	time-reversal
UT	uninformed transmitter
VA	Viterbi algorithm

Chapter 1

INTRODUCTION

The demand for capacity in cellular and wireless local area networks has grown in a literally explosive manner during the last decade. In particular, the need for wireless Internet access and multimedia applications require an increase in information throughput with orders of magnitude compared to the data rates made available by today's technology. One major technological breakthrough that will make this increase in data rate possible is the use of *multiple antennas* at the transmitters and receivers in the system. A system with multiple transmit and receive antennas is often called a multiple-input multiple-output (MIMO) system. The feasibility of implementing MIMO systems and the associated signal processing algorithms is enabled by the corresponding increase of computational power of integrated circuits, which is generally believed to grow with time in an exponential fashion.

1.1 Why Space-Time Diversity?

Depending on the surrounding environment, a transmitted radio signal usually propagates through several different paths before it reaches the receiver antenna. This phenomenon is often referred to as *multipath* propagation. The radio signal received by the receiver antenna consists of the superposition of the various multipaths. If there is no line-of-sight between the transmitter and the receiver, the attenuation coefficients corresponding to different paths are often assumed to be independent and identically distributed, in which case the central limit theorem [PAPOULIS, 2002, CH. 7] applies and the resulting path gain can be modelled as a complex Gaussian random variable (which has a uniformly distributed phase and a Rayleigh distributed magnitude). In such a situation, the channel is said to be *Rayleigh fading*.

Since the propagation environment usually varies with time, the fading is time-variant and owing to the Rayleigh distribution of the received amplitude, the channel gain can sometimes be so small that the channel becomes useless. One way to mitigate this problem is to employ *diversity*, which amounts to transmitting the same information over multiple channels which fade independently of each other. Some common diversity techniques include time diversity and frequency diversity, where the same information is transmitted at different time instants or in different frequency bands, as well as antenna diversity, where one exploits the fact that the fading is (at least partly) independent between different points in space.

One way of exploiting antenna diversity is to equip a communication system with *multiple antennas at the receiver.* Doing so usually leads to a considerable performance gain, both in terms of a better link budget and in terms of tolerance to co-channel interference. The signals from the multiple receive antennas are typically combined in digital hardware, and the so-obtained performance gain is related to the diversity effect obtained from the independence of the fading of the signal paths corresponding to the different antennas. Many established communication systems today use receive diversity at the base station. For instance, a base station in the Global System for Mobile communications (GSM) [MOULY AND PAUTET, 1992] typically has two receive antennas. Clearly, a base station that employs receive diversity can improve the quality of the *uplink* (from the mobile to the base station) without adding any cost, size or power consumption to the mobile. See, for example, [WINTERS ET AL., 1994] for a general discussion on the use of receive diversity in cellular systems and its impacts on the system capacity.

In recent years it has been realized that many of the benefits as well as a substantial amount of the performance gain of receive diversity can be reproduced by using *multiple antennas at the transmitter* to achieve *transmit diversity.* The development of transmit diversity techniques started in the early 1990's and since then the interest in the topic has grown in a rapid fashion. In fact, the potential increase in data rates and performance of wireless links offered by transmit diversity and MIMO technology has proven to be so promising that we can expect MIMO technology to be a cornerstone of many future wireless communication systems. The use of transmit diversity at the base stations in a cellular or wireless local area network has attracted a special interest; this is so primarily because a performance increase is possible without adding extra antennas, power consumption or significant complexity to the mobile. Also, the cost of the extra transmit antenna at the base station can be shared among all users.

1.2 Space-Time Coding

Perhaps one of the first forms of transmit diversity was antenna hopping. In a system using antenna hopping, two or more transmit antennas are used interchangeably to achieve a diversity effect. For instance, in a burst or packet-based system with coding across the bursts, every other burst can be transmitted via the first antenna and the remaining bursts through the second antenna. Antenna hopping attracted some attention during the early 1990's as a comparatively inexpensive way of achieving a transmit diversity gain in systems such as GSM. More recently there has been a strong interest in *systematic* transmission techniques that can use multiple transmit antennas in an *optimal* manner. See [Paulraj and Kailath, 1993], [Wittneben, 1991], [Alamouti, 1998], [Foschini, Jr., 1996], [Yang and Roy, 1993], [Telatar, 1999], [Raleigh and Cioffi, 1998], [Tarokh et al., 1998], [Guey et al., 1999] for some articles that are often cited as pioneering work or that present fundamental contributions. The review papers [Ottersten, 1996], [Paulraj and Papadias, 1997], [Naguib et al., 2000], [Liu et al., 2001b], [Liew and Hanzo, 2002] also contain a large number of relevant references to earlier work (both on space-time coding and antenna array processing for wireless communications in general). However, despite the rather large body of literature on space-time coding, the current knowledge on optimal signal processing and coding for MIMO systems is probably still only the tip of the iceberg.

Space-time coding finds its applications in cellular communications as well as in wireless local area networks. Some of the work on space-time coding focuses on explicitly improving the performance of existing systems (in terms of the probability of incorrectly detected data packets) by employing extra transmit antennas, and other research capitalizes on the promises of information theory to use the extra antennas for increasing the throughput. Speaking in very general terms, the design of space-time codes amounts to finding a constellation of matrices that satisfy certain optimality criteria. In particular, the construction of space-time coding schemes is to a large extent a trade-off between the three conflicting goals of maintaining a simple decoding (i.e., limit the complexity of the receiver), maximizing the error performance, and maximizing the information rate.

1.3 An Introductory Example

The purpose of this book is to explain the concepts of antenna diversity and space-time coding in a systematic way. However, before we introduce the necessary formalism and notation for doing so, we will illustrate the fundamentals of receive and transmit diversity by studying a simple example.

1.3.1 One Transmit Antenna and Two Receive Antennas

Let us consider a communication system with one transmit antenna and two receive antennas (see Figure 1.1), and suppose that a complex symbol s is transmitted. If the fading is frequency flat, the two received samples can then be written:

$$\begin{aligned} y_1 &= h_1 s + e_1 \\ y_2 &= h_2 s + e_2 \end{aligned} \tag{1.3.1}$$

where h_1 and h_2 are the channel gains between the transmit antenna and the two receive antennas, and e_1, e_2 are mutually uncorrelated noise terms. Suppose that given y_1 and y_2, we attempt to recover s by the following *linear combination*:

$$\hat{s} = w_1^* y_1 + w_2^* y_2 = (w_1^* h_1 + w_2^* h_2)s + w_1^* e_1 + w_2^* e_2 \tag{1.3.2}$$

where w_1 and w_2 are weights (to be chosen appropriately). The SNR in $\hat{s}$ is given by:

$$\text{SNR} = \frac{|w_1^* h_1 + w_2^* h_2|^2}{(|w_1|^2 + |w_2|^2) \cdot \sigma^2} \cdot E\left[|s|^2\right] \tag{1.3.3}$$

where σ^2 is the power of the noise. We can choose w_1 and w_2 that maximize this SNR. A useful tool towards this end is the Cauchy-Schwarz inequality [HORN AND JOHNSON, 1985, TH. 5.1.4], the application of which yields:

$$\text{SNR} = \frac{|w_1^* h_1 + w_2^* h_2|^2}{(|w_1|^2 + |w_2|^2) \cdot \sigma^2} \cdot E\left[|s|^2\right] \leq \frac{|h_1|^2 + |h_2|^2}{\sigma^2} \cdot E\left[|s|^2\right] \tag{1.3.4}$$

where equality holds whenever w_1 and w_2 are chosen proportional to h_1 and h_2:

$$\begin{aligned} w_1 &= \alpha \cdot h_1 \\ w_2 &= \alpha \cdot h_2 \end{aligned} \tag{1.3.5}$$

for some (complex) scalar α. The resulting SNR in (1.3.4) is proportional to $|h_1|^2 + |h_2|^2$. Therefore, loosely speaking, even if one of h_1 or h_2 is equal to zero, s can still be detected from $\hat{s}$. More precisely, if the fading is Rayleigh, then $|h_1|^2 + |h_2|^2$ is χ^2-distributed, and we can show that the error probability of detecting s decays as SNR_a^{-2} when $\text{SNR}_a \to \infty$ (by SNR_a here we mean the average channel SNR). This must be contrasted to the error rate for transmission

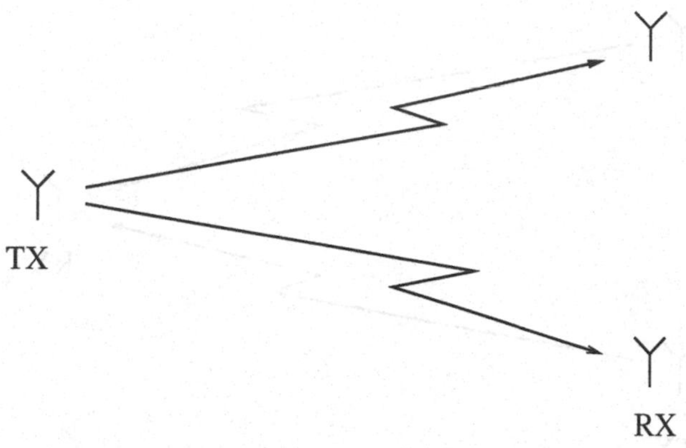

Figure 1.1. A system with one transmit antenna and two receive antennas.

and reception with a single antenna in Rayleigh fading, which typically behaves as SNR_a^{-1}.

In loose words, the *diversity order* of a system is the slope of the BER curve if plotted versus the average SNR on a log-log scale (a more formal definition is given in Chapter 4). Hence, we can say that the above considered system, provided that w_1 and w_2 are chosen optimally, achieves a diversity of order two.

1.3.2 Two Transmit Antennas and One Receive Antenna

Let us now study the "dual" case, namely a system with two transmit antennas and one receive antenna (see Figure 1.2). At a given time instant, let us transmit a symbol s, that is pre-weighted with two weights w_1 and w_2. The received sample can be written:

$$y = h_1 w_1 s + h_2 w_2 s + e \tag{1.3.6}$$

where e is a noise sample and h_1, h_2 are the channel gains. The SNR in y is:

$$\text{SNR} = \frac{|h_1 w_1 + h_2 w_2|^2}{\sigma^2} \cdot E\left[|s|^2\right] \tag{1.3.7}$$

If w_1 and w_2 are fixed, this SNR has the same statistical distribution (to within a scaling factor) as $|h_1|^2$ (or $|h_2|^2$). Therefore, if the weights w_1 and w_2 are not allowed to depend on h_1 and h_2 it is impossible to achieve a diversity of order two. However, it turns out that if we assume that the transmitter knows the channel, and w_1 and w_2 are chosen to be functions of h_1 and h_2, it is possible to achieve

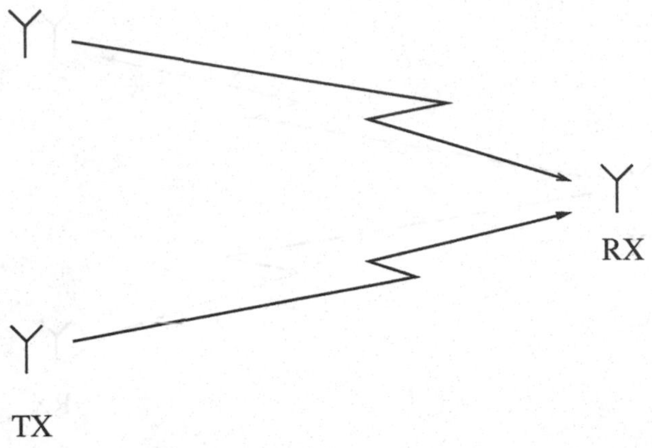

Figure 1.2. A system with two transmit antennas and one receive antenna.

an error probability that behaves as SNR_a^{-2}. We defer a deeper discussion of this aspect to Section 6.1.

We have seen that without channel knowledge at the transmitter, diversity cannot be achieved. However, if we are allowed to use more than one time interval for the transmission, we can achieve a diversity of order two rather easily. To illustrate this, suppose that we use *two* time intervals to transmit a single symbol s, where in the first interval only the first antenna is used and where during the second time interval only the second antenna is used. We get the following two received samples:

$$\begin{aligned} y_1 &= h_1 s + e_1 \\ y_2 &= h_2 s + e_2 \end{aligned} \tag{1.3.8}$$

Equation (1.3.8) is of the same form as (1.3.1) and hence the error rate associated with this method is equal to that for the case where we had one transmit and two receive antennas. However, *the data rate* is halved.

This simple example shows that transmit diversity is easy to achieve, if a sacrifice in information rate is acceptable. Space-time coding is concerned with the harder and more interesting topic: how can we maximize the transmitted information rate, at the same time as the error probability is minimized? This book will present some of the major ideas and results from the last decade's research on this topic.

1.4 Outline of the Book

Our book is organized as follows. We begin in Chapter 2 by introducing a formal model for the MIMO channel, along with appropriate notation. In Chapter 3, we study the promises of the MIMO channels from an information theoretical point of view. Chapter 4 is devoted to the analysis of error probabilities for transmission over a fading MIMO channel. In Chapter 5, we study a "classical" receive diversity system with an arbitrary number of receive antennas. This discussion sets, in some sense, the goal for transmit diversity techniques. In Chapter 6, we go on to discuss how transmit diversity can be achieved and also review some space-time coding methods that achieve such diversity. Chapter 7 studies a large and interesting class of space-time coding methods, namely linear space-time block coding (STBC) for the case of frequency flat fading. The case of frequency selective fading is treated in the subsequent Chapter 8. In Chapter 9 we discuss receiver structures for linear STBC, both for the coherent and the noncoherent case. Finally, Chapters 10 and 11 treat two special topics: space-time coding for transmitters with partial channel knowledge, and space-time coding in a multiuser environment.

1.5 Problems

1. Prove (1.3.3).

2. In (1.3.5), find the value of α such that

$$\hat{s} = s + e \tag{1.5.1}$$

where e is a zero-mean noise term. What is the variance of e? Can you interpret the quantity $\hat{s}$?

3. In Section 1.3.2, suppose that the transmitter *knows the channel* and that it can use this knowledge to choose w_1 and w_2 in an "adaptive" fashion. What is the SNR-optimal choice of w_1 and w_2 (as a function of h_1 and h_2)? Prove that by a proper choice of w_1 and w_2, we can achieve an error rate that behaves as SNR_a^{-2}.

4. In Section 1.3.2, can you suggest a method to transmit *two* symbols during two time intervals such that transmit diversity is achieved, without knowledge of h_1 and h_2 at the transmitter, and without sacrificing the transmission rate?

Chapter 2

THE TIME-INVARIANT LINEAR MIMO CHANNEL

If we adopt the standard complex baseband representation of narrowband signals (see Appendix A), the input-output relation associated with a linear and time-invariant MIMO communication channel can easily be expressed in a matrix-algebraic framework. In this chapter, we will discuss both the frequency flat and the frequency-selective case. Models for single-input single-output (SISO), single-input multiple-output (SIMO) and multiple-input single-output (MISO) channels follow as special cases.

2.1 The Frequency Flat MIMO Channel

We consider a system where the transmitter has n_t antennas, and the receiver has n_r antennas (see Figure 2.1). In the current section we also assume that the bandwidth of the transmitted signal is so small that no intersymbol interference (ISI) occurs, or equivalently, that each signal path can be represented by a complex gain factor. For practical purposes, it is common to model the channel as frequency flat whenever the bandwidth of the system is smaller than the inverse of the delay spread of the channel; hence a wideband system operating where the delay spread is fairly small (for instance, indoors) may sometimes also be considered as frequency flat. Models for frequency-selective multiantenna channels, i.e., MIMO channels with non-negligible ISI, will be presented in Section 2.2.

Let $h_{m,n}$ be a complex number corresponding to the channel gain between transmit antenna n and receive antenna m. If at a certain time instant the complex signals $\{x_1, \ldots, x_{n_t}\}$ are transmitted via the n_t antennas, respectively, the received signal at antenna m can be expressed as

$$y_m = \sum_{n=1}^{n_t} h_{m,n} x_n + e_m \tag{2.1.1}$$

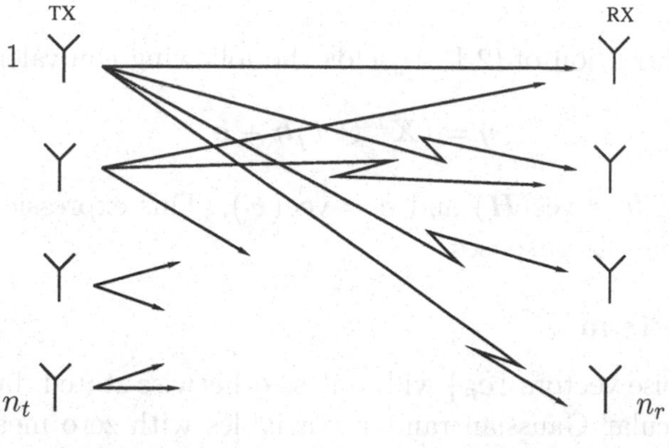

Figure 2.1. A MIMO channel with n_t transmit and n_r receive antennas.

where e_m is a noise term (to be discussed later). The relation (2.1.1) is easily expressed in a matrix framework. Let $\boldsymbol{x}$ and $\boldsymbol{y}$ be n_t and n_r vectors containing the transmitted and received data, respectively. Define the following $n_r \times n_t$ channel gain matrix:

$$\boldsymbol{H} = \begin{bmatrix} h_{1,1} & \cdots & h_{1,n_t} \\ \vdots & & \vdots \\ h_{n_r,1} & \cdots & h_{n_r,n_t} \end{bmatrix} \tag{2.1.2}$$

Then we have

$$\boldsymbol{y} = \boldsymbol{H}\boldsymbol{x} + \boldsymbol{e} \tag{2.1.3}$$

where $\boldsymbol{e} = [e_1 \ \cdots \ e_{n_r}]^T$ is a vector of noise samples. If several consecutive vectors $\{\boldsymbol{x}_1, \ldots, \boldsymbol{x}_N\}$ are transmitted, the corresponding received data can be arranged in a matrix

$$\boldsymbol{Y} = [\boldsymbol{y}_1 \ \cdots \ \boldsymbol{y}_N] \tag{2.1.4}$$

and written as follows:

$$\boldsymbol{Y} = \boldsymbol{H}\boldsymbol{X} + \boldsymbol{E} \tag{2.1.5}$$

where

$$\boldsymbol{X} = [\boldsymbol{x}_1 \ \cdots \ \boldsymbol{x}_N] \tag{2.1.6}$$

and

$$\boldsymbol{E} = [\boldsymbol{e}_1 \ \cdots \ \boldsymbol{e}_N] \tag{2.1.7}$$

Note that vectorization of (2.1.5) yields the following equivalent model:

$$\boldsymbol{y} = \left(\boldsymbol{X}^T \otimes \boldsymbol{I}\right)\boldsymbol{h} + \boldsymbol{e} \tag{2.1.8}$$

where $\boldsymbol{y} = \text{vec}(\boldsymbol{Y})$, $\boldsymbol{h} = \text{vec}(\boldsymbol{H})$ and $\boldsymbol{e} = \text{vec}(\boldsymbol{E})$. This expression will be useful for performance analysis purposes.

2.1.1 The Noise Term

In this text, the noise vectors $\{\boldsymbol{e}_n\}$ will, unless otherwise stated, be assumed to be spatially white circular Gaussian random variables with zero mean and variance σ^2:

$$\boldsymbol{e}_n \sim N_C(\boldsymbol{0}, \sigma^2 \boldsymbol{I}) \tag{2.1.9}$$

Such noise is called additive white Gaussian noise (AWGN).

The Gaussian assumption is customary, as there are at least two strong reasons for making it. First, Gaussian distributions tend to yield mathematical expressions that are relatively easy to deal with. Second, a Gaussian distribution of a disturbance term can often be motivated via the central limit theorem.

Unless otherwise stated we will also assume throughout this book that the noise is temporally white. Although such an assumption is customary, it is clearly an approximation. In particular, the $\boldsymbol{E}$ term may contain interference consisting of modulated signals that are not perfectly white.

To summarize, the set of complex Gaussian vectors $\{\boldsymbol{e}_n\}$ has the following statistical properties:

$$\begin{aligned} E[\boldsymbol{e}_n \boldsymbol{e}_n^H] &= \sigma^2 \boldsymbol{I} \\ E[\boldsymbol{e}_n \boldsymbol{e}_k^H] &= \boldsymbol{0}, \quad n \neq k \\ E[\boldsymbol{e}_n \boldsymbol{e}_k^T] &= \boldsymbol{0}, \quad \text{for all } n, k \end{aligned} \tag{2.1.10}$$

2.1.2 Fading Assumptions

The elements of the matrix $\boldsymbol{H}$ correspond to the (complex) channel gains between the transmit and receive antennas. For the purpose of assessing and predicting the performance of a communication system, it is necessary to postulate a statistical distribution of these elements. This is true to some degree also for receiver design, in the sense that knowledge of the statistical behavior of $\boldsymbol{H}$ could potentially be used to improve the performance of the receiver.

Throughout this book we will assume that the elements of the channel matrix $\boldsymbol{H}$ are complex Gaussian random variables with zero mean. This assumption is made to model the fading effects induced by local scattering in the absence of a line-of-sight component [STÜBER, 2001, CH. 2], [RAPPAPORT, 2002, CH. 5], [MEYR ET AL., 1998, CH. 11], [PROAKIS, 2001, CH. 14]. Consequently the magnitudes of the channel gains $\{|h_{m,n}|\}$ have a Rayleigh distribution, or equivalently expressed, $\{|h_{m,n}|^2\}$ are exponentially distributed. The presence of a line-of-sight component can be modelled by letting $\{h_{m,n}\}$ have a Gaussian distribution with a non-zero mean (this is called Rice fading in the literature), but for notational simplicity we will not do this in this text. In addition to using the Rayleigh and Rice distributions, there are many other ways of modeling the statistical behavior of a fading channel (see, e.g., [STÜBER, 2001, CH. 2] and [SIMON AND ALOUINI, 2000, CH. 2]).

Another commonly made assumption on $\boldsymbol{H}$ is that its elements are statistically independent of one another. Although we will make this assumption in this book unless otherwise explicitly stated, it is usually a rough approximation. That $\{h_{m,n}\}$ are not independent in practice can easily be understood intuitively: if two electromagnetic waves, originating from two different antennas, are reflected by the same object, the propagation coefficients associated with each of these waves will be correlated. In general, the elements of $\boldsymbol{H}$ are correlated by an amount that depends on the propagation environment as well as the polarization of the antenna elements and the spacing between them [ERTEL ET AL., 1998].

One possible model for $\boldsymbol{H}$ that takes the fading correlation into account splits the fading correlation into two independent components called *receive correlation* and *transmit correlation*, respectively (see, e.g., [KERMOAL ET AL., 2002], [GESBERT ET AL., 2002]). This amounts to modeling $\boldsymbol{H}$ as follows:

$$\boldsymbol{H} = \boldsymbol{R}_r^{1/2} \boldsymbol{H}_w \boldsymbol{R}_t^{1/2} \tag{2.1.11}$$

where $\boldsymbol{H}_w$ is a matrix whose elements are Gaussian and *independent* with unity variance and where $(\cdot)^{1/2}$ stands for the Hermitian square root of a matrix. The matrix $\boldsymbol{R}_r$ determines the correlation between the rows of $\boldsymbol{H}$, and is used to model the correlation between the receive antennas; note that this correlation is the same regardless of which transmit antenna that is considered. The matrix $\boldsymbol{R}_t$ is called transmit correlation matrix and models the covariance of the columns of $\boldsymbol{H}$ in a corresponding manner; according to the model (2.1.11), this correlation is equal for all receive antennas under consideration. Let $\boldsymbol{h}_w \triangleq \operatorname{vec}(\boldsymbol{H}_w)$. Then

$$\boldsymbol{h} \triangleq \operatorname{vec}(\boldsymbol{H}) = \left(\boldsymbol{R}_t^{T/2} \otimes \boldsymbol{R}_r^{1/2}\right) \boldsymbol{h}_w \tag{2.1.12}$$

and hence

$$h \sim N_C(\mathbf{0}, \boldsymbol{R}_t^T \otimes \boldsymbol{R}_r) \tag{2.1.13}$$

Of course, one could envision the use of a channel model with a more general covariance structure than that in (2.1.11) but such a matrix would lose the interpretation in terms of a separate transmit and receive correlation.

The correlation coefficients in $\boldsymbol{R}_r$ and $\boldsymbol{R}_t$ in (2.1.11) can be obtained analytically by making the assumption that the transmitter and receiver are surrounded by a cluster of scatterers with a certain spatial distribution (see, e.g., [ZETTERBERG, 1996]). The corresponding correlation coefficients will in general decrease as the size of the scatterer cluster (relative to the distance between the transmitter and receiver) increases. If both the transmitter and the receiver are located in the middle of one large and dense isotropic distribution of scatterers, the correlation coefficients can become virtually zero.

The relations between the scattering environment and the properties of $\boldsymbol{R}_t$ and $\boldsymbol{R}_r$ are illustrated in Figure 2.2. In Figure 2.2(a), both the transmitter and the receiver are located within a homogenous field of scatterers and we can expect both $\boldsymbol{R}_r$ and $\boldsymbol{R}_t$ to be proportional to identity matrices. In Figure 2.2(b), only the transmitter is located within a field of scatterers, and from the receiver this field is seen within a relatively small angle. The receiver is not surrounded by any significant amount of scatterers. In this case, we can expect $\boldsymbol{R}_r$ to have a low effective rank (in other words, the ratio between the largest eigenvalue and the smallest eigenvalues is fairly large), whereas $\boldsymbol{R}_t$ should be proportional to an identity matrix. Finally, Figure 2.2(c) illustrates the opposite case in which only the receiver is located within a scatterer field that is visible within a small angle from the transmitter (which is not surrounded by local scatterers). In this case, $\boldsymbol{R}_r$ should be proportional to an identity matrix whereas $\boldsymbol{R}_t$ may be of low rank. This last case can occur, for instance, in the downlink in a cellular communication system when the base station antennas are located on the top of a radio mast and the mobile user is in an urban area.

In practice, the amount of correlation between the elements of $\boldsymbol{H}$ can be assessed by measurements. A recent experimental study [KYRITSI AND COX, 2001] considers an indoor scenario with the transmitter and receiver situated in two different rooms, and reports correlation coefficients with magnitudes between 0.25 and 0.75, depending on the polarization and the distance between the transmitter and receiver. Other related results have been reported in [SWINDLEHURST ET AL., 2001], [STRIDH AND OTTERSTEN, 2000], [MOLISCH ET AL., 2002], [YU ET AL., 2001]. Also note that the model in (2.1.11) is only one possible

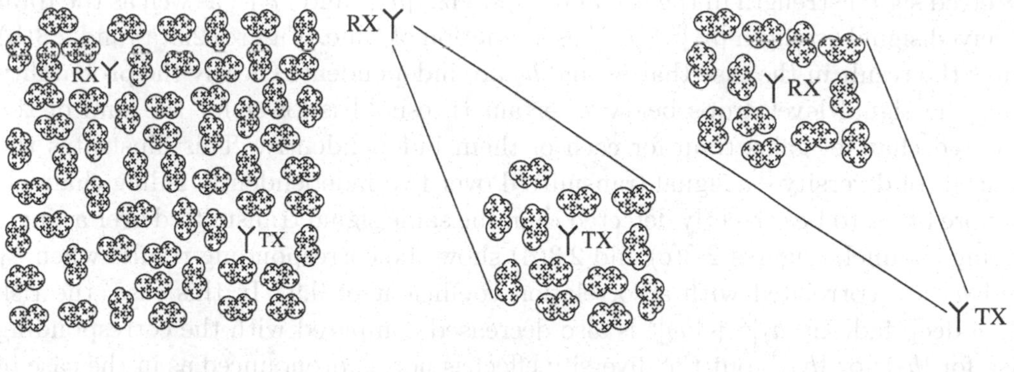

(a) $R_t \propto I$; $R_r \propto I$ (b) $R_t \propto I$; R_r rank deficient (c) $R_r \propto I$; R_t rank deficient

Figure 2.2. Illustration of local scattering.

way of modeling the correlation between the elements of $\boldsymbol{H}$. See, for example, [PAULRAJ ET AL., 2003], [DURGIN, 2002], [ERTEL ET AL., 1998], [GESBERT ET AL., 2002], [ZETTERBERG, 1996], [YU AND OTTERSTEN, 2002], [CORREIA, 2001] for a more detailed discussion of propagation modeling for wireless MIMO channels, and [BLISS ET AL., 2002] for an analysis of how the statistics of $\boldsymbol{H}$ influence quantities such as channel capacity. The case of polarization diversity is studied in, for example, [KYRITSI ET AL., 2002], [NABAR ET AL., 2002] and the references therein.

As we will show later in the text, the transmit and receive correlation can affect the achievable system performance heavily. The ideal case is when both $\boldsymbol{R}_r$ and $\boldsymbol{R}_t$ are proportional to identity matrices, whereas in the case when both $\boldsymbol{R}_r$ and $\boldsymbol{R}_t$ have rank one, the achievable radio link performance will effectively be reduced to that of a conventional SISO system. This is easy to understand intuitively: if the magnitudes of all propagation coefficients $\{|h_{m,n}|\}$ fade independently the risk that all will be in a deep fade at the same time is significantly reduced compared to the case where they fade in a correlated fashion. We illustrate these facts in the following example.

Example 2.1: Correlated and Uncorrelated Fading.

Consider a system with one transmit antenna and two receive antennas. Let the two corresponding channel gains be denoted by h_1 and h_2, and assume that they are complex Gaussian random variables as above. In Figure 2.3 we illustrate the received signal strength in the two antennas, viz., $|h_1|^2$ and $|h_2|^2$, as well as the total received signal strength $|h_1|^2+|h_2|^2$ as a function of time. Figures 2.3(a) and 2.3(b) show the result in the case that h_1 and h_2 are independent. Clearly the probability that the signal level drops below a certain threshold is lower for the sum of the squared channel gains than for each of them independently. This illustrates the concept of diversity – a signal transmitted over two independently fading channels is more likely to be correctly detected than the same signal transmitted over a single fading channel. Figures 2.3(c) and 2.3(d) show the corresponding results when h_1 and h_2 are correlated with a correlation coefficient of 0.9. In this case, the risk for a deep fade in $|h_1|^2 + |h_2|^2$ is also decreased compared with the corresponding risk for $|h_1|^2$ or $|h_2|^2$, but the diversity effect is not as pronounced as in the case of independently fading channel gains. For comparison, the average signal strengths are also shown in each figure as a horizontal line. ■

2.2 The Frequency-Selective MIMO Channel

A frequency-selective multipath MIMO channel can be modeled as a causal continuous-time linear system with an impulse response $\boldsymbol{H}_{\text{cont}}(t)$, $0 \le t < \infty$, consisting of a (virtually infinite) number of impulses with coefficient matrices $\{\boldsymbol{A}_k\}$ and relative timing $\{\tau_k\}$ (see, e.g., [Rappaport, 2002, Chap. 5] for a general discussion of modeling of multipath channels, and [Paulraj et al., 2003] for a discussion of MIMO channel modeling):

$$\boldsymbol{H}_{\text{cont}}(t) = \sum_k \boldsymbol{A}_k \delta(t - \tau_k) \tag{2.2.1}$$

(As all quantities in this text will later be discrete-time, we emphasize the continuous-time nature of $\boldsymbol{H}_{\text{cont}}(t)$ by the subscript $(\cdot)_{\text{cont}}$.) The transfer function of the channel is:

$$\boldsymbol{H}_{\text{cont}}(\omega) = \int_{-\infty}^{\infty} \boldsymbol{H}_{\text{cont}}(t) e^{-i\omega t} dt \tag{2.2.2}$$

To obtain a model suitable for analysis, it is convenient to use a sampled representation; also, envisioning a digital implementation of receiver algorithms, a discrete-time model is necessary. Let τ be the sampling period, and suppose that the transmitted signal is ideally bandlimited to the Nyquist frequency $1/\tau$; hence we can assume that the channel is bandlimited to $1/\tau$ as well. Then assuming that the sampling jitter is chosen such that a sample is taken at $t = 0$, the sampled

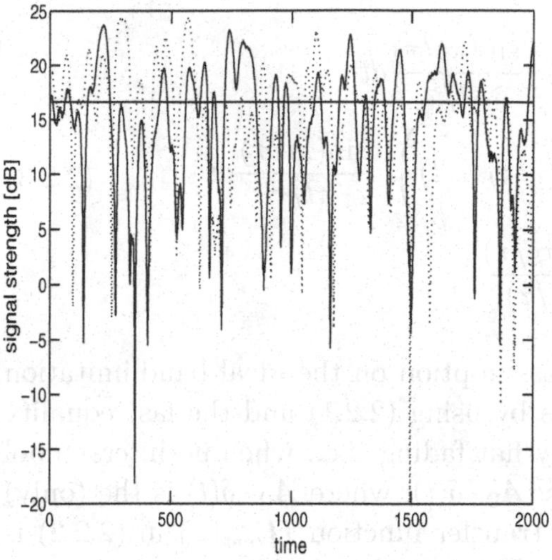

(a) Plot of $|h_1|^2$ (solid) and $|h_2|^2$ (dotted). Here h_1 and h_2 are uncorrelated.

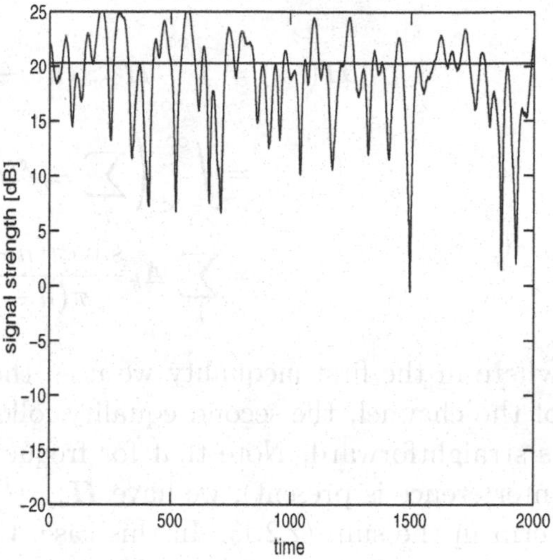

(b) Plot of $|h_1|^2 + |h_2|^2$. Here h_1 and h_2 are uncorrelated.

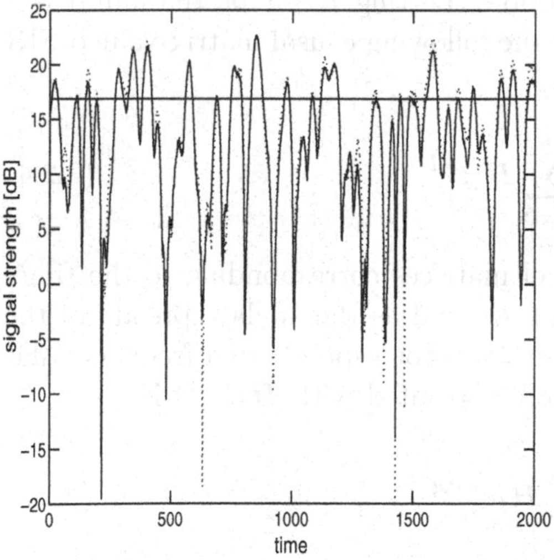

(c) Plot of $|h_1|^2$ (solid) and $|h_2|^2$ (dashed). The correlation between h_1 and h_2 is 0.9.

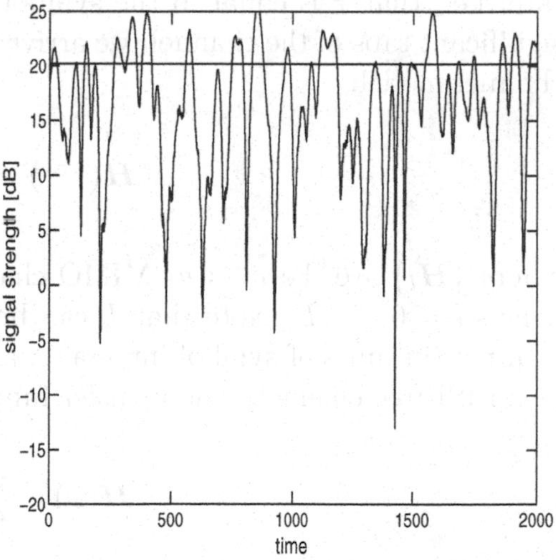

(d) Plot of $|h_1|^2 + |h_2|^2$. The correlation between h_1 and h_2 is 0.9.

Figure 2.3. Illustration of fading and diversity. In all figures, the horizontal line shows the time average.

impulse response of the channel becomes:

$$\begin{aligned}
\boldsymbol{H}(n\tau) &= \int_{-\infty}^{\infty} \boldsymbol{H}_{\text{cont}}(n\tau - t) \cdot \frac{\sin(\pi t/\tau)}{\pi t/\tau} dt \\
&= \int_{-\infty}^{\infty} \left(\sum_k \boldsymbol{A}_k \delta(n\tau - \tau_k - t) \right) \cdot \frac{\sin(\pi t/\tau)}{\pi t/\tau} dt \\
&= \sum_k \boldsymbol{A}_k \frac{\sin(\pi(n - \tau_k/\tau))}{\pi(n - \tau_k/\tau)}
\end{aligned} \tag{2.2.3}$$

where in the first inequality we used the assumption on the ideal bandlimitation of the channel, the second equality follows by using (2.2.1) and the last equality is straightforward. Note that for frequency flat fading (i.e., when no intersymbol interference is present), we have $\boldsymbol{H}_{\text{cont}}(t) \approx \boldsymbol{A}_0 \cdot \delta(t)$, where $\boldsymbol{A}_0 \cdot \delta(t)$ is the (only) term in the sum (2.2.1). In this case, the transfer function $\boldsymbol{H}_{\text{cont}}(\omega)$ in (2.2.2) is approximately independent of ω within the frequency band of interest.

In general the sampled impulse response $\boldsymbol{H}(n\tau)$ in (2.2.3) has an infinite length, but we will truncate it to an "effective length," i.e., a number of taps which contain most of the power. We will also assume that all signals are sampled at the symbol rate, i.e., that τ is equal to the symbol period. Letting $L+1$ be the number of significant taps of the channel, we arrive at the following causal matrix-valued FIR channel model:

$$\boldsymbol{H}(z^{-1}) = \sum_{l=0}^{L} \boldsymbol{H}_l z^{-l} \tag{2.2.4}$$

where $\{\boldsymbol{H}_l\}$ are the $n_r \times n_t$ MIMO channel matrices corresponding to the time-delays $l = 0, \ldots, L$. Note that L can be interpreted as the "delay-spread" of the channel (in units of symbol intervals); also, $L = 0$ corresponds to a frequency-flat or an ISI-free channel. The transfer function associated with $\boldsymbol{H}(z^{-1})$ is:

$$\boldsymbol{H}(\omega) = \sum_{l=0}^{L} \boldsymbol{H}_l e^{-i\omega l} \tag{2.2.5}$$

2.2.1 Block Transmission

Since the channel has memory, sequentially transmitted symbols will interfere with each other at the receiver, and therefore we need to consider a sequence, or block, of symbols $\{s(0), s(1), \ldots\}$, instead of a single symbol at a time as before. In general, we will assume that the transmitted sequence is preceded by a preamble

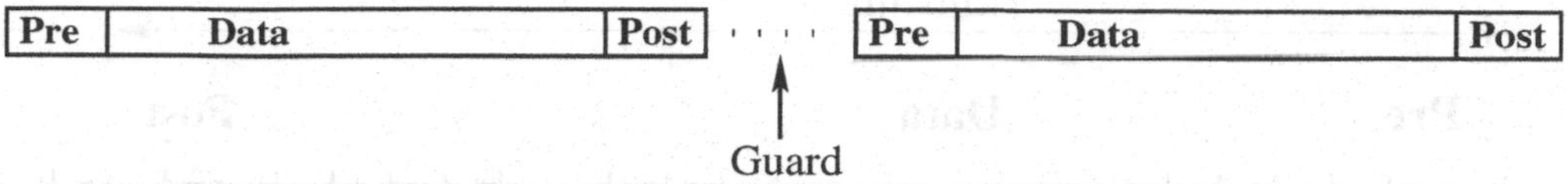

(a) Transmission of blocks with a separating guard interval.

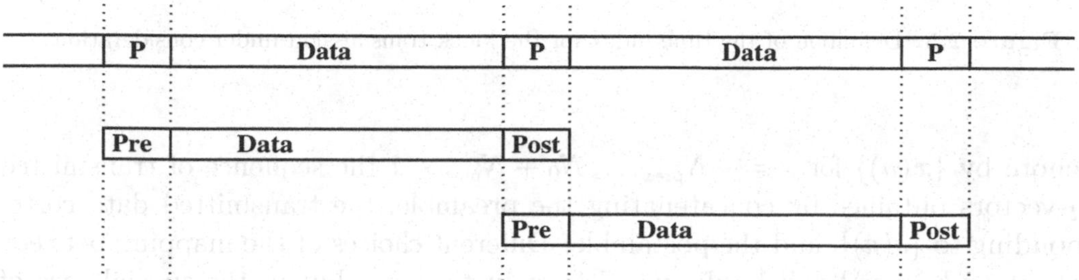

(b) Continuous transmission of blocks; the preamble and postamble of consecutive blocks coincide.

Figure 2.4. Illustration of block-transmission.

and followed by a postamble. The preamble and postamble play the important role of preventing subsequent blocks from interfering with each other and, as we will see later in the book, they can also be essential for providing diversity. If several blocks follow each other, the postamble of one block may also function as the preamble of the following block, and so on. This is the case for example in some wireless local area network standards [VAN NEE ET AL., 1999]. However, it is also possible that the blocks are separated with a guard interval; this is the case for instance in the GSM cellular phone system [MOULY AND PAUTET, 1992]. These two possibilities are illustrated in Figure 2.4.

Let us consider the transmission of a given block (see Figure 2.5). Assume that the preamble and postamble are of the length N_{pre} and N_{post}, respectively. Note that unless there is an additional guard interval between the blocks (as in Figure 2.4(a)), we must have $N_{\text{pre}} \geq L$ to prevent the data (or postamble) of a preceding burst from interfering with the data part of the block under consideration. We let the preamble of the block under consideration be transmitted during the time intervals $n = -N_{\text{pre}}, \ldots, -1$, we let the data be transmitted at $n = 0, \ldots, N_0 - 1$ and we assume that the postamble is transmitted at $n = N_0, \ldots, N_0 + N_{\text{post}} - 1$. We

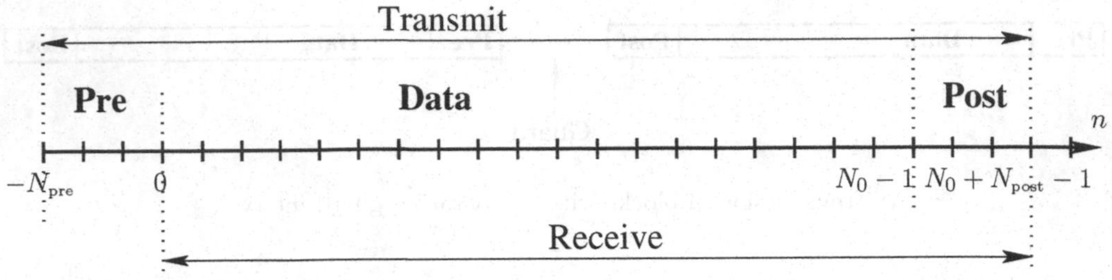

Figure 2.5. Definition of the time index for the block transmission under consideration.

denote by $\{\boldsymbol{x}(n)\}$ for $n = -N_{\text{pre}}, \ldots, N_0 + N_{\text{post}} - 1$ the sequence of transmitted n_t-vectors obtained by concatenating the preamble, the transmitted data corresponding to $\{s(n)\}$, and the postamble. Different choices of the mapping between $\{s(n)\}$ and $\{\boldsymbol{x}(n)\}$ will be discussed later in the text, but in the special case of one transmit antenna ($n_t = 1$), $\boldsymbol{x}(n)$ becomes a scalar $x(n)$ and in that case no transmit encoding is necessary so we can simply set $x(n) = s(n)$.

The received signal is well-defined for $n = L - N_{\text{pre}}, \ldots, N_0 + N_{\text{post}} - 1$, but it depends on the transmitted data only for $n = 0, \ldots, N_0 + L - 1$; in the discussion we assume that $N_{\text{post}} \leq L$. The samples $n = 0, \ldots, N_0 + L - 1$ that depend on the data can be written as:

$$\boldsymbol{y}(n) = \sum_{l=0}^{L} \boldsymbol{H}_l \boldsymbol{x}(n-l) + \boldsymbol{e}(n) = \boldsymbol{H}(z^{-1})\boldsymbol{x}(n) + \boldsymbol{e}(n) \tag{2.2.6}$$

where $\boldsymbol{e}(n)$ is a noise vector. Note that parts of the signal received during the preamble can in general depend on the data and the postamble in the previous burst. Also note that the received samples during the preamble and postamble, which do *not* depend on the unknown data, can be used for other tasks such as channel estimation.

2.2.2 Matrix Formulations

We can rewrite the model (2.2.6) to express the received signal of the form (2.1.3) as follows. Let

$$\boldsymbol{G} = \begin{bmatrix} \boldsymbol{H}_L & \boldsymbol{H}_{L-1} & \cdots & \boldsymbol{H}_1 & \boldsymbol{H}_0 & \boldsymbol{0} & \cdots & \boldsymbol{0} \\ \boldsymbol{0} & \boldsymbol{H}_L & \ddots & & \ddots & \boldsymbol{H}_0 & \ddots & \vdots \\ \vdots & \ddots & \ddots & & & & \ddots & \boldsymbol{0} \\ \boldsymbol{0} & \cdots & \boldsymbol{0} & \boldsymbol{H}_L & \boldsymbol{H}_{L-1} & \cdots & \boldsymbol{H}_1 & \boldsymbol{H}_0 \end{bmatrix} \tag{2.2.7}$$

and

$$\begin{aligned} \boldsymbol{x} &= \left[\boldsymbol{x}^T(-L) \ \cdots \ \boldsymbol{x}^T(N_0+L-1)\right]^T \\ \boldsymbol{y} &= \left[\boldsymbol{y}^T(0) \ \cdots \ \boldsymbol{y}^T(N_0+L-1)\right]^T \\ \boldsymbol{e} &= \left[\boldsymbol{e}^T(0) \ \cdots \ \boldsymbol{e}^T(N_0+L-1)\right]^T \end{aligned} \tag{2.2.8}$$

Then we have that:

$$\boldsymbol{y} = \boldsymbol{G}\boldsymbol{x} + \boldsymbol{e} \tag{2.2.9}$$

Although (2.2.9) has the same form as the relation (2.1.3) that describes the input-output relation associated with a flat MIMO channel, the matrix $\boldsymbol{G}$ in (2.2.9) has a very special structure, as opposed to the matrix $\boldsymbol{H}$ in Section 2.1. However, we can rearrange the model (2.2.9) to obtain an expression of the form (2.1.8), where $\boldsymbol{h}$ is an unstructured vector, and hence flat and frequency-selective channels can be analyzed in a common matrix-algebraic framework. First note that

$$\begin{aligned} &\left[\boldsymbol{y}(0) \ \cdots \ \boldsymbol{y}(N_0+L-1)\right] \\ &= \left[\boldsymbol{H}_0 \ \cdots \ \boldsymbol{H}_L\right]\boldsymbol{X} + \left[\boldsymbol{e}(0) \ \cdots \ \boldsymbol{e}(N_0+L-1)\right] \end{aligned} \tag{2.2.10}$$

where $\boldsymbol{X}$ is a block-Toeplitz matrix built from the transmitted data $\{\boldsymbol{x}(n)\}_{n=-L}^{N_0+L-1}$ according to:

$$\boldsymbol{X} = \begin{bmatrix} \boldsymbol{x}(0) & & \cdots & \cdots & \cdots & \boldsymbol{x}(N_0+L-1) \\ \boldsymbol{x}(-1) & \ddots & & & & \boldsymbol{x}(N_0+L-2) \\ \vdots & & \ddots & & & \vdots \\ \boldsymbol{x}(-L) & \cdots & \cdots & \boldsymbol{x}(0) & \cdots & \boldsymbol{x}(N_0-1) \end{bmatrix} \tag{2.2.11}$$

If we let

$$\boldsymbol{h} = \mathrm{vec}\left(\left[\boldsymbol{H}_0 \ \cdots \ \boldsymbol{H}_L\right]\right) \tag{2.2.12}$$

then by vectorizing both sides of (2.2.10) we find that:

$$\boldsymbol{y} = (\boldsymbol{X}^T \otimes \boldsymbol{I}_{n_r})\boldsymbol{h} + \boldsymbol{e} \tag{2.2.13}$$

which has the form (2.1.8).

Although the FIR-filtering representation in (2.2.6) may be easier to interpret intuitively than (2.2.13), the latter formulation will be more useful for performance analysis.

2.3 Summary and Discussion

In this chapter we have introduced a unified matrix-algebraic framework for the MIMO propagation channel. Relying on the standard complex baseband representation of narrowband signals, we found in Section 2.1 that a frequency flat MIMO channel can be described by a matrix equation of the form $\boldsymbol{Y} = \boldsymbol{H}\boldsymbol{X} + \boldsymbol{E}$ (where all quantities are defined in Section 2.1). For frequency-selective channels, we modeled the channel via a symbol-spaced tapped-delay-line with coefficient matrices $\{\boldsymbol{H}_l\}_{l=0}^{L}$ where $L + 1$ is the number of taps (see Section 2.2). Finally, we found that the input-output relation associated with both frequency flat and frequency-selective channels can be cast in the form $\boldsymbol{y} = (\boldsymbol{X}^T \otimes \boldsymbol{I})\boldsymbol{h} + \boldsymbol{e}$ where $\boldsymbol{y}$ is a vector of received data, $\boldsymbol{h}$ is a vector of the channel, $\boldsymbol{X}$ is a (structured) matrix containing transmitted data and $\boldsymbol{e}$ is noise. While the results of this chapter will be used in the entire remaining text, the latter general model will be particularly useful in Chapters 4, 5 and 8 when we analyze the error performance of transmission schemes for both flat and frequency-selective channels.

While the model for flat MIMO channels does not involve any significant approximation, concerning frequency-selective channels we should stress that the models used in Section 2.2 are idealized descriptions of reality. In particular, a sampled channel impulse response has usually an infinite length even when the continuous-time delay spread is finite. However, in practice some of the channel taps may be very small; hence for analysis purposes it may make sense to let L be the *effective* length of the channel, i.e., the number of significant taps in the channel impulse response. This means that the effective rank of $E[\boldsymbol{h}\boldsymbol{h}^H]$ will be (at most) equal to $n_r n_t(L + 1)$. Also, note that in some receivers it is common practice to use an oversampled version of the received signal. In this case, the received signal can still be modelled as in (2.2.6), but using a FIR filter of length $\hat{L} > L$ where L is the length of the channel (sampled at symbol rate). The same may be true in the presence of symbol synchronization errors, in which case it may be necessary to use a longer FIR filter ($\hat{L} > L$) to accurately represent the propagation channel. However, in both these cases some elements of the vector $\boldsymbol{h}$ will be highly correlated so that they have a rank deficient covariance matrix; therefore the rank of $E[\boldsymbol{h}\boldsymbol{h}^H]$ is strictly less than $n_r n_t(\hat{L} + 1)$. More details on the transition from continuous to discrete time can be found in [MEYR ET AL., 1998]; a discussion on sufficiency of the statistics obtained by sampling a continuous-time model, in the context of maximum-likelihood equalization, can also be found in [CHUGG AND POLYDOROS, 1996].

2.4 Problems

1. Give some examples of practical communication systems where the propagation channel is

   (a) frequency flat

   (b) frequency-selective

   (c) time-invariant

   (d) time-variant

2. Discuss some tradeoffs involved in the choice of the length N_0 of the data block, and the lengths N_{pre} and N_{post} of the preambles and postambles (see Figure 2.4).

3. Discuss the validity of the assumptions in Section 2.1.1. Can you give an example of a situation where $\boldsymbol{E}$ is *not* statistically white? Can you give an example of a case when it is not Gaussian?

4. Discuss the correlation between the matrices $\boldsymbol{H}_l$ in (2.2.4). Are they independent in general? If not, what factors influence the correlation between them?

5. Study the scenarios in Figure 2.2.

   (a) Consider two antennas, spaced several wavelengths apart, located in the middle of an area with uniformly distributed scatterers (this is the situation for both the transmitter and receiver in Figure 2.2(a)). Use a central limit theorem-type of reasoning to argue that the fading associated with the two antennas is (practically) independent.

   (b) Consider two antennas located in free space, and suppose that a signal impinges from a direction in space with a very small angular spread (this is the situation for the receiver in Figure 2.2(b) and for the transmitter in Figure 2.2(c)). Explain why we can expect the fading associated with these two antennas to be highly correlated.

6. Give an example of a covariance matrix $\boldsymbol{R}$ that does *not* have the Kronecker structure in (2.1.13).

Chapter 3

MIMO INFORMATION THEORY

The primary goal of this chapter is to review a few basic concepts from information theory and see how these can be applied to the analysis of a MIMO system. We will obtain expressions for the capacity of a MIMO system, and study how this capacity behaves in certain situations. Such capacity results are important because they express bounds on the maximal achievable data rate, and they will also be used for the analysis of different transmission schemes later in the text. While many results in this chapter are classical, some of them are more recent. Our intention is to offer a brief, but as consistent as possible, treatment of the topic. Some of the material in the present chapter is rather mathematical; the reader may skim through this chapter and return to its details when needed.

3.1 Entropy and Mutual Information

Let x be a random variable with a discrete probability mass function $p_x(x)$. Then the quantity

$$H(x) = -E_x\left[\log_2\left(p_x(x)\right)\right] = -\sum_k \log_2\left(p_x(x_k)\right)p_x(x_k) \tag{3.1.1}$$

is called the *entropy* of x. The entropy is essentially a measure of how many bits of information are needed, on the average, to convey the information contained in x provided that an optimal source coding algorithm is used [Cover and Thomas, 1991, Ch. 5]. In a similar way, for a continuous random variable (in general, a random vector), we can define:

$$H(\boldsymbol{x}) = -E_{\boldsymbol{x}}\left[\log_2\left(p_{\boldsymbol{x}}(\boldsymbol{x})\right)\right] = -\int d\boldsymbol{x}\; p_{\boldsymbol{x}}(\boldsymbol{x})\log_2\left(p_{\boldsymbol{x}}(\boldsymbol{x})\right) \tag{3.1.2}$$

For continuous probability distributions, the quantity $H(\boldsymbol{x})$ is often called the *differential entropy* in the information theory literature.

If $\boldsymbol{x}$ is a real-valued Gaussian random vector of length n:

$$\boldsymbol{x} \sim N(\boldsymbol{\mu}, \boldsymbol{P}) \tag{3.1.3}$$

then we leave as an exercise to the reader to verify that

$$\begin{aligned} H(\boldsymbol{x}) &= -\int d\boldsymbol{x}\, \frac{\exp\left(-\frac{1}{2}\boldsymbol{x}^T\boldsymbol{P}^{-1}\boldsymbol{x}\right)}{(2\pi)^{n/2}\cdot|\boldsymbol{P}|^{1/2}} \log_2\left(\frac{\exp\left(-\frac{1}{2}\boldsymbol{x}^T\boldsymbol{P}^{-1}\boldsymbol{x}\right)}{(2\pi)^{n/2}\cdot|\boldsymbol{P}|^{1/2}}\right) \\ &= \frac{n}{2}\cdot\log_2(2\pi e) + \frac{1}{2}\log_2|\boldsymbol{P}| \end{aligned} \tag{3.1.4}$$

Note that the entropy $H(\boldsymbol{x})$ in (3.1.4) does not depend on $\boldsymbol{\mu}$. In the same way, if $\boldsymbol{x}$ is a circular Gaussian random vector (see Appendix A.3) of length n:

$$\boldsymbol{x} \sim N_C(\boldsymbol{\mu}, \boldsymbol{P}) \tag{3.1.5}$$

then

$$H(\boldsymbol{x}) = n\log_2(e\pi) + \log_2|\boldsymbol{P}| \tag{3.1.6}$$

(independent of $\boldsymbol{\mu}$).

Gaussian vectors have a very special property: among all random vectors $\boldsymbol{x}$ with zero mean and covariance matrix $\boldsymbol{P}$, the entropy $H(\boldsymbol{x})$ is maximized if $\boldsymbol{x}$ is Gaussian. We state this classical result in the following theorem.

Theorem 3.1: Entropy-maximizing property of a Gaussian random variable.
Let $\boldsymbol{x}$ be a real-valued random vector with

$$\begin{aligned} E[\boldsymbol{x}] &= \boldsymbol{0} \\ E[\boldsymbol{x}\boldsymbol{x}^T] &= \boldsymbol{P} \end{aligned} \tag{3.1.7}$$

Then $H(\boldsymbol{x})$ is maximized if $\boldsymbol{x} \sim N(\boldsymbol{0}, \boldsymbol{P})$.

Proof: This result is well-known (see, e.g., [Cover and Thomas, 1991, Th. 9.6.5] and [Telatar, 1999]). For completeness, we present a short proof in Section 3.8. ■

The above result extends directly to complex random vectors, in which case the entropy is maximized for a circular Gaussian vector.

The entropy can be evaluated conditioned on another variable $\boldsymbol{y}$, in which case we write $H(\boldsymbol{x}|\boldsymbol{y})$. The so-obtained quantity, which is called conditional entropy, is defined as:

$$H(\boldsymbol{x}|\boldsymbol{y}) = -E\big[\log_2\left(p(\boldsymbol{x}|\boldsymbol{y})\right)\big] \tag{3.1.8}$$

For two random vectors $\boldsymbol{x}$ and $\boldsymbol{y}$, we can associate a quantity called *mutual information* defined as follows:

$$M(\boldsymbol{y},\boldsymbol{x}) = H(\boldsymbol{y}) - H(\boldsymbol{y}|\boldsymbol{x}) \tag{3.1.9}$$

The mutual information is essentially a measure of how much information about the random vector $\boldsymbol{y}$ is contained in the vector $\boldsymbol{x}$.

The concept of mutual information is particularly relevant in the context of the following linear model:

$$\boldsymbol{y} = \boldsymbol{H}\boldsymbol{x} + \boldsymbol{e} \tag{3.1.10}$$

where $\boldsymbol{y}$, $\boldsymbol{x}$ and $\boldsymbol{e}$ are random vectors of length n, m, and n, respectively. Let us first assume that all quantities in (3.1.10) are real-valued, and that $\boldsymbol{e}$ is zero-mean Gaussian:

$$\boldsymbol{e} \sim N(\boldsymbol{0}, \sigma^2 \cdot \boldsymbol{I}) \tag{3.1.11}$$

Then by (3.1.4):

$$\begin{aligned} M_{\text{real}}(\boldsymbol{y},\boldsymbol{x}|\boldsymbol{H}) &= H(\boldsymbol{y}|\boldsymbol{H}) - H(\boldsymbol{y}|\boldsymbol{x},\boldsymbol{H}) \\ &= H(\boldsymbol{y}|\boldsymbol{H}) - \frac{n}{2}\cdot\log_2(2\pi e) - \frac{1}{2}\log_2|\sigma^2\boldsymbol{I}| \end{aligned} \tag{3.1.12}$$

Clearly, if $\boldsymbol{x}$ has zero mean and covariance matrix $\boldsymbol{P}$, then $\boldsymbol{y}$ has zero mean and covariance matrix $\boldsymbol{H}\boldsymbol{P}\boldsymbol{H}^T + \sigma^2\boldsymbol{I}$. By Theorem 3.1, it follows that among all zero-mean vectors $\boldsymbol{x}$ with covariance matrix $\boldsymbol{P}$, the mutual information in (3.1.12) is maximized when $\boldsymbol{y}$ is Gaussian, which in turn holds if $\boldsymbol{x}$ is Gaussian. For such a Gaussian choice of $\boldsymbol{x}$, we have that (cf. (3.1.4)):

$$H(\boldsymbol{y}|\boldsymbol{H}) = \frac{n}{2}\cdot\log_2(2\pi e) + \frac{1}{2}\log_2|\sigma^2\boldsymbol{I} + \boldsymbol{H}\boldsymbol{P}\boldsymbol{H}^T| \tag{3.1.13}$$

and consequently

$$\begin{aligned} M_{\text{real}}(\boldsymbol{y},\boldsymbol{x}|\boldsymbol{H}) &= \frac{1}{2}\log_2|\sigma^2\boldsymbol{I} + \boldsymbol{H}\boldsymbol{P}\boldsymbol{H}^T| - \frac{1}{2}\log_2|\sigma^2\boldsymbol{I}| \\ &= \frac{1}{2}\log_2\left|\boldsymbol{I} + \frac{1}{\sigma^2}\boldsymbol{H}\boldsymbol{P}\boldsymbol{H}^T\right| \end{aligned} \tag{3.1.14}$$

If instead all quantities in (3.1.10) are complex-valued, and $\boldsymbol{e}$ is circular Gaussian:

$$\boldsymbol{e} \sim N_C(\mathbf{0}, \sigma^2 \cdot \boldsymbol{I}) \tag{3.1.15}$$

then we can show that

$$M_{\text{complex}}(\boldsymbol{y}, \boldsymbol{x}|\boldsymbol{H}) = \log_2 \left| \boldsymbol{I} + \frac{1}{\sigma^2}\boldsymbol{H}\boldsymbol{P}\boldsymbol{H}^H \right| \tag{3.1.16}$$

3.2 Capacity of the MIMO Channel

In the previous section we have outlined the steps that lead to the following central result.

Theorem 3.2: Capacity of a constant MIMO channel with Gaussian noise.
For a given $\boldsymbol{H}$, the Shannon capacity of a MIMO channel is given by the maximum mutual information between the received vectors $\{\boldsymbol{y}_n\}$ and the transmitted vectors $\{\boldsymbol{x}_n\}$. Among all zero-mean vectors $\{\boldsymbol{x}\}$ with covariance matrix $\boldsymbol{P}$, the largest mutual information is achieved when $\{\boldsymbol{x}\}$ is Gaussian. Then the total transmission power is equal to $P = Tr\{\boldsymbol{P}\}$ and the resulting channel capacity is equal to:

$$C(\boldsymbol{H}) = B \cdot \log_2 \left| \boldsymbol{I} + \frac{1}{\sigma^2}\boldsymbol{H}\boldsymbol{P}\boldsymbol{H}^H \right| \tag{3.2.1}$$

where B is the bandwidth (in Hz) of the channel.

Proof: From Shannon's information theory [Cover and Thomas, 1991] it follows that the capacity of a channel $\boldsymbol{x} \to \boldsymbol{y}$ is equal to $B \cdot M(\boldsymbol{y}, \boldsymbol{x})$. Since (3.1.10) is essentially equivalent to the linear MIMO channel model model (2.1.3) the result follows. More rigorous arguments can be found in [Telatar, 1999]. ∎

The channel capacity $C(\boldsymbol{H})$ in Theorem 3.2 gives a bound on the maximum achievable information rate (in bits/second) for a constant channel. This bound is achievable only under certain (quite restrictive) conditions. For instance, a rate of $C(\boldsymbol{H})$ is not achievable unless the length of the data block under consideration goes to infinity [Cover and Thomas, 1991].

Note that for a SISO system that transmits with power P, for which the channel gain is constant and equal to h (in the SISO case, the matrix $\boldsymbol{H}$ reduces to a scalar

h), and which is affected by AWGN noise such that the SNR is equal to $P|h|^2/\sigma^2$, the channel capacity in Theorem 3.2 simplifies to:

$$C(h) = B \cdot \log_2 \left(1 + \frac{P}{\sigma^2}|h|^2\right) \tag{3.2.2}$$

which is the classical Shannon capacity.

The next two examples show that the expression (3.2.1) can be specialized to many situations.

Example 3.1: Capacity of parallel and independent channels.
If $\boldsymbol{H}$ and $\boldsymbol{P}$ are diagonal matrices, then the system is equivalent to n_t parallel and independent SISO systems. The aggregate capacity of n_t such systems is equal to:

$$\begin{aligned} C(\boldsymbol{H}) &= B \cdot \log_2 \left| \boldsymbol{I} + \frac{1}{\sigma^2} \boldsymbol{H}\boldsymbol{P}\boldsymbol{H}^H \right| \\ &= B \cdot \sum_{n=1}^{n_t} \log_2 \left(1 + \frac{|[\boldsymbol{H}]_{n,n}|^2 [\boldsymbol{P}]_{n,n}}{\sigma^2}\right) \end{aligned} \tag{3.2.3}$$

■

Example 3.2: Capacity of a frequency-selective channel.
If $\{\boldsymbol{H}_l\}$ is the impulse response of a frequency-selective MIMO channel, its channel capacity is obtained by integrating (3.2.1) (omitting the factor B) over the frequency band Ω that is available for transmission:

$$C(\boldsymbol{H}(z^{-1})) = \int_\Omega d\omega \, \log_2 \left| \boldsymbol{I} + \frac{1}{\sigma^2} \boldsymbol{H}(\omega)\boldsymbol{P}(\omega)\boldsymbol{H}^H(\omega) \right| \tag{3.2.4}$$

where $\boldsymbol{H}(\omega)$ is the transfer function of the channel defined in (2.2.5):

$$\boldsymbol{H}(\omega) = \sum_{l=0}^{L} \boldsymbol{H}_l e^{-i\omega l} \tag{3.2.5}$$

and $\boldsymbol{P}(\omega)$ is the matrix spectral density function of the transmitted vectors. ■

The channel capacity is a quantity that is not always easy to interpret, and which does not express the entire truth about the achievable throughput. However, it is a quantity that is often relevant to study in order to gain insight into the possibilities and the fundamental limitations of a system. In particular, since the channel capacity of an AWGN SISO channel is achievable to within less than one dB using state-of-the-art coding and signal processing [Sklar, 1997], the corresponding MIMO capacity figure will often provide an indication of the ultimately achievable throughput.

3.3 Channel Capacity for Informed Transmitters

If the channel $\boldsymbol{H}$ is known to the transmitter, the transmit correlation matrix $\boldsymbol{P}$ can be chosen to maximize the channel capacity for a given realization of the channel. The main tool for performing this maximization is the following result, which is commonly referred to as "water-filling," and the resulting maximal channel capacity $C_{\text{IT}}(\boldsymbol{H})$ is called the *informed transmitter* (IT) capacity. Since this result is also used in other contexts than maximizing the mutual information, we state it in a general form as follows.

Theorem 3.3: Water-filling.
Let $\boldsymbol{Q}$ and $\boldsymbol{P}$ be positive semi-definite matrices of dimension $m \times m$ and define the following function:

$$f(\boldsymbol{P}) = |\boldsymbol{I} + \boldsymbol{P}\boldsymbol{Q}| \tag{3.3.1}$$

Assume that $\boldsymbol{Q}$ has $m_+ \leq m$ nonzero eigenvalues and let

$$\boldsymbol{Q}^{1/2} = \boldsymbol{U}\boldsymbol{\Gamma}\boldsymbol{U}^H = \begin{bmatrix}\boldsymbol{U}_+ & \boldsymbol{U}_0\end{bmatrix}\begin{bmatrix}\boldsymbol{\Gamma}_+ & \boldsymbol{0}\\ \boldsymbol{0} & \boldsymbol{0}\end{bmatrix}\begin{bmatrix}\boldsymbol{U}_+^H\\ \boldsymbol{U}_0^H\end{bmatrix} \tag{3.3.2}$$

be the eigen-decomposition of the Hermitian square root of $\boldsymbol{Q}$. Here $\boldsymbol{U}$ and $\boldsymbol{\Gamma}$ are of dimension $m \times m$, $\boldsymbol{U}_+$ is of dimension $m \times m_+$, $\boldsymbol{U}_0$ is of dimension $m \times (m - m_+)$ and

$$\boldsymbol{\Gamma}_+ = \begin{bmatrix}\gamma_1 & 0 & \cdots & 0\\ 0 & \gamma_2 & \ddots & \vdots\\ \vdots & \ddots & \ddots & 0\\ 0 & \cdots & 0 & \gamma_{m_+}\end{bmatrix} \tag{3.3.3}$$

is an $m_+ \times m_+$ diagonal matrix such that $\gamma_1 \geq \cdots \geq \gamma_{m_+}$.

Consider the maximization of (3.3.1) with respect to $\boldsymbol{P}$ for a fixed $\boldsymbol{Q}$, and subject to the constraint that $Tr\{\boldsymbol{P}\} \leq P$. Then it holds that

$$f(\boldsymbol{P}) \leq |\boldsymbol{\Gamma}_-^2| \cdot \left(\frac{Tr\{\boldsymbol{\Gamma}_-^{-2}\} + P}{m_-}\right)^{m_-} \tag{3.3.4}$$

with equality if $\boldsymbol{P}$ is chosen as follows:

$$\boldsymbol{P} = \boldsymbol{U}\boldsymbol{\Phi}\boldsymbol{U}^H \tag{3.3.5}$$

where

$$\boldsymbol{\Gamma}_- = \begin{bmatrix} \gamma_1 & 0 & \cdots & 0 \\ 0 & \gamma_2 & \ddots & \vdots \\ \vdots & \ddots & \ddots & 0 \\ 0 & \cdots & 0 & \gamma_{m_-} \end{bmatrix} \tag{3.3.6}$$

and

$$\boldsymbol{\Phi} = \text{diag}\{\phi_1, \ldots, \phi_{m_-}, 0, \ldots, 0\} \tag{3.3.7}$$

Also,

$$\phi_m = \begin{cases} \frac{Tr\{\boldsymbol{\Gamma}_-^{-2}\}+P}{m_-} - \frac{1}{\gamma_m^2}, & \frac{Tr\{\boldsymbol{\Gamma}_-^{-2}\}+P}{m_-} > \frac{1}{\gamma_m^2} \\ 0, & otherwise \end{cases} \tag{3.3.8}$$

for $m = 1, \ldots, m_-$. The number of nonzero ϕ_m is equal to m_- and is defined implicitly via (3.3.8).

Proof: This result is well-known from information theory (see, e.g., [Cover and Thomas, 1991, Th. 13.3.3] using a slightly different formulation). For completeness, we outline a proof in Section 3.8. ■

By applying the theorem with

$$\boldsymbol{Q} = \frac{1}{\sigma^2}\boldsymbol{H}^H\boldsymbol{H} \tag{3.3.9}$$

(recall the fact that $|\boldsymbol{I}+\boldsymbol{AB}| = |\boldsymbol{I}+\boldsymbol{BA}|$ for any matrices $\boldsymbol{A}$ and $\boldsymbol{B}$ of compatible dimension) we can easily compute the transmit correlation matrix $\boldsymbol{P}$ that maximizes the mutual information between the transmitter and receiver for a given $\boldsymbol{H}$. According to Theorem 3.3, the available transmit power should be distributed on the dominant eigenvectors of the MIMO channel. Only a subset of the eigenvectors is used; depending on the SNR, between one and all of the eigen-modes of the channel are exploited.

3.4 Ergodic Channel Capacity

For a fading channel, the matrix $\boldsymbol{H}$ is a random quantity and hence the associated channel capacity $C(\boldsymbol{H})$ is also a random variable. The average of (3.2.1) over the distribution of $\boldsymbol{H}$:

$$B \cdot E_{\boldsymbol{H}}\Big[\log_2\Big|\boldsymbol{I} + \frac{1}{\sigma^2}\boldsymbol{HPH}^H\Big|\Big] \tag{3.4.1}$$

is sometimes called *average capacity*, or *ergodic capacity* and it cannot be achieved unless coding is employed across an infinite number of independently fading blocks. We can choose $\boldsymbol{P}$ to maximize the average capacity (subject to a power constraint). The following classical theorem, which we present without proof, shows that the maximal ergodic capacity is achieved when $\boldsymbol{P}$ is proportional to an identity matrix.

Theorem 3.4: Maximizing the ergodic capacity.
Let $\boldsymbol{H}$ be a Gaussian random matrix with independent and identically distributed elements. Then the average capacity

$$B \cdot E_{\boldsymbol{H}}\left[\log_2 \left|\boldsymbol{I} + \frac{1}{\sigma^2}\boldsymbol{H}\boldsymbol{P}\boldsymbol{H}^H\right|\right] \tag{3.4.2}$$

is maximized subject to the power constraint

$$Tr\{\boldsymbol{P}\} \leq P \tag{3.4.3}$$

when

$$\boldsymbol{P} = \frac{P}{n_t}\boldsymbol{I} \tag{3.4.4}$$

Proof: See [TELATAR, 1999]. ■

Theorem 3.4 shows that to maximize the average capacity, the antennas should transmit uncorrelated streams with the same average power. This observation is intuitively appealing. In a more stringent setting, it also led to the concept of *unitary space-time modulation* [HOCHWALD AND MARZETTA, 2000], [HOCHWALD ET AL., 2000].

With $\boldsymbol{P}$ chosen according to (3.4.4), the resulting capacity (for a fixed $\boldsymbol{H}$) is given by the formula:

$$C_{\mathrm{UT}}(\boldsymbol{H}) = B \cdot \log_2 \left|\boldsymbol{I} + \frac{P}{n_t\sigma^2}\boldsymbol{H}\boldsymbol{H}^H\right| \tag{3.4.5}$$

The quantity $C_{\mathrm{UT}}(\boldsymbol{H})$ in (3.4.5) is relevant in the case when the transmitter does not have channel knowledge (and hence chooses $\boldsymbol{P}$ according to (3.4.4)) and it is called the *uninformed-transmitter channel capacity* in this text. We stress the difference between the informed capacity $C_{\mathrm{IT}}(\boldsymbol{H})$ and the uninformed capacity $C_{\mathrm{UT}}(\boldsymbol{H})$. The former is the capacity (for a fixed $\boldsymbol{H}$) obtained by using channel knowledge at the transmitter to optimize $\boldsymbol{P}$ via Theorem 3.3, whereas the latter is the capacity that results if $\boldsymbol{P}$ is chosen according to (3.4.4).

Example 3.3: Average MIMO channel capacity.
Assume that $\boldsymbol{H}$ has independent and Gaussian elements with unit variance and that the total (accumulated over all antennas) transmit power is equal to $P = 1$. The noise is white and Gaussian and has a power of σ^2 per antenna. Figure 3.1 shows the average capacity (i.e., $E[C(\boldsymbol{H})]$ using $\boldsymbol{P} = (1/n_t)\boldsymbol{I}$) of the system, obtained by numerical averaging over the distribution of the channel, for some different values of n_r and n_t. From this figure, we can see that the gain in capacity obtained by employing an extra receive antenna is around 3 dB relative to a SISO system ($n_t = 1, n_r = 1$); this gain can be viewed as a consequence of the fact that the extra receive antenna effectively doubles the received power. The gain of a system with ($n_t = 2, n_r = 1$) relative to the SISO system is small; as far as the average capacity is concerned there is practically no benefit in adding an extra transmit antenna to the SISO system. Finally, the capacity of a system with $n_t = 2$, $n_r = 2$ is around a factor two higher than that of the SISO system, which is a promising observation. ■

3.5 The Ratio Between IT and UT Channel Capacities

We can expect that the IT capacity given by Theorem 3.3 will be higher than the UT capacity (3.4.5), because an informed transmitter can optimize the transmit covariance matrix for a given channel. This is indeed the case, yet the IT capacity converges to the UT capacity as the SNR grows without bound. The following result (from [Bliss et al., 2002]) quantifies the relation between $C_{\mathrm{IT}}(\boldsymbol{H})$ and $C_{\mathrm{UT}}(\boldsymbol{H})$.

Theorem 3.5: The Ratio between UT and IT capacities.

(i) The informed and uninformed capacities coincide for high SNR:

$$\frac{C_{IT}(\boldsymbol{H})}{C_{UT}(\boldsymbol{H})} \to 1 \tag{3.5.1}$$

as $P/\sigma^2 \to \infty$.

(ii) For low SNR, the ratio between the IT and UT capacities behaves as:

$$\frac{C_{IT}(\boldsymbol{H})}{C_{UT}(\boldsymbol{H})} \approx n_t \cdot \frac{\lambda_{max}(\boldsymbol{H}\boldsymbol{H}^H)}{Tr\{\boldsymbol{H}\boldsymbol{H}^H\}} \tag{3.5.2}$$

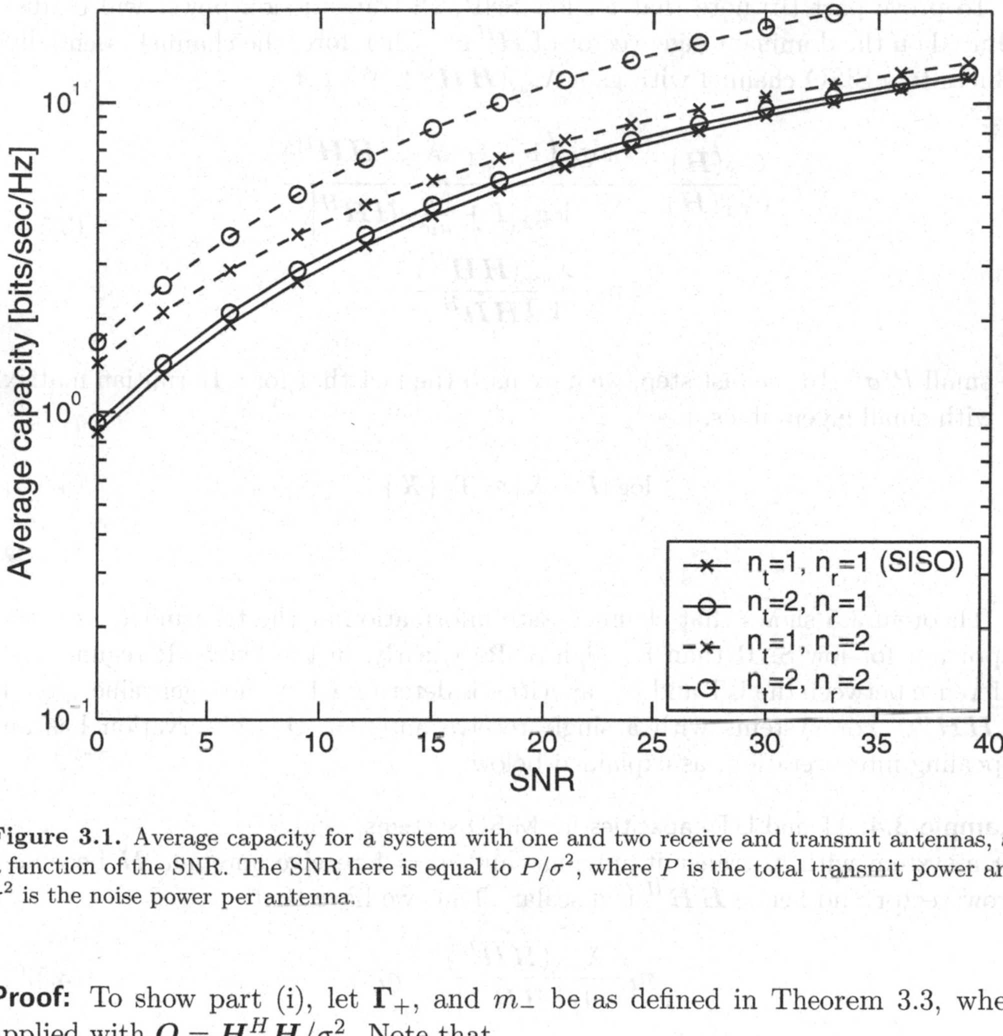

Figure 3.1. Average capacity for a system with one and two receive and transmit antennas, as a function of the SNR. The SNR here is equal to P/σ^2, where P is the total transmit power and σ^2 is the noise power per antenna.

Proof: To show part (i), let $\mathbf{\Gamma}_+$, and m_- be as defined in Theorem 3.3, when applied with $\boldsymbol{Q} = \boldsymbol{H}^H\boldsymbol{H}/\sigma^2$. Note that

$$\frac{C_{\mathrm{IT}}(\boldsymbol{H})}{C_{\mathrm{UT}}(\boldsymbol{H})} \to \frac{\log_2\left(|\mathbf{\Gamma}_+|\cdot\left(\frac{\mathrm{Tr}\{\mathbf{\Gamma}_+^{-2}\}+P}{m_-}\right)^{m_-}\right)}{\log_2\left|\boldsymbol{I}+\frac{P}{n_t}\mathbf{\Gamma}_+\right|} \to \frac{m_-\cdot\log_2\left(\frac{P}{m_-}\right)+\log_2|\mathbf{\Gamma}_+|}{m_-\cdot\log_2\left(\frac{P}{n_t}\right)+\log_2|\mathbf{\Gamma}_+|} \to 1 \tag{3.5.3}$$

when $P/\sigma^2 \to \infty$.

To prove part (ii) note that for low SNR, all transmission power will be distributed on the dominant eigenvector of $\boldsymbol{H}^H\boldsymbol{H}$. Therefore, the channel essentially reduces to a SISO channel with gain $\lambda_{\max}(\boldsymbol{H}\boldsymbol{H}^H)$. We get:

$$\begin{aligned} \frac{C_{\mathrm{IT}}(\boldsymbol{H})}{C_{\mathrm{UT}}(\boldsymbol{H})} &\to \frac{\log_2\left(1+\frac{P}{\sigma^2}\cdot\lambda_{\max}(\boldsymbol{H}\boldsymbol{H}^H)\right)}{\log_2\left|\boldsymbol{I}+\frac{P}{n_t\sigma^2}\boldsymbol{H}\boldsymbol{H}^H\right|} \\ &\approx n_t\cdot\frac{\lambda_{\max}(\boldsymbol{H}\boldsymbol{H}^H)}{\mathrm{Tr}\{\boldsymbol{H}\boldsymbol{H}^H\}} \end{aligned} \tag{3.5.4}$$

for small P/σ^2. In the last step, we have used the fact that for a Hermitian matrix $\boldsymbol{X}$ with small eigenvalues,

$$\log|\boldsymbol{I}+\boldsymbol{X}| \approx \mathrm{Tr}\{\boldsymbol{X}\} \tag{3.5.5}$$

■

Theorem 3.5 shows that channel state information at the transmitter is more important for low SNR than for high SNR. Clearly, in the low-SNR regime, the difference between the UT and IT capacities is determined by the eigenvalue spread of $\boldsymbol{H}\boldsymbol{H}^H$. For systems with a single receive antenna, this observation has an appealing interpretation, as explained below.

Example 3.4: IT and UT capacities for MISO systems.
For a system with n_t transmit antennas and $n_r = 1$ receive antenna, $\boldsymbol{H}$ becomes a row vector and hence $\boldsymbol{H}\boldsymbol{H}^H$ is a scalar. Thus we have that

$$n_t\cdot\frac{\lambda_{\max}(\boldsymbol{H}\boldsymbol{H}^H)}{\mathrm{Tr}\{\boldsymbol{H}\boldsymbol{H}^H\}} = n_t \tag{3.5.6}$$

Therefore, for such a system, the asymptotic (for low SNR) ratio between the IT and UT channel capacities is equal to n_t. Since for low values of SNR

$$\alpha\cdot\log_2\left|\boldsymbol{I}+\frac{P}{n_t\sigma^2}\boldsymbol{H}\boldsymbol{H}^H\right| \approx \log_2\left|\boldsymbol{I}+\frac{\alpha\cdot P}{n_t\sigma^2}\boldsymbol{H}\boldsymbol{H}^H\right| \tag{3.5.7}$$

the capacity of an informed transmitter can be reached by an uninformed transmitter only if the transmission power is increased by a factor n_t. This factor of n_t is closely related to what we will call "array gain" in Chapter 6. ■

3.6 Outage Capacity

Since the channel capacity is a random variable, it is meaningful to consider its statistical distribution. A particularly useful measure of its statistical behavior is the so-called *outage* capacity. The α-outage capacity C_α is the capacity in (3.2.1) that can be surpassed with probability α:

$$P(C(\boldsymbol{H}) > C_\alpha) = \alpha \tag{3.6.1}$$

The outage capacity is often a more relevant measure than the average capacity, especially for systems based on packet, or burst, transmission where the performance is specified in terms of a maximal frame-error-rate (FER), and where the channel is assumed to remain constant during the transmission of a packet. Note that the outage capacity cannot be achieved unless the length of the blocks (during which the channel remains constant) goes to infinity. By imposing delay constraints it is possible to define measures such as *delay-limited* capacity [Caire et al., 1999, Def. 5], which is a quantity that is relevant yet somewhat outside the scope of our discussion.

Example 3.5: 99%-Outage MIMO channel capacity.
Figure 3.2 shows the 99%-outage channel capacity (3.6.1), for the same systems as in Example 3.3. This capacity measure is relevant for obtaining a bound on the maximal throughput in a packet-based system that allows a packet error rate of 1%. Considering the outage capacity, a significant gain is obtained by employing an extra receive or transmit antenna (compared to the SISO system). This gain is much larger than the corresponding gain in average capacity (cf. Example 3.3). The increase in outage capacity of the ($n_t = 2, n_r = 2$) system compared to the SISO system is around four; hence the theory predicts that a fourfold increase in data rate can be possible by employing two extra antennas. ■

Example 3.6: Bound on the Frame-Error Rate.
The previous capacity results can be illustrated in a variety of ways, but a particularly interesting comparison is obtained when the outage probability is plotted as a function of SNR for a given rate. For a system transmitting with a higher rate than the channel capacity, error-free communication is impossible. Hence, for a system with unity bandwidth transmitting packets with a bit rate R, the probability of a frame, or burst error, can be lower bounded as follows:

$$P(\text{frame error}) \geq P(C(\boldsymbol{H}) < R) = P\Big(\log_2 \Big|\boldsymbol{I} + \frac{P}{n_t\sigma^2}\boldsymbol{H}\boldsymbol{H}^H\Big| < R\Big) \tag{3.6.2}$$

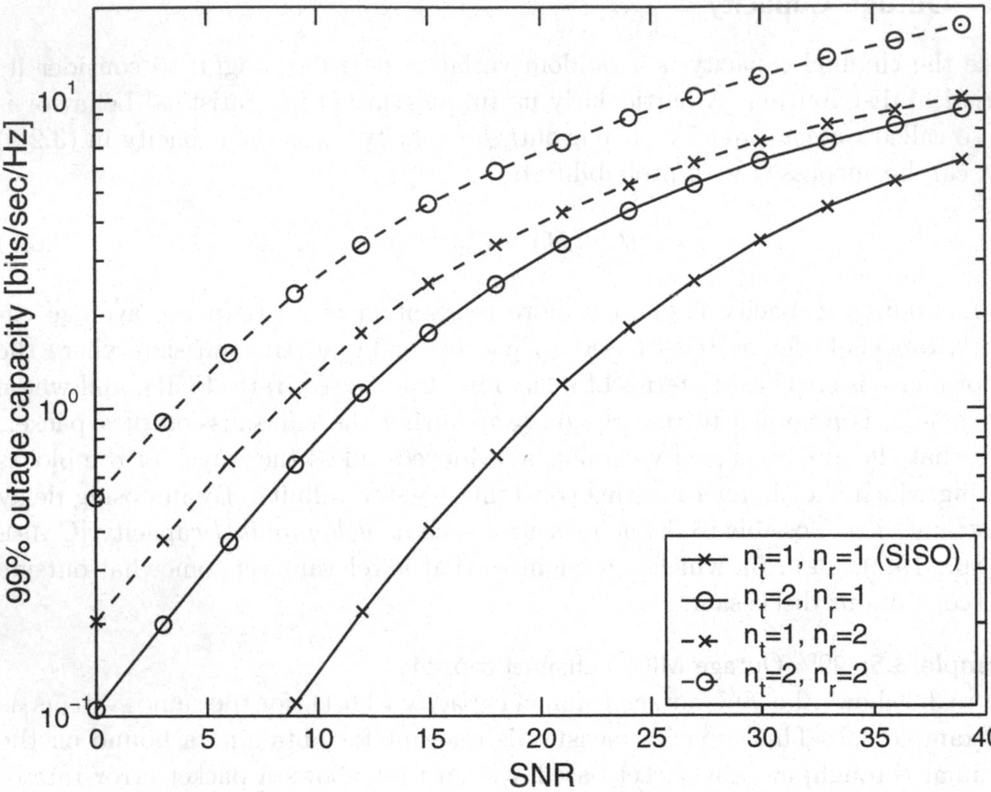

Figure 3.2. 99%-outage capacity for a system with one or two receive and transmit antennas, as a function of the SNR.

Figure 3.3 shows the lower bound on the frame-error-rate obtained via (3.6.2) for some different numbers of transmit and receive antennas, and for a rate of $R = 2$ bits/s/Hz. ■

3.7 Summary and Discussion

The goal of this chapter has been to quantify the promise of the MIMO channel from an information theoretic point of view. Following the introductory discussion in Section 3.1, we provided a heuristic motivation for the channel capacity formula for a constant (i.e., time-invariant) and frequency-flat MIMO channel $\boldsymbol{H}$ with

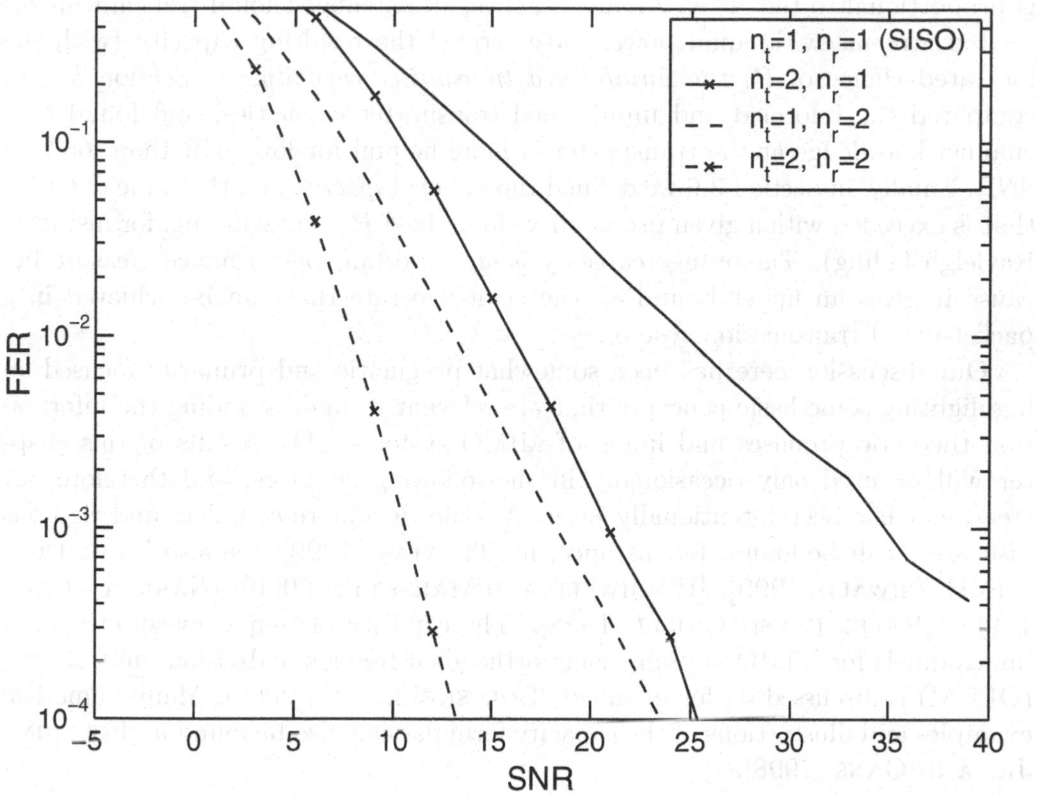

Figure 3.3. Lower bound on the frame-error-rate for some different numbers of antennas and for a rate of $R = 2$ bits/s/Hz.

bandwidth B and subject to additive white Gaussian noise with variance σ^2:

$$C(\boldsymbol{H}) = B \cdot \log_2 \left| \boldsymbol{I} + \frac{1}{\sigma^2} \boldsymbol{HPH}^H \right| \tag{3.7.1}$$

where $\boldsymbol{P}$ is the covariance matrix of the transmitted signals (see Section 3.2). Next, in Section 3.3 we observed that for a fixed $\boldsymbol{H}$, and with channel knowledge at the transmitter, $\boldsymbol{P}$ can be chosen to maximize $C(\boldsymbol{H})$ via a result referred to as water-filling (see Theorem 3.3 on page 27); we called the resulting capacity the *informed transmitter capacity.* On the other hand, for fading (i.e., randomly time-varying) channels, $C(\boldsymbol{H})$ is a random variable. For this case, we stated in Section 3.4 without proof the result that in Rayleigh fading and in the absence of channel knowledge at the transmitter, the matrix $\boldsymbol{P}$ that maximizes $E_{\boldsymbol{H}}[C(\boldsymbol{H})]$

is proportional to the identity matrix; hence, all antennas should transmit uncorrelated streams with equal power. We termed the resulting capacity (with the indicated choice of $\boldsymbol{P}$) the *uninformed transmitter capacity.* In Section 3.5 we compared the informed and uninformed transmitter capacities, and found that channel knowledge at the transmitter is more helpful for low SNR than for high SNR. Finally, in Section 3.6 we defined the *outage capacity,* i.e., the value of $C(\boldsymbol{H})$ that is exceeded with a given probability (for a fixed $\boldsymbol{P}$ and assuming, for instance, Rayleigh fading). The outage capacity is an important performance measure because it gives an upper bound on the error-free rate that can be achieved in a packet-based transmission system.

Our discussion here has been somewhat pragmatic and primarily focused on highlighting some basic concepts that are relevant to understanding the information theoretic promises and limits of MIMO systems. The results of this chapter will be used only occasionally in the following chapters, and therefore our treatment has been intentionally brief. A somewhat more complete and rigorous discussion can be found, for instance, in [TELATAR, 1999]; see also [MARZETTA AND HOCHWALD, 1999], [HOCHWALD AND MARZETTA, 2000], [NARULA ET AL., 1999], [RALEIGH AND CIOFFI, 1998]. The capacity of frequency-selective fading channels for MIMO systems using orthogonal frequency division multiplexing (OFDM) is discussed in, for instance, [BÖLCSKEI ET AL., 2002]. Many numerical examples and illustrations of the capacity formulas can also be found in [FOSCHINI, JR. AND GANS, 1998].

3.8 Proofs

Proof of Theorem 3.1

Let $\boldsymbol{x}_G$ be a real-valued Gaussian random vector with zero mean and covariance matrix $\boldsymbol{P}$ and let $p_{\boldsymbol{x}_G}(\boldsymbol{x})$ be its p.d.f. Also, let $\boldsymbol{x}$ be any other random vector with zero mean and covariance matrix $\boldsymbol{P}$ and denote its p.d.f. by $p_{\boldsymbol{x}}(\boldsymbol{x})$. We want to show that

$$H(\boldsymbol{x}) \leq H(\boldsymbol{x}_G) \tag{3.8.1}$$

Since by assumption, $\boldsymbol{x}$ and $\boldsymbol{x}_G$ have the same first and second order moments, we have that

$$\int d\boldsymbol{x}\ p_{\boldsymbol{x}_G}(\boldsymbol{x}) x_k x_l = \int d\boldsymbol{x}\ p_{\boldsymbol{x}}(\boldsymbol{x}) x_k x_l \tag{3.8.2}$$

for any k, l. Next, note that

$$p_{\boldsymbol{x}_G}(\boldsymbol{x}) = \frac{1}{(2\pi)^{n/2} \cdot |\boldsymbol{P}|^{1/2}} \exp\Big(-\frac{1}{2}\boldsymbol{x}^T \boldsymbol{P}^{-1}\boldsymbol{x}\Big) \tag{3.8.3}$$

Hence $\log_2\big(p_{\boldsymbol{x}_G}(\boldsymbol{x})\big)$ is a quadratic function of $\boldsymbol{x}$. It follows from (3.8.2) that

$$\int d\boldsymbol{x}\; p_{\boldsymbol{x}_G}(\boldsymbol{x}) \log_2\left(p_{\boldsymbol{x}_G}(\boldsymbol{x})\right) = \int d\boldsymbol{x}\; p_{\boldsymbol{x}}(\boldsymbol{x}) \log_2\left(p_{\boldsymbol{x}_G}(\boldsymbol{x})\right) \tag{3.8.4}$$

Therefore,

$$\begin{aligned} H(\boldsymbol{x}) - H(\boldsymbol{x}_G) &= \int d\boldsymbol{x}\; p_{\boldsymbol{x}_G}(\boldsymbol{x}) \log_2\left(p_{\boldsymbol{x}_G}(\boldsymbol{x})\right) - \int d\boldsymbol{x}\; p_{\boldsymbol{x}}(\boldsymbol{x}) \log_2\left(p_{\boldsymbol{x}}(\boldsymbol{x})\right) \\ &= \int d\boldsymbol{x}\; p_{\boldsymbol{x}}(\boldsymbol{x}) \log_2\left(p_{\boldsymbol{x}_G}(\boldsymbol{x})\right) - \int d\boldsymbol{x}\; p_{\boldsymbol{x}}(\boldsymbol{x}) \log_2\left(p_{\boldsymbol{x}}(\boldsymbol{x})\right) \\ &= \int d\boldsymbol{x}\; p_{\boldsymbol{x}}(\boldsymbol{x}) \log_2\left(\frac{p_{\boldsymbol{x}_G}(\boldsymbol{x})}{p_{\boldsymbol{x}}(\boldsymbol{x})}\right) \end{aligned} \tag{3.8.5}$$

By Jensen's inequality [Cover and Thomas, 1991, Th. 2.6.2] and the concavity of the logarithm it follows that

$$\begin{aligned} \int d\boldsymbol{x}\; p_{\boldsymbol{x}}(\boldsymbol{x}) \log_2\left(\frac{p_{\boldsymbol{x}_G}(\boldsymbol{x})}{p_{\boldsymbol{x}}(\boldsymbol{x})}\right) &\leq \log_2\left(\int d\boldsymbol{x}\; p_{\boldsymbol{x}}(\boldsymbol{x}) \cdot \frac{p_{\boldsymbol{x}_G}(\boldsymbol{x})}{p_{\boldsymbol{x}}(\boldsymbol{x})}\right) \\ &= \log_2\left(\int d\boldsymbol{x}\; p_{\boldsymbol{x}_G}(\boldsymbol{x})\right) \\ &= \log_2(1) = 0 \end{aligned} \tag{3.8.6}$$

which proves (3.8.1).

Proof of Theorem 3.3

First, let us define $\boldsymbol{\Phi} = \boldsymbol{U}^H\boldsymbol{P}\boldsymbol{U}$ as a general (not necessarily diagonal) Hermitian matrix; we will shortly show that the optimum matrix $\boldsymbol{P}$ is such that the above $\boldsymbol{\Phi}$ is diagonal. With this notation and the definitions made in the theorem, it holds (by the Hadamard inequality [Horn and Johnson, 1985, Th. 7.8.1]) that

$$f(\boldsymbol{P}) = \left| \boldsymbol{I} + [\boldsymbol{\Gamma}_+ \;\; \boldsymbol{0}]\boldsymbol{U}^H\boldsymbol{P}\boldsymbol{U}\begin{bmatrix}\boldsymbol{\Gamma}_+ \\ \boldsymbol{0}\end{bmatrix} \right| \leq \prod_{k=1}^{m_+}(1 + \gamma_k^2 \Phi_{k,k}) \tag{3.8.7}$$

The equality in (3.8.7) is achieved if $\boldsymbol{\Phi}$ is diagonal as in (3.3.7), which we assume in what follows. Consequently:

$$f(\boldsymbol{P}) = \prod_{k=1}^{m_+} \gamma_k^2 \cdot \prod_{k=1}^{m_+} \left(\frac{1}{\gamma_k^2} + \Phi_{k,k} \right) \tag{3.8.8}$$

Next we note that, depending on the value of P, not all $\{\Phi_{k,k}\}_{k=1}^{m_+}$ can be strictly positive, i.e., some $\Phi_{k,k} = 0$. As $\gamma_1^2 \geq \gamma_2^2 \geq \cdots$, it is evident from (3.8.7) and the constraint

$$\operatorname{Tr}\{\boldsymbol{P}\} = \sum_{k=1}^{m_+} \Phi_{k,k} \leq P \tag{3.8.9}$$

that if some $\Phi_{k,k}$ corresponding to the optimal solution are zero, then they must be $\Phi_{m_+,m_+}, \Phi_{m_++1,m_++1}, \ldots$ and not $\Phi_{1,1}, \ldots$. Let m_- be the number of $\Phi_{k,k} > 0$; the value of m_- is yet to be determined, see below. Then it follows from the previous observation and (3.8.8) that for the $\boldsymbol{\Phi}$ in (3.3.7) we get:

$$\begin{aligned} f(\boldsymbol{P}) &= \prod_{k=1}^{m_+} \gamma_k^2 \cdot \prod_{k=1}^{m_-} \left(\frac{1}{\gamma_k^2} + \Phi_{k,k} \right) \cdot \prod_{k=m_-+1}^{m_+} \frac{1}{\gamma_k^2} \\ &\leq |\boldsymbol{\Gamma}_-^2| \cdot \left(\frac{1}{m_-} \sum_{k=1}^{m_-} \left(\frac{1}{\gamma_k^2} + \Phi_{k,k} \right) \right)^{m_-} \\ &\leq |\boldsymbol{\Gamma}_-^2| \cdot \left(\frac{\operatorname{Tr}\{\boldsymbol{\Gamma}_-^{-2}\} + P}{m_-} \right)^{m_-} \end{aligned} \tag{3.8.10}$$

The first inequality above follows from the relation between the arithmetic and geometric means; the equalities in (3.8.10) hold if

$$\frac{1}{\gamma_k^2} + \Phi_{k,k} = \frac{\operatorname{Tr}\{\boldsymbol{\Gamma}_-^{-2}\} + P}{m_-} \tag{3.8.11}$$

for $k = 1, \ldots, m_-$ which proves the theorem. Note that (3.3.8) *implicitly* defines m_-.

3.9 Problems

1. Prove Equations (3.1.4), (3.1.6) and (3.1.16).

Hint: If $\boldsymbol{x}$ is an n-dimensional complex Gaussian vector with zero mean and covariance matrix $\boldsymbol{Q}$, then

$$\begin{aligned}\int d\boldsymbol{x}\; p_{\boldsymbol{x}}(\boldsymbol{x})\boldsymbol{x}^H\boldsymbol{Q}^{-1}\boldsymbol{x} =& \operatorname{Tr}\left\{\boldsymbol{Q}^{-1}\cdot\int d\boldsymbol{x}\; \boldsymbol{x}\boldsymbol{x}^H p_{\boldsymbol{x}}(\boldsymbol{x})\right\} \\ =& \operatorname{Tr}\left\{\boldsymbol{Q}^{-1}\boldsymbol{Q}\right\} = n\end{aligned} \tag{3.9.1}$$

2. Prove (3.5.5).

Hint: use an eigen-decomposition of $\boldsymbol{X}$ along with a Taylor series approximation.

3. Write a computer program that reproduces Figures 3.1 and 3.2. Also plot the corresponding curves for larger values of n_r and n_t. How do the average and outage capacities behave as functions of n_r and n_t for a fixed SNR?

4. Consider a scenario with two receive and two transmit antennas and where the transmit and receive correlation matrices (see Section 2.1.2) are given by

$$\boldsymbol{R}_t = \begin{bmatrix} 1 & \rho_t \\ \rho_t & 1 \end{bmatrix} \tag{3.9.2}$$

and

$$\boldsymbol{R}_r = \begin{bmatrix} 1 & \rho_r \\ \rho_r & 1 \end{bmatrix} \tag{3.9.3}$$

Use Monte-Carlo simulation to compute and plot the corresponding 99%−outage capacity versus ρ_t and ρ_r for some different SNR. Discuss the result.

5. Plot the IT capacity and the UT capacity for some different numbers of antennas and for some different SNR. How do the IT and UT capacities behave for low and high SNR, respectively? Discuss the result.

Chapter 4

ERROR PROBABILITY ANALYSIS

In the high-SNR region, the average probability of a detection error (assuming maximum-likelihood detection or maximum-ratio combining) for communication over a fading channel usually behaves as:

$$P(\text{error}) \sim G_c \cdot \text{SNR}^{-G_d}$$

The constant G_c is referred to as the *coding advantage* and G_d is called the *diversity order* of the system. If the error rate is plotted versus the SNR on a log-log scale the diversity order can be interpreted as the slope of the so-obtained curve, whereas the coding advantage corresponds to the horizontal position of the curve. This chapter presents some results on error performance analysis for SISO and MIMO systems, that will be essential for the analysis of different transmission schemes later in the book. Like for Chapter 3, some of the material in this chapter is of a fairly mathematical nature, and therefore the time reader may wish to read the material briefly before studying its details. While this chapter focuses on the analysis of the detection error probability, Chapter 9 will discuss practical detection algorithms.

4.1 Error Probability Analysis for SISO Channels

We start by studying the average probability of a detection error for a SISO channel. Our treatment is quite brief; for a more in-depth discussion the reader is referred to, for instance, [Simon and Alouini, 2000] which is an extensive source of exact and approximate error probability expressions.

Assume that a symbol s_0, taken from a finite constellation $\mathcal{S}$, is transmitted, and suppose that we observe:

$$y = hs_0 + e \tag{4.1.1}$$

where h is a complex number and e is zero-mean complex Gaussian noise with variance σ^2. In general, (4.1.1) arises in the analysis of a SISO channel, but it is also relevant in the context of *beamforming* (see Chapter 6) and when several received signals are combined coherently in a ML or maximum-SNR sense. Suppose that we employ the ML detection rule to detect s_0:

$$\hat{s} = \underset{s \in \mathcal{S}}{\operatorname{argmin}} |y - hs| \tag{4.1.2}$$

Let ρ_b^2 be the transmitted energy per information bit. Hence if the constellation $\mathcal{S}$ has M elements, the average transmitted energy per information-bearing symbol is

$$\rho_s^2 = E[|s_0|^2] = \log_2(M) \cdot \rho_b^2 \tag{4.1.3}$$

For many signalling schemes, the SER (P_s) or BER (P_b) can be bounded or exactly expressed via an expression that contains the Gaussian Q-function [Proakis, 2001, Ch. 5]:

$$P_0 = c \cdot Q\left(\sqrt{g \cdot \rho_b^2 \frac{|h|^2}{\sigma^2}}\right) \tag{4.1.4}$$

where c and g are constants and $\rho_b^2|h|^2/\sigma^2$ is a measure of the SNR. Typically, the value of g is related to the minimum distance in the constellation, and c is related to the number of constellation points that achieve this minimum distance.

Binary Phase Shift Keying (BPSK): For BPSK, the SER is equal to the BER and can be found exactly from (4.1.4) as: $P_b = P_s = P_0$ with $c = 1$ and $g = 2$.

Quadrature Phase Shift Keying (QPSK): For QPSK with Gray coding, the exact BER can be found as $P_b = P_0$ with $c = 1$ and $g = 2$; an expression for the associated SER is derived in [Simon and Alouini, 2000, Sec. 8.1.1.3]. As is well-known, QPSK doubles the bit rate of BPSK at the same BER and transmitted energy per bit.

M-ary Quadrature Amplitude Modulation (M-QAM): For M-QAM signalling with an average transmitted power equal to one, a tight bound of the SER is $P_s \leq P_0$ with $c = 4$ and $g = 3/(M-1)$. An accurate (for high SNR) and often useful approximation of the BER is obtained [Simon and Alouini, 2000, Sec. 8.1.1.2] by setting

$$\begin{aligned} c &= 4\frac{\sqrt{M}-1}{\sqrt{M} \cdot \log_2 M} \\ g &= \frac{3}{M-1} \log_2 M \end{aligned} \tag{4.1.5}$$

in (4.1.4).

M-ary Phase Shift Keying (M-PSK): For M-PSK and high SNR, the BER can be approximated by (4.1.4) with $c = 2/\log_2 M$ and $g = 2\sin^2(\pi/M)\log_2 M$; see [Simon and Alouini, 2000, Sec. 8.1.1.3].

Better approximations than those listed above, as well as other related useful expressions, can be found in the cited book [Simon and Alouini, 2000]. However, for our purpose the stated expressions are sufficient.

For transmission over an AWGN channel, the received complex sample is a realization of (4.1.1) where h is a constant channel gain. For a fading channel, h is a random variable (for instance, for a Rayleigh fading channel, h is zero-mean complex Gaussian), and it is useful to compute the BER or SER averaged over h. We next present a theorem that is useful towards this end.

Theorem 4.1: Averaging the Gaussian Q-function.
Consider, for a fixed h, the general error probability in (4.1.4):

$$P_0 = c \cdot Q\left(\sqrt{g \cdot \frac{\gamma^2}{\sigma^2}}\right) \tag{4.1.6}$$

where c and g are constants, and $\gamma^2 = |h|^2 \rho_b^2$. Suppose that h is a random variable such that γ^2 can be written as:

$$\gamma^2 = \boldsymbol{h}^H \boldsymbol{h} = \|\boldsymbol{h}\|^2 \tag{4.1.7}$$

where

$$\boldsymbol{h} \sim N_C(\boldsymbol{0}, \boldsymbol{\Upsilon}) \tag{4.1.8}$$

and $\boldsymbol{\Upsilon}$ is a positive semi-definite matrix. Let m be the dimension of $\boldsymbol{\Upsilon}$ and let $n = \text{rank}\{\boldsymbol{\Upsilon}\}$. Also, let $\{\varepsilon_1, \ldots, \varepsilon_n\}$ be the nonzero eigenvalues of $\boldsymbol{\Upsilon}$. Then the following statements hold.

(i) The error probability, averaged over $\boldsymbol{h}$, is bounded by

$$E_{\boldsymbol{h}}[P_0] \le c \cdot \left|\boldsymbol{I} + \frac{g}{2\sigma^2}\boldsymbol{\Upsilon}\right|^{-1} \le c \cdot \left(\frac{g}{2\sigma^2}\right)^{-n} \cdot \prod_{k=1}^{n} \varepsilon_k^{-1} \tag{4.1.9}$$

(ii) Suppose that all nonzero eigenvalues of $\boldsymbol{\Upsilon}$ are equal to a constant ρ^2:

$$\varepsilon_k = \frac{Tr\{\boldsymbol{\Upsilon}\}}{n} \triangleq \rho^2, \quad k = 1, \ldots, n \tag{4.1.10}$$

Then the average of (4.1.6) can be computed exactly:

$$\begin{aligned} E_{\boldsymbol{h}}[P_0] =& c \cdot 2^{-n} \cdot \left(1 - \sqrt{\frac{g\rho^2}{2\sigma^2 + g\rho^2}}\right)^n \\ & \cdot \sum_{k=0}^{n-1} \left[2^{-k} \binom{n-1+k}{k} \cdot \left(1 + \sqrt{\frac{g\rho^2}{2\sigma^2 + g\rho^2}}\right)^k \right] \end{aligned} \tag{4.1.11}$$

Proof: See Section 4.4. ■

The reason behind the model for γ^2 in (4.1.7) will become clear in later chapters when we study the performance of receive and transmit diversity systems. However, already at this point we should note that the SNR for the system studied in the introductory example in Section 1.3 takes on the form (4.1.7); hence Theorem 4.1 can be applied to study the BER performance of this system.

Note from part (i) of Theorem 4.1 that the slope of the average BER curve, i.e., the diversity order, is equal to n, whereas the relative shift of this curve is determined by the eigenvalues of $\boldsymbol{\Upsilon}$. Also note that to obtain a more convenient formula than (4.1.11) in part (ii) of Theorem 4.1, the equation referred to can be approximated by

$$E_{\boldsymbol{h}}[P_0] \approx c \cdot (2g)^{-n} \binom{2n-1}{n} \left(\frac{\rho^2}{\sigma^2}\right)^{-n} \tag{4.1.12}$$

In practice, for example when applied to QPSK, this approximation is quite accurate; in particular, it indicates that the diversity order is equal to n and hence it reinforces part (i) of the theorem.

4.2 Error Probability Analysis for MIMO Channels

We saw in Chapter 2 that both the flat and the frequency-selective MIMO channel exhibit a matrix-algebraic representation of the following general form:

$$\boldsymbol{y} = \boldsymbol{X}\boldsymbol{h} + \boldsymbol{e} \tag{4.2.1}$$

where the transmitted data matrix $\boldsymbol{X}$ is of dimension $r \times m$ and belongs to a finite constellation of (in general non-square) matrices $\mathcal{X}$. The model (4.2.1) is applicable both to the analysis of a frequency flat MIMO channel (in which case $r = Nn_r$ and $m = n_r n_t$; cf. (2.1.8)) and a frequency selective MIMO channel (in which case $r = n_r(N_0 + L)$ and $m = n_r n_t(L+1)$; cf. (2.2.13)).

4.2.1 Pairwise Error Probability and Union Bound

Most of the results to follow are concerned with the so-called pairwise error probability, i.e., the probability that a certain "true" transmitted code matrix $\boldsymbol{X}_0$ is mistaken for a different matrix $\boldsymbol{X} \neq \boldsymbol{X}_0$ at the receiver, $P(\boldsymbol{X}_0 \to \boldsymbol{X})$ (assuming that these two matrices are the only two matrices in the code). Clearly, this error measure depends on which particular matrices $\boldsymbol{X}$ and $\boldsymbol{X}_0$ one chooses to study. In general, the system performance is dominated by the error events where $\boldsymbol{X}$ and $\boldsymbol{X}_0$ are close (measured in some appropriate metric) and hence the pairwise error probability is usually a meaningful quantity.

An upper bound on the *total* error probability of the system can be obtained via the *union bound* [PROAKIS, 2001, SEC. 5.2]. Assuming that all code matrices $\boldsymbol{X}$ are equally likely to be transmitted, the average error rate is upper bounded by

$$P_{\text{tot}} \leq \frac{1}{|\mathcal{X}|} \sum_{\boldsymbol{X}_n \in \mathcal{X}} \sum_{\boldsymbol{X}_k \in \mathcal{X}, k \neq n} P(\boldsymbol{X}_n \to \boldsymbol{X}_k) \tag{4.2.2}$$

where $P(\boldsymbol{X}_n \to \boldsymbol{X}_k)$ denotes the probability that a transmitted code matrix $\boldsymbol{X}_n$ is mistaken for a different code matrix $\boldsymbol{X}_k$, and $|\mathcal{X}|$ is the number of elements in the matrix constellation $\mathcal{X}$.

4.2.2 Coherent Maximum-Likelihood Detection

We next study the error probability for coherent maximum-likelihood (ML) detection, i.e., assuming that $\boldsymbol{h}$ is known to the receiver. Under the assumption that $\boldsymbol{h}$ is known and that the elements of $\boldsymbol{e}$ are independent zero-mean Gaussian random variables with variance σ^2, $\boldsymbol{y}$ is a complex Gaussian random vector with mean $E[\boldsymbol{y}] = \boldsymbol{X}\boldsymbol{h}$ and covariance matrix

$$E\left[(\boldsymbol{y} - \boldsymbol{X}\boldsymbol{h})(\boldsymbol{y} - \boldsymbol{X}\boldsymbol{h})^H\right] = \sigma^2 \boldsymbol{I} \tag{4.2.3}$$

It follows that the likelihood function of the received data $\boldsymbol{y}$, conditioned on the transmitted matrix $\boldsymbol{X}$, can be written:

$$p(\boldsymbol{y}|\boldsymbol{X}; \boldsymbol{h}, \sigma^2) = \pi^{-r}\sigma^{-2r} \exp\left(-\frac{1}{\sigma^2}\|\boldsymbol{y} - \boldsymbol{X}\boldsymbol{h}\|^2\right) \tag{4.2.4}$$

where r is the dimension of $\boldsymbol{y}$. Hence, detecting $\boldsymbol{X}$ in an ML sense given $\boldsymbol{y}$, which amounts to maximizing the conditional probability density function (or likelihood function) in (4.2.4), is equivalent to minimizing the following *ML metric*:

$$\hat{\boldsymbol{X}} = \underset{\boldsymbol{X} \in \mathcal{X}}{\operatorname{argmin}} \|\boldsymbol{y} - \boldsymbol{X}\boldsymbol{h}\|^2 \tag{4.2.5}$$

The following theorem gives an upper bound on the probability that the coherent ML detector in (4.2.5) takes an incorrect decision.

Theorem 4.2: Pairwise error probability of coherent detection for MIMO channel. *Let us study the linear model in (4.2.1), where* $\boldsymbol{X}$ *is of dimension* $r \times m$, *and consider event that the ML detector (4.2.5) decides for a matrix* $\boldsymbol{X} \neq \boldsymbol{X}_0$ *in favor of* $\boldsymbol{X}_0$. *Then the following results hold.*

(i) For a given $\boldsymbol{h}$, *the probability of error is*

$$P(\boldsymbol{X}_0 \to \boldsymbol{X}) = Q\left(\sqrt{\frac{\|(\boldsymbol{X}_0 - \boldsymbol{X})\boldsymbol{h}\|^2}{2\sigma^2}}\right) \tag{4.2.6}$$

(ii) Suppose $\boldsymbol{h}$ *is random as follows:*

$$\boldsymbol{h} \sim N_C(\boldsymbol{0}, \boldsymbol{\Upsilon}) \tag{4.2.7}$$

where $\boldsymbol{\Upsilon}$ *is an* $m \times m$ *positive semi-definite matrix, and let*

$$n = \operatorname{rank}\{(\boldsymbol{X}_0 - \boldsymbol{X})\boldsymbol{\Upsilon}(\boldsymbol{X}_0 - \boldsymbol{X})^H\} \tag{4.2.8}$$

Then the probability of an incorrect decision (averaged over $\boldsymbol{h}$*) can be bounded as follows:*

$$\begin{aligned} E_{\boldsymbol{h}}[P(\boldsymbol{X}_0 \to \boldsymbol{X})] &\leq \left|\boldsymbol{I} + \frac{1}{4\sigma^2}(\boldsymbol{X}_0 - \boldsymbol{X})\boldsymbol{\Upsilon}(\boldsymbol{X}_0 - \boldsymbol{X})^H\right|^{-1} \\ &\leq \left(\frac{1}{4\sigma^2}\right)^{-n} \cdot \prod_{k=1}^{n} \varepsilon_k^{-1} \end{aligned} \tag{4.2.9}$$

where $\{\varepsilon_1, \ldots, \varepsilon_n\}$ *are the nonzero eigenvalues of* $(\boldsymbol{X}_0 - \boldsymbol{X})\boldsymbol{\Upsilon}(\boldsymbol{X}_0 - \boldsymbol{X})^H$.

Proof: See Section 4.4. ■

It follows from Theorem 4.2 that the slope of the BER curve, which corresponds to the diversity order, is at least

$$G_d = n \tag{4.2.10}$$

and that the coding gain is essentially determined by the product of the nonzero eigenvalues of $(\boldsymbol{X}_0 - \boldsymbol{X})\boldsymbol{\Upsilon}(\boldsymbol{X}_0 - \boldsymbol{X})^H$. Note also that in the special case of

$\boldsymbol{\Upsilon} = \rho^2 \boldsymbol{I}$, the average error probability is bounded by (provided that $r > m$, i.e., that $\boldsymbol{X}$ is a tall matrix):

$$E_{\boldsymbol{h}}[P(\boldsymbol{X}_0 \to \boldsymbol{X})] \leq \left|(\boldsymbol{X}_0 - \boldsymbol{X})^H (\boldsymbol{X}_0 - \boldsymbol{X})\right|^{-1} \cdot \left(\frac{\rho^2}{4\sigma^2}\right)^{-n} \tag{4.2.11}$$

and hence a diversity order equal to n can be achieved.

From Theorem 4.2 we can also see that to minimize the probability of an error, given $\boldsymbol{h}$, we should choose $\boldsymbol{X}$ and $\boldsymbol{X}_0$ that maximize the Euclidean distance $\|(\boldsymbol{X}_0 - \boldsymbol{X})\boldsymbol{h}\|^2$. This is an intuitively appealing result. On the other hand, to minimize the *average* error probability when $\boldsymbol{h}$ is random we shall maximize the determinant of the matrix $(\boldsymbol{X}_0 - \boldsymbol{X})\boldsymbol{\Upsilon}(\boldsymbol{X}_0 - \boldsymbol{X})^H$. If $\boldsymbol{\Upsilon} = \rho^2 \boldsymbol{I}$, this determinant is proportional to the product of the eigenvalues of the matrix $(\boldsymbol{X}_0 - \boldsymbol{X})^H(\boldsymbol{X}_0 - \boldsymbol{X})$; such a product is sometimes referred to as *product distance.*

The following example illustrates some subtle aspects of diversity as a performance measure. As we will see, even though the diversity gain is a well-defined quantity, it has to be carefully interpreted.

Example 4.1: Diversity and Under-Determined Equation Systems.
Consider a system with $n_t = 2$ transmit antennas and $n_r = 1$ receive antenna, and assume that we transmit a complex symbol s_1 via the first antenna and at the same time another complex symbol s_2 via the second antenna. If $\boldsymbol{h}^T = [h_1 \ h_2]$ is a vector that contains the two channel gains between the transmit and receive antennas, the receiver observes the following scalar signal:

$$y = \boldsymbol{h}^T \boldsymbol{s} + e = \boldsymbol{s}^T \boldsymbol{h} + e \tag{4.2.12}$$

where $\boldsymbol{s} = [s_1 \ s_2]^T$ and e is noise.

Assume that h_1 and h_2 are independent zero-mean complex Gaussian random variables with covariance matrix $\boldsymbol{\Upsilon} = \rho^2 \boldsymbol{I}$. Clearly, the system is under-determined in the sense that we have one scalar measurement (y) and two scalar unknowns (s_1 and s_2). However, if s_1 and s_2 belong to a certain finite constellation $\mathcal{S}$, application of Theorem 4.2 shows that minimizing the ML metric

$$|y - h_1 s_1 - h_2 s_2|^2 \tag{4.2.13}$$

with respect to $\{s_1, s_2\} \in \mathcal{S}$ gives an average error rate that is bounded by:

$$E_{\boldsymbol{h}}\big[P((s_1^0, s_2^0) \to (s_1, s_2))\big] \leq \left| 1 + \left(\frac{\rho^2}{4\sigma^2}\right) \cdot \begin{bmatrix} s_1^0 - s_1 & s_2^0 - s_2 \end{bmatrix} \begin{bmatrix} (s_1^0 - s_1)^* \\ (s_2^0 - s_2)^* \end{bmatrix} \right|^{-1}$$
$$\leq \left(\frac{\rho^2}{4\sigma^2}\right)^{-1} \left(|s_1^0 - s_1|^2 + |s_2^0 - s_2|^2 \right)^{-1} \tag{4.2.14}$$

Hence the probability of making an error in detecting $\boldsymbol{s}$ goes to zero as σ^2 when $\sigma^2 \to 0$ (i.e., when the SNR goes to infinity). This shows that *even though the system is under-determined we can detect both symbols consistently as SNR $\to \infty$.* However, since given one sample we must determine *two* symbols (compare with a conventional system where one sample is typically used to determine *one* symbol only), we can expect the BER performance for a finite SNR (or the coding gain) to be poor. Of course the fact that the transmitted symbols belong to a finite constellation is essential in arriving at the conclusion that the detection rule is consistent as SNR $\to \infty$. ■

The next example shows how diversity is related to the concept of *observability*.

Example 4.2: Diversity and "observability."
We say that a code is observable (or identifiable) if, in the noise-free case, two distinct transmitted codewords (of dimension $r \times m$, where $r \geq m$) $\boldsymbol{X}_1$ and $\boldsymbol{X}_2$ give rise to two distinct received codewords as long as the channel vector $\boldsymbol{h}$ has at least one nonzero element. Interestingly enough, a code is observable exactly when it provides maximal diversity: $\boldsymbol{X}_1\boldsymbol{h} \neq \boldsymbol{X}_2\boldsymbol{h}$ for all $\boldsymbol{h} \neq \boldsymbol{0}$ if and only if the determinant of $(\boldsymbol{X}_1 - \boldsymbol{X}_2)^H(\boldsymbol{X}_1 - \boldsymbol{X}_2)$ is nonzero. To see this, note first that if $(\boldsymbol{X}_1 - \boldsymbol{X}_2)^H(\boldsymbol{X}_1 - \boldsymbol{X}_2)$ is nonsingular, then $\boldsymbol{X}_1\boldsymbol{h} \neq \boldsymbol{X}_2\boldsymbol{h}$ for all $\boldsymbol{h} \neq \boldsymbol{0}$ and hence the code is observable. Conversely, if $(\boldsymbol{X}_1 - \boldsymbol{X}_2)^H(\boldsymbol{X}_1 - \boldsymbol{X}_2)$ is singular, then for any vector $\boldsymbol{h}$ in the null space of this matrix we have $\boldsymbol{X}_1\boldsymbol{h} = \boldsymbol{X}_2\boldsymbol{h}$ and the code is not observable. ■

If we consider the model for flat MIMO channels in (2.1.5) instead of the (more general) linear model (4.2.1), we can specialize the error bound for coherent detection and obtain a very useful formula. This is illustrated in the next example.

Example 4.3: Error bounds for flat MIMO channels.
Study the model for a flat MIMO channel (2.1.5):

$$\boldsymbol{Y} = \boldsymbol{H}\boldsymbol{X} + \boldsymbol{E} \tag{4.2.15}$$

By rewriting (4.2.15) on the form (2.1.8):

$$y = \mathrm{vec}(\boldsymbol{Y}) = \mathrm{vec}(\boldsymbol{H}\boldsymbol{X} + \boldsymbol{E}) = (\boldsymbol{X}^T \otimes \boldsymbol{I})\boldsymbol{h} + \boldsymbol{e} \tag{4.2.16}$$

with obvious implicit definitions of $\boldsymbol{h}$ and $\boldsymbol{e}$, we can use Theorem 4.2(a) to bound the error rate. We get:

$$P(\boldsymbol{X}_0 \to \boldsymbol{X}) = Q\left(\sqrt{\frac{\|(\boldsymbol{X}_0^T \otimes \boldsymbol{I} - \boldsymbol{X}^T \otimes \boldsymbol{I})\boldsymbol{h}\|^2}{2\sigma^2}}\right) = Q\left(\sqrt{\frac{\|\boldsymbol{H}(\boldsymbol{X}_0 - \boldsymbol{X})\|^2}{2\sigma^2}}\right) \tag{4.2.17}$$

Assuming that $E\left[\boldsymbol{h}\boldsymbol{h}^H\right] = \rho^2 \boldsymbol{I}$ and that $N \geq n_t$, we get from Theorem 4.2(b):

$$\begin{aligned} &E\left[P(\boldsymbol{X}_0 \to \boldsymbol{X})\right] \\ &\leq \left|\boldsymbol{I} + \frac{\rho^2}{4\sigma^2}(\boldsymbol{X}_0^T \otimes \boldsymbol{I} - \boldsymbol{X}^T \otimes \boldsymbol{I})(\boldsymbol{X}_0^T \otimes \boldsymbol{I} - \boldsymbol{X}^T \otimes \boldsymbol{I})^H\right|^{-1} \\ &= \left|\boldsymbol{I} + \frac{\rho^2}{4\sigma^2}(\boldsymbol{X}_0 - \boldsymbol{X})^H(\boldsymbol{X}_0 - \boldsymbol{X})\right|^{-n_r} \\ &= \left|\boldsymbol{I} + \frac{\rho^2}{4\sigma^2}(\boldsymbol{X}_0 - \boldsymbol{X})(\boldsymbol{X}_0 - \boldsymbol{X})^H\right|^{-n_r} \\ &\leq \left|(\boldsymbol{X}_0 - \boldsymbol{X})(\boldsymbol{X}_0 - \boldsymbol{X})^H\right|^{-n_r} \cdot \left(\frac{\rho^2}{4\sigma^2}\right)^{-n_r n_t} \end{aligned} \tag{4.2.18}$$

■

4.2.3 Detection with Imperfect Channel Knowledge

The following result is concerned with the error probability in the case in which an estimate of $\boldsymbol{h}$ is used in lieu of the true $\boldsymbol{h}$ in the coherent detector in (4.2.5).

Theorem 4.3: Diversity with imperfect channel knowledge.
Assume that $\boldsymbol{X}_0 - \boldsymbol{X}$ has full column rank for all $\boldsymbol{X} \neq \boldsymbol{X}_0$ and that the following detection rule is employed in lieu of (4.2.5):

$$\hat{\boldsymbol{X}} = \underset{\boldsymbol{X} \in \mathcal{X}}{\mathrm{argmin}} \|\boldsymbol{y} - \boldsymbol{X}\hat{\boldsymbol{h}}\|^2 \tag{4.2.19}$$

where

$$\hat{\boldsymbol{h}} \sim N_C(\boldsymbol{h}, \sigma^2 \boldsymbol{\Sigma}) \tag{4.2.20}$$

and $\boldsymbol{\Sigma}$ is a constant positive semi-definite matrix, and where the constant σ^2 in (4.2.20) is the same as the noise power σ^2. Then the system provides a diversity of order n; in other words, the diversity order remains unchanged if $\boldsymbol{h}$ is replaced by a perturbed value of itself in the ML detection rule (yet, the reader should note that $\hat{\boldsymbol{h}} \to \boldsymbol{h}$ as $\sigma^2 \to 0$).

Proof: See Section 4.4. ■

Replacing the true channel $\boldsymbol{h}$ by an estimate $\hat{\boldsymbol{h}}$ in the coherent detector is common practice in training-based detectors (see, e.g., Section 9.4 in this text). In this context, the result in Theorem 4.3 is encouraging as it indicates that although channel estimation errors in general decrease the error performance, they do not destroy the diversity properties of the system. Also, the code design criteria from Theorem 4.2 remain valid also in the case when an estimated channel is used in lieu of the true one. Similar conclusions were also drawn in [TAROKH ET AL., 1999B].

4.2.4 Joint ML Estimation/Detection

The following result quantifies the diversity and coding advantage of a system if a joint detection/estimation approach is used at the receiver.

Theorem 4.4: Bound on the error rate of noncoherent detection.
Assume that $\boldsymbol{h}$ is unknown and that $\boldsymbol{X}$ is detected via a generalized likelihood-ratio test (GLRT), i.e., by minimizing $\|\boldsymbol{y} - \boldsymbol{X}\boldsymbol{h}\|^2$ jointly with respect to $\boldsymbol{h}$ and $\boldsymbol{X}$:

$$\min_{\boldsymbol{h}, \boldsymbol{X} \in \mathcal{X}} \|\boldsymbol{y} - \boldsymbol{X}\boldsymbol{h}\|^2 = \min_{\boldsymbol{X} \in \mathcal{X}} \left\{ \min_{\boldsymbol{h}} \|\boldsymbol{y} - \boldsymbol{X}\boldsymbol{h}\|^2 \right\} \tag{4.2.21}$$

Suppose also that the code matrices have full column rank. Then, neglecting a higher-order term, the average error probability is upper bounded by:

$$E_{\boldsymbol{h}}[P(\boldsymbol{X}_0 \to \boldsymbol{X})] \leq \left| \boldsymbol{I} + \frac{1}{4\sigma^2} \Pi_{\boldsymbol{X}}^{\perp} \boldsymbol{X}_0 \boldsymbol{\Upsilon} \boldsymbol{X}_0^H \right|^{-1} \leq \left(\frac{1}{4\sigma^2} \right)^{-n} \cdot \prod_{k=1}^{n} \varepsilon_k^{-1} \tag{4.2.22}$$

where $\{\varepsilon_1, \ldots, \varepsilon_n\}$ are the nonzero eigenvalues of $\Pi_{\boldsymbol{X}}^{\perp} \boldsymbol{X}_0 \boldsymbol{\Upsilon} \boldsymbol{X}_0^H$.

Proof: See Section 4.4. ■

The formulation in (4.2.21) above assumes that the receiver implements a generalized likelihood ratio test (GLRT) [KAY, 1998, SEC. 6.4.2]. Another possibility is to consider $\boldsymbol{h}$ random (for instance, with a Gaussian distribution), and maximize (with respect to $\boldsymbol{X}$) the likelihood function *averaged over the distribution of* $\boldsymbol{h}$. It turns out that these two approaches give the same result under certain conditions [LARSSON ET AL., 2002B] (see also Exercise 9.7 on page 209). See also [BREHLER AND VARANASI, 2001] for a more detailed discussion of error probabilities for noncoherent detection.

Theorem 4.4 is useful for the design of space-time codes for noncoherent systems, with differential encoding as an important special case (see Section 9.6). In essence, the theorem shows that provided that the matrix $\Pi_{\boldsymbol{X}}^{\perp}\boldsymbol{X}_0\boldsymbol{\Upsilon}\boldsymbol{X}_0^H$ has full rank, maximal diversity is achievable even in total absence of channel knowledge at the receiver. In a way, this is a stronger conclusion than that of Theorems 4.2 and 4.3, which showed that full diversity can be achieved when the code matrices are detected coherently and via a training-based approach, respectively.

Example 4.4: Geometrical interpretation of the error bound.
Suppose that $\boldsymbol{\Upsilon} = \rho^2\boldsymbol{I}$. Since

$$\left|\boldsymbol{I} + \frac{1}{4\sigma^2}\Pi_{\boldsymbol{X}}^{\perp}\boldsymbol{X}_0\boldsymbol{\Upsilon}\boldsymbol{X}_0^H\right| = \left|\boldsymbol{I} + \frac{\rho^2}{4\sigma^2}\boldsymbol{X}_0^H\Pi_{\boldsymbol{X}}^{\perp}\boldsymbol{X}_0\right| \quad (4.2.23)$$

it is clear from Theorem 4.4 that the asymptotic error rate is inversely proportional to the product of the eigenvalues of the matrix $\boldsymbol{X}_0^H\Pi_{\boldsymbol{X}}^{\perp}\boldsymbol{X}_0$. Therefore, to minimize the detection error probability, the eigenvalues of this matrix, which give the "angles" between the range space of $\boldsymbol{X}_0$ and the range space of $\boldsymbol{X}$, should be maximized. ■

Example 4.5: Comparison between coherent and noncoherent ML.
Let $\boldsymbol{\Upsilon} = \rho^2\boldsymbol{I}$. If we compare the error bound for noncoherent detection in Theorem 4.4 with that for coherent detection in Theorem 4.2, we discover that for large SNR$=\rho^2/\sigma^2$:

$$\begin{aligned} E_{\boldsymbol{h}}[P(\boldsymbol{X}_0 \to \boldsymbol{X})]\Big|_{\text{coherent}} &\leq \left|(\boldsymbol{X}_0 - \boldsymbol{X})^H(\boldsymbol{X}_0 - \boldsymbol{X})\right|^{-1}\left(\frac{\rho^2}{4\sigma^2}\right)^{-n} \\ E_{\boldsymbol{h}}[P(\boldsymbol{X}_0 \to \boldsymbol{X})]\Big|_{\text{noncoherent}} &\leq \left|\boldsymbol{X}_0^H\Pi_{\boldsymbol{X}}^{\perp}\boldsymbol{X}_0\right|^{-1}\left(\frac{\rho^2}{4\sigma^2}\right)^{-n} \end{aligned} \quad (4.2.24)$$

The ratio between these error probabilities is:

$$\frac{\left|\boldsymbol{X}_0^H\Pi_{\boldsymbol{X}}^{\perp}\boldsymbol{X}_0\right|}{\left|(\boldsymbol{X}_0 - \boldsymbol{X})^H(\boldsymbol{X}_0 - \boldsymbol{X})\right|} \quad (4.2.25)$$

Since $\Pi^{\perp}_{\boldsymbol{X}}\boldsymbol{X} = 0$, we have that:

$$\begin{aligned}&(\boldsymbol{X}-\boldsymbol{X}_0)^H(\boldsymbol{X}-\boldsymbol{X}_0) - \boldsymbol{X}_0^H\Pi^{\perp}_{\boldsymbol{X}}\boldsymbol{X}_0\\ &= (\boldsymbol{X}-\boldsymbol{X}_0)^H(\boldsymbol{X}-\boldsymbol{X}_0) - (\boldsymbol{X}-\boldsymbol{X}_0)^H\Pi^{\perp}_{\boldsymbol{X}}(\boldsymbol{X}-\boldsymbol{X}_0)\\ &= (\boldsymbol{X}-\boldsymbol{X}_0)^H\Pi_{\boldsymbol{X}}(\boldsymbol{X}-\boldsymbol{X}_0)\end{aligned} \tag{4.2.26}$$

The matrix in (4.2.26) is always positive semi-definite. Hence

$$\left|\boldsymbol{X}_0^H\Pi^{\perp}_{\boldsymbol{X}}\boldsymbol{X}_0\right| \le \left|(\boldsymbol{X}-\boldsymbol{X}_0)^H(\boldsymbol{X}-\boldsymbol{X}_0)\right| \tag{4.2.27}$$

and therefore the ratio in (4.2.25) is always less than one.

Using the union bound, (4.2.2), we find that the detection performance loss (i.e., the increase in SNR that is necessary for a noncoherent ML detector to achieve the performance of the coherent ML detector) can be approximated by:

$$\Delta_{\mathrm{SNR}} \approx \left(\frac{\sum_{\boldsymbol{X}_n\in\mathcal{X}}\sum_{\boldsymbol{X}_k\in\mathcal{X},k\neq n}\left|(\boldsymbol{X}_n-\boldsymbol{X}_k)^H(\boldsymbol{X}_n-\boldsymbol{X}_k)\right|}{\sum_{\boldsymbol{X}_n\in\mathcal{X}}\sum_{\boldsymbol{X}_k\in\mathcal{X},k\neq n}\left|\boldsymbol{X}_n^H\Pi^{\perp}_{\boldsymbol{X}_k}\boldsymbol{X}_n\right|}\right)^{1/n} \tag{4.2.28}$$

Note that the above bounds for coherent and non-coherent detection may not be tight, but we can expect them to be off by a similar amount; hence their relative comparison should make sense.

■

4.3 Summary and Discussion

The purpose of this chapter has been to derive exact expressions for, and bounds on, the detection error probability for transmission over a MIMO channel. The expressions derived in this chapter are useful both for the analysis of given space-time codes, as well as for the design of new coding schemes, and they will be used in most of the following chapters.

We started in Section 4.1 by establishing a formula (see Theorem 4.1) for the average of the Gaussian Q-function when its argument is a sum of exponentially distributed random variables; this is usually the case for systems with diversity combining. For the case when these exponentially distributed random variables are i.i.d., we derived an exact expression, whereas in the general case we resorted to a Chernoff bound.

Next in Section 4.2 we established formulas for the pairwise error probability in a MIMO system assuming coherent detection (Theorem 4.2), detection with imperfect channel knowledge (Theorem 4.3) and noncoherent detection (Theorem 4.4).

Considering a general data model of the form $\boldsymbol{y} = \boldsymbol{X}\boldsymbol{h} + \boldsymbol{e}$, we showed that for Rayleigh fading the probability (averaged over the channel distribution) that a transmitted matrix $\boldsymbol{X}_0$ is mistaken for another matrix $\boldsymbol{X}$ essentially is inversely proportional to the quantity $|\boldsymbol{I} + \rho^2/(4\sigma^2)(\boldsymbol{X}_0 - \boldsymbol{X})^H(\boldsymbol{X}_0 - \boldsymbol{X})|$ (for coherent detection), and on $|\boldsymbol{I} + \rho^2/(4\sigma^2)\boldsymbol{X}_0^H\Pi_{\boldsymbol{X}}^{\perp}\boldsymbol{X}_0|$ (for noncoherent detection), where ρ^2/σ^2 is the SNR. Hence the error rate will decay as $(\rho^2/\sigma^2)^{-\operatorname{rank}\{\boldsymbol{X}_0 - \boldsymbol{X}\}}$ and $(\rho^2/\sigma^2)^{-\operatorname{rank}\{\Pi_{\boldsymbol{X}}^{\perp}\boldsymbol{X}_0\}}$, for coherent and noncoherent detection, respectively, when the SNR goes to infinity; these formulas can be used to determine the diversity order of a given MIMO system.

Although the error bounds presented in this chapter may not be tight in all cases (in particular for a large number of transmit or receive antennas, see [Biglieri et al., 2002]), they provide a systematic way of tackling the problem of designing space-time constellations and therefore their use has become one of the major design criteria for space-time codes. Note that under certain circumstances, a *randomly* chosen (finite) constellation of matrices will yield a code that gives maximal diversity with probability one, provided that the receiver uses maximum-likelihood decoding. However, it is in general difficult to find constellations that (in addition to maximal diversity) also provide high coding gains and at the same time a receiver structure with tractable complexity. This issue is further complicated by the fact that the matrix $\boldsymbol{X}$ above may be forced to have a special structure; cf. (2.1.8) and (2.2.13). We conclude that the problem of finding an "optimal" constellation of space-time matrices by using the above criteria is in general a difficult problem.

4.4 Proofs

Proof of Theorem 4.1

Consider the following eigen-decomposition:

$$\boldsymbol{\Upsilon} = \begin{bmatrix} \boldsymbol{U} & \boldsymbol{U}_0 \end{bmatrix} \begin{bmatrix} \boldsymbol{\Delta} & \boldsymbol{0} \\ \boldsymbol{0} & \boldsymbol{0} \end{bmatrix} \begin{bmatrix} \boldsymbol{U}^H \\ \boldsymbol{U}_0^H \end{bmatrix} \tag{4.4.1}$$

where

$$\boldsymbol{\Delta} = \{\varepsilon_1, \ldots, \varepsilon_n\} \tag{4.4.2}$$

is a positive definite diagonal matrix of dimension $n \times n$, $\boldsymbol{U}$ is an $m \times n$ matrix, $\boldsymbol{U}_0$ is of dimension $m \times (m - n)$ and

$$\begin{bmatrix} \boldsymbol{U} & \boldsymbol{U}_0 \end{bmatrix} \begin{bmatrix} \boldsymbol{U}^H \\ \boldsymbol{U}_0^H \end{bmatrix} = \boldsymbol{I} \tag{4.4.3}$$

Then

$$\begin{bmatrix} \boldsymbol{U}^H \\ \boldsymbol{U}_0^H \end{bmatrix} \boldsymbol{\Upsilon} \begin{bmatrix} \boldsymbol{U} & \boldsymbol{U}_0 \end{bmatrix} = \begin{bmatrix} \boldsymbol{\Delta} & \boldsymbol{0} \\ \boldsymbol{0} & \boldsymbol{0} \end{bmatrix} \tag{4.4.4}$$

and consequently

$$\begin{bmatrix} \boldsymbol{U}^H \\ \boldsymbol{U}_0^H \end{bmatrix} \boldsymbol{h} \sim N_C\left(\boldsymbol{0}, \begin{bmatrix} \boldsymbol{\Delta} & \boldsymbol{0} \\ \boldsymbol{0} & \boldsymbol{0} \end{bmatrix}\right) \tag{4.4.5}$$

It follows that

$$\|\boldsymbol{h}\|^2 = \boldsymbol{h}^H \boldsymbol{h} = \sum_{l=1}^{n} \varepsilon_l \chi_l^2 \tag{4.4.6}$$

where $\{\chi_l^2\}_{l=1}^n$ are independent and exponentially distributed random variables with the same probability density function

$$p_{\chi_l^2}(x) = e^{-x} \tag{4.4.7}$$

The average error probability is

$$\begin{aligned} E_{\boldsymbol{h}}[P_0] =& c \cdot \int_0^\infty \cdots \int_0^\infty \mathrm{d}s_1^2 \cdots \mathrm{d}s_n^2 \\ & \cdot \left(\prod_{l=1}^{n} \frac{1}{\varepsilon_l} \exp\left(-\frac{s_l^2}{\varepsilon_l} \right) \right) \cdot Q\left(\sqrt{g \cdot \frac{\sum_{k=1}^n s_k^2}{\sigma^2}} \right) \end{aligned} \tag{4.4.8}$$

If $\varepsilon_1 = \ldots = \varepsilon_n = \rho^2$, the integral in (4.4.8) can be evaluated exactly [SIMON AND ALOUINI, 2000, SEC. 9.2.2]. The result is (4.1.11). In the general case, an exact expression is hard to obtain (a closed-form expression involving an indefinite integral can be found in [SIMON AND ALOUINI, 2000, SEC. 9.2.3]). However, we can easily obtain a bound on the error probability as well as establish that the diversity order is equal to rank$\{\boldsymbol{\Upsilon}\}$. By the Chernoff bound [PROAKIS, 2001, SEC. 2.1.5]:

$$Q\left(\sqrt{g \cdot \frac{\sum_{k=1}^n s_k^2}{\sigma^2}} \right) \leq \exp\left(-g \cdot \frac{\sum_{k=1}^n s_k^2}{2\sigma^2} \right) \tag{4.4.9}$$

Hence we get

$$\begin{aligned}
E_{\boldsymbol{h}}[P_0] &\leq c \cdot \int_0^\infty \cdots \int_0^\infty \mathrm{d}s_1^2 \cdots \mathrm{d}s_n^2 \prod_{l=1}^{n} \left(\frac{1}{\varepsilon_l} \exp\left(-g \cdot \frac{s_l^2}{2\sigma^2} - \frac{s_l^2}{\varepsilon_l} \right) \right) \\
&\leq c \cdot \prod_{l=1}^{n} \left(\frac{1}{\varepsilon_l} \int_0^\infty \mathrm{d}s^2 \exp\left(-g \cdot \frac{s^2}{2\sigma^2} - \frac{s^2}{\varepsilon_l} \right) \right) \\
&= c \cdot \prod_{l=1}^{n} \left(g \cdot \frac{\varepsilon_l}{2\sigma^2} + 1 \right)^{-1} \\
&= c \cdot \left| \boldsymbol{I} + \frac{g}{2\sigma^2} \boldsymbol{\Delta} \right|^{-1} \\
&= c \cdot \left| \boldsymbol{I} + \frac{g}{2\sigma^2} \boldsymbol{\Upsilon} \right|^{-1}
\end{aligned} \tag{4.4.10}$$

Obviously, we also have

$$\prod_{l=1}^{n} \left(g \cdot \frac{\varepsilon_l}{2\sigma^2} + 1 \right)^{-1} \leq \left(\frac{g}{2\sigma^2} \right)^{-n} \cdot \prod_{l=1}^{n} \varepsilon_l^{-1} \tag{4.4.11}$$

and the proof is completed.

Proof of Theorem 4.2

Let $\breve{\boldsymbol{X}} = \boldsymbol{X}_0 - \boldsymbol{X}$. To show part (i), note first that an incorrect decision is taken in favor of $\boldsymbol{X} \neq \boldsymbol{X}_0$ if and only if

$$\|\boldsymbol{y} - \boldsymbol{X}\boldsymbol{h}\|^2 < \|\boldsymbol{y} - \boldsymbol{X}_0\boldsymbol{h}\|^2 \tag{4.4.12}$$

or equivalently

$$2\mathrm{Re}\{\boldsymbol{e}^H \breve{\boldsymbol{X}} \boldsymbol{h}\} < -\boldsymbol{h}^H \breve{\boldsymbol{X}}^H \breve{\boldsymbol{X}} \boldsymbol{h} \tag{4.4.13}$$

For a given $\breve{\boldsymbol{X}}$ and a given $\boldsymbol{h}$, it is easy to see that

$$2\mathrm{Re}\{\boldsymbol{e}^H \breve{\boldsymbol{X}} \boldsymbol{h}\} \sim N(0, 2\sigma^2 \boldsymbol{h}^H \breve{\boldsymbol{X}}^H \breve{\boldsymbol{X}} \boldsymbol{h}) \tag{4.4.14}$$

and hence

$$\begin{aligned}
P(\boldsymbol{X}_0 \to \boldsymbol{X}) &= P\left(2\mathrm{Re}\{\boldsymbol{e}^H \breve{\boldsymbol{X}} \boldsymbol{h}\} < -\boldsymbol{h}^H \breve{\boldsymbol{X}}^H \breve{\boldsymbol{X}} \boldsymbol{h} \right) \\
&= Q\left(\sqrt{\frac{\|\breve{\boldsymbol{X}}\boldsymbol{h}\|^2}{2\sigma^2}} \right)
\end{aligned} \tag{4.4.15}$$

Part (ii) follows by applying Theorem 4.1 to (4.4.15).

Proof of Theorem 4.3

Let $\breve{\boldsymbol{X}} = \boldsymbol{X}_0 - \boldsymbol{X}$. A calculation similar to that in the proof of Theorem 4.2 shows that the probability of error is

$$\begin{aligned} P(\boldsymbol{X}_0 \to \boldsymbol{X}) =& P\Big(\|\boldsymbol{y} - \boldsymbol{X}\hat{\boldsymbol{h}}\|^2 < \|\boldsymbol{y} - \boldsymbol{X}_0\hat{\boldsymbol{h}}\|^2\Big) \\ =& P\Big(2\mathrm{Re}\{\boldsymbol{e}^H \breve{\boldsymbol{X}}\boldsymbol{h}\} - 2\mathrm{Re}\{\boldsymbol{\delta}^H \boldsymbol{X}^H \breve{\boldsymbol{X}}\boldsymbol{h}\} \\ & < -\boldsymbol{h}^H \breve{\boldsymbol{X}}^H \breve{\boldsymbol{X}}\boldsymbol{h} - 2\mathrm{Re}\{\boldsymbol{e}^H \breve{\boldsymbol{X}}\boldsymbol{\delta}\} \\ & - \boldsymbol{\delta}^H \boldsymbol{X}^H \boldsymbol{X}\boldsymbol{\delta} + \boldsymbol{\delta}^H \boldsymbol{X}_0^H \boldsymbol{X}_0\boldsymbol{\delta}\Big) \end{aligned} \tag{4.4.16}$$

where $\boldsymbol{\delta} = \hat{\boldsymbol{h}} - \boldsymbol{h}$. From the Chernoff bound and the fact that

$$\frac{\boldsymbol{h}^H \breve{\boldsymbol{X}}^H \boldsymbol{X}\boldsymbol{\Sigma}\boldsymbol{X}^H \breve{\boldsymbol{X}}\boldsymbol{h}}{\boldsymbol{h}^H \breve{\boldsymbol{X}}^H \breve{\boldsymbol{X}}\boldsymbol{h}} \le \lambda_{\max}(\boldsymbol{X}\boldsymbol{\Sigma}\boldsymbol{X}^H) \triangleq \alpha \tag{4.4.17}$$

where α does not depend on $\boldsymbol{h}$, it follows that (also see the argument following (4.4.19)):

$$\begin{aligned} P(\boldsymbol{X}_0 \to \boldsymbol{X}) =& P\Big(2\mathrm{Re}\{\boldsymbol{e}^H \breve{\boldsymbol{X}}\boldsymbol{h}\} - 2\mathrm{Re}\{\boldsymbol{\delta}^H \boldsymbol{X}^H \breve{\boldsymbol{X}}\boldsymbol{h}\} < -\boldsymbol{h}^H \breve{\boldsymbol{X}}^H \breve{\boldsymbol{X}}\boldsymbol{h}\Big) \\ \le& \exp\left(-\frac{1}{4\sigma^2} \cdot \frac{(\boldsymbol{h}^H \breve{\boldsymbol{X}}^H \breve{\boldsymbol{X}}\boldsymbol{h})^2}{\boldsymbol{h}^H \breve{\boldsymbol{X}}^H \breve{\boldsymbol{X}}\boldsymbol{h} + \boldsymbol{h}^H \breve{\boldsymbol{X}}^H \boldsymbol{X}\boldsymbol{\Sigma}\boldsymbol{X}^H \breve{\boldsymbol{X}}\boldsymbol{h}}\right) \\ \le& \exp\left(-\frac{\boldsymbol{h}^H \breve{\boldsymbol{X}}^H \breve{\boldsymbol{X}}\boldsymbol{h}}{4(1+\alpha)\sigma^2}\right) \end{aligned} \tag{4.4.18}$$

The remaining three terms in (4.4.16),

$$2\mathrm{Re}\{\boldsymbol{e}^H \breve{\boldsymbol{X}}\boldsymbol{\delta}\} + \boldsymbol{\delta}^H \boldsymbol{X}^H \boldsymbol{X}\boldsymbol{\delta} - \boldsymbol{\delta}^H \boldsymbol{X}_0^H \boldsymbol{X}_0\boldsymbol{\delta} \tag{4.4.19}$$

are higher-order terms and can be neglected in an asymptotic analysis. In particular, they have a variance of order σ^4 whereas the variance of the other terms in (4.4.16) is of the order σ^2. A more rigorous analysis would involve verifying the known fact that the tail of the probability density function for indefinite quadratic forms of Gaussian random variables decays exponentially.

Applying the averaging technique used at the end of the proof of Theorem 4.1 to (4.4.18) we find that the error probability decays as $\sigma^{2\cdot\mathrm{rank}\{\boldsymbol{\Upsilon}\}}$ also in this case.

Proof of Theorem 4.4

Minimization of $\|\boldsymbol{y} - \boldsymbol{X}\boldsymbol{h}\|$ with respect to $\boldsymbol{h}$ yields (provided that $\boldsymbol{X}$ has full column rank):

$$\hat{\boldsymbol{h}} = (\boldsymbol{X}^H\boldsymbol{X})^{-1}\boldsymbol{X}^H\boldsymbol{y} \tag{4.4.20}$$

Inserting this expression into the original cost function shows that the joint estimation/detection problem is equivalent to maximizing

$$\boldsymbol{y}^H\boldsymbol{X}(\boldsymbol{X}^H\boldsymbol{X})^{-1}\boldsymbol{X}^H\boldsymbol{y} = \|\Pi_{\boldsymbol{X}}\boldsymbol{y}\|^2 \tag{4.4.21}$$

with respect to $\boldsymbol{X} \in \mathcal{X}$. Let

$$\boldsymbol{\Delta}_\Pi = \Pi_{\boldsymbol{X}_0} - \Pi_{\boldsymbol{X}} \tag{4.4.22}$$

Note that the matrix $\boldsymbol{\Delta}_\Pi$ is Hermitian, but *not* necessarily positive definite. Also note that

$$\boldsymbol{\Delta}_\Pi\boldsymbol{X}_0 = \Pi_{\boldsymbol{X}}^\perp\boldsymbol{X}_0 \tag{4.4.23}$$

The probability that the ML detector makes a mistake is:

$$\begin{aligned}
&P(\boldsymbol{X}_0 \to \boldsymbol{X}) = P\big(\|\Pi_{\boldsymbol{X}}\boldsymbol{y}\|^2 > \|\Pi_{\boldsymbol{X}_0}\boldsymbol{y}\|^2\big)\\
=&P\Big((\boldsymbol{X}_0\boldsymbol{h}+\boldsymbol{e})^H\boldsymbol{\Delta}_\Pi(\boldsymbol{X}_0\boldsymbol{h}+\boldsymbol{e}) < 0\Big)\\
=&P\Big(2\mathrm{Re}\{\boldsymbol{h}^H\boldsymbol{X}_0^H\boldsymbol{\Delta}_\Pi\boldsymbol{e}\} + \boldsymbol{e}^H\boldsymbol{\Delta}_\Pi\boldsymbol{e} < -\boldsymbol{h}^H\boldsymbol{X}_0^H\boldsymbol{\Delta}_\Pi\boldsymbol{X}_0\boldsymbol{h}\Big)\\
=&P\Big(2\mathrm{Re}\{\boldsymbol{h}^H\boldsymbol{X}_0^H\Pi_{\boldsymbol{X}}^\perp\boldsymbol{e}\} + \boldsymbol{e}^H\boldsymbol{\Delta}_\Pi\boldsymbol{e} < -\boldsymbol{h}^H\boldsymbol{X}_0^H\Pi_{\boldsymbol{X}}^\perp\boldsymbol{X}_0\boldsymbol{h}\Big)
\end{aligned} \tag{4.4.24}$$

The term $\boldsymbol{e}^H\boldsymbol{\Delta}_\Pi\boldsymbol{e}$ is of higher order and may be neglected in an asymptotic analysis (cf. the remark in the previous proof). By the Chernoff bound we have that:

$$\begin{aligned}
&P\Big(2\mathrm{Re}\{\boldsymbol{h}^H\boldsymbol{X}_0^H\Pi_{\boldsymbol{X}}^\perp\boldsymbol{e}\} < -\boldsymbol{h}^H\boldsymbol{X}_0^H\Pi_{\boldsymbol{X}}^\perp\boldsymbol{X}_0\boldsymbol{h}\Big)\\
&\leq \exp\left(-\frac{1}{4\sigma^2}\cdot\frac{\Big(\boldsymbol{h}^H\boldsymbol{X}_0^H\Pi_{\boldsymbol{X}}^\perp\boldsymbol{X}_0\boldsymbol{h}\Big)^2}{\boldsymbol{h}^H\boldsymbol{X}_0^H\Pi_{\boldsymbol{X}}^\perp\boldsymbol{X}_0\boldsymbol{h}}\right)\\
&= \exp\Big(-\frac{\boldsymbol{h}^H\boldsymbol{X}_0^H\Pi_{\boldsymbol{X}}^\perp\boldsymbol{X}_0\boldsymbol{h}}{4\sigma^2}\Big)
\end{aligned} \tag{4.4.25}$$

Hence, by averaging in a way similar to the proof of Theorem 4.1, and observing that

$$\Pi_{\boldsymbol{X}}^\perp\Pi_{\boldsymbol{X}}^\perp = \Pi_{\boldsymbol{X}}^\perp \tag{4.4.26}$$

we find that:

$$\begin{aligned} E_{\boldsymbol{h}}[P(\boldsymbol{X}_0 \to \boldsymbol{X})] &\leq \left| \boldsymbol{I} + \frac{1}{4\sigma^2} \Pi_{\boldsymbol{X}}^{\perp} \boldsymbol{X}_0 \boldsymbol{\Upsilon} \boldsymbol{X}_0^H \Pi_{\boldsymbol{X}}^{\perp} \right|^{-1} \\ &= \left| \boldsymbol{I} + \frac{1}{4\sigma^2} \Pi_{\boldsymbol{X}}^{\perp} \boldsymbol{X}_0 \boldsymbol{\Upsilon} \boldsymbol{X}_0^H \right|^{-1} \end{aligned} \tag{4.4.27}$$

and the result follows. The second inequality in (4.2.22) is trivial.

4.5 Problems

1. Consider the linear model in (4.2.1) and study a space-time code consisting of three matrices as follows:

$$\boldsymbol{X}_1 = \begin{bmatrix} 1 & 0 \\ 0 & 1 \end{bmatrix}, \quad \boldsymbol{X}_2 = \begin{bmatrix} 1 & 0 \\ 0 & -1 \end{bmatrix}, \quad \boldsymbol{X}_3 = \begin{bmatrix} 0 & 1 \\ 1 & 0 \end{bmatrix} \tag{4.5.1}$$

Suppose that the channel $\boldsymbol{h}$ is of dimension 2×1 and that $\boldsymbol{h} \sim N_C(\boldsymbol{0}, \rho^2 \boldsymbol{I})$ where σ^2 is the power (variance) of the noise. Hence ρ^2/σ^2 is effectively a measure of the signal-to-noise ratio (SNR).

(a) Use Theorem 4.2 to derive a bound of the average error probabilities $P(\boldsymbol{X}_1 \to \boldsymbol{X}_2), P(\boldsymbol{X}_2 \to \boldsymbol{X}_3)$ and $P(\boldsymbol{X}_1 \to \boldsymbol{X}_3)$. Simplify the result as much as possible.

(b) Which pair of two matrices do you think are the most likely to be "mistaken for each other" at the receiver? Discuss the result.

(c) Does the code provide full diversity order? If not, suggest a modification of it to give maximal diversity.

2. Let $h \sim N_C(0, 1)$.

(a) Compute (via numerical averaging or "Monte-Carlo simulation") the following expectation (over h):

$$E\left[Q\left(\sqrt{\alpha \cdot |h|^2}\right)\right] \tag{4.5.2}$$

Plot the result as a function of α on a dB-log scale (i.e., vertical axis logarithmic and horizontal axis in dB). (We can think of this as the average BER for a system on a Rayleigh fading channel without diversity, where α is the average SNR and h is a channel gain.) Discuss the result.

(b) Next, use the Theorem 4.1(b) to compute the average *exactly*. Compare with your numerical average.

(c) Finally, use Theorem 4.1(a) to find a (Chernoff) bound on the average error probability. Plot this bound in the same figure. How tight is the bound?

3. Prove (4.2.18). Also, derive a counterpart to (4.2.18) for that case when $E\left[\boldsymbol{h}\boldsymbol{h}^H\right] = \boldsymbol{R}_t^T \otimes \boldsymbol{R}_r$ (see Section 2.1.2). Finally, show that for noncoherent detection (see Section 4.2.4), the corresponding error bound becomes:

$$E_{\boldsymbol{H}}[P(\boldsymbol{X}_0 \to \boldsymbol{X})]\Big|_{\text{noncoherent}} \leq |\boldsymbol{X}_0 \Pi^{\perp}_{\boldsymbol{X}^H} \boldsymbol{X}_0^H|^{-n_r} \left(\frac{\rho^2}{4\sigma^2}\right)^{-n_r n_t} \tag{4.5.3}$$

where n_r and n_t are the number of receive and transmit antennas respectively.

Hint: for square matrices $\boldsymbol{A}$ and $\boldsymbol{B}$ of dimensions $m \times m$ and $n \times n$, respectively, it holds that $|\boldsymbol{A} \otimes \boldsymbol{B}| = |\boldsymbol{A}|^n \cdot |\boldsymbol{B}|^m$.

4. In Example 4.1, suppose that s_1 and s_2 are BPSK symbols. Implement the ML detector (4.2.13) and plot its empirical error rate as a function of SNR. Compare the result to the union bound obtained via (4.2.2) and (4.2.14), as well as to the the exact error probability for BPSK over a Rayleigh fading SISO channel.

5. State and prove the extensions of Theorems 4.1, 4.2, 4.3 and 4.4 to the case of Rice fading:

$$\boldsymbol{h} \sim N_C(\boldsymbol{\mu_h}, \boldsymbol{\Upsilon}) \tag{4.5.4}$$

6. Extend Theorems 4.2, 4.3 and 4.4 to the case of colored noise:

$$\boldsymbol{e} \sim N_C(\boldsymbol{0}, \boldsymbol{\Lambda}) \tag{4.5.5}$$

for some positive definite (known) matrix $\boldsymbol{\Lambda}$.

Hint: Note that $\boldsymbol{e}$ can be written

$$\mathbf{e} = \boldsymbol{\Lambda}^{1/2} \mathbf{e}_w \tag{4.5.6}$$

where

$$\mathbf{e}_w \sim N_C(\mathbf{0}, \mathbf{I}) \tag{4.5.7}$$

7. Formally prove that neglecting the higher-order terms in (4.4.16) has no effect on the asymptotic slope of the error probability curve.

Hint: First show that for any two random variables x and y that satisfy

$$\begin{aligned} P(|x| \geq t) &\leq M \cdot \exp(-at^{\gamma}) \\ P(|y| \geq t) &\leq M \cdot \exp(-at^{\gamma}) \end{aligned} \tag{4.5.8}$$

for some positive constants a and γ, it holds that

$$\begin{aligned} P(|x+y| \geq t) &\leq 2M \cdot \exp\big(-a(t/2)^{\gamma}\big) \\ P(|xy| \geq t) &\leq 2M \cdot \exp\left(-at^{\gamma/2}\right) \end{aligned} \tag{4.5.9}$$

Chapter 5

RECEIVE DIVERSITY

In this chapter we study some basic properties of a system with a single transmit antenna ($n_t = 1$), but an arbitrary number $n_r \geq 1$ of receive antennas (see Figure 5.1). Such systems play an important role in practice: receive diversity has been used extensively during the 1990's for improving the uplink performance in cellular systems. The discussion in the chapter serves as important background material, and it will also help in setting the framework and the goals for the rest of the book.

5.1 Flat Channels

For a system with $n_t = 1$ transmit antenna and $n_r \geq 1$ receive antennas, the transmitted matrix $\boldsymbol{X}$ in Section 2.1 reduces to a scalar x (hence $N = 1$). Since no transmit encoding is necessary, we will simply set $x = s$; the sequence of symbols s is typically obtained by encoding a bitstream. Also, in this case, the MIMO channel matrix $\boldsymbol{H}$ in (2.1.2) reduces to a column vector $\boldsymbol{h}$ of length n_r that contains the n_r channel gains between the transmit antenna and the n_r receive antennas.

Assume that at a certain time instant, a complex symbol s is transmitted. Then the receiver observes

$$\boldsymbol{y} = \boldsymbol{h}s + \boldsymbol{e} \tag{5.1.1}$$

where $\boldsymbol{e}$ is an n_r-vector of noise. We assume, as before, that $\boldsymbol{e}$ is spatially white Gaussian noise:

$$\boldsymbol{e} \sim N_C(\boldsymbol{0}, \sigma^2 \boldsymbol{I}) \tag{5.1.2}$$

If $\boldsymbol{h}$ is a Gaussian random variable satisfying

$$\boldsymbol{h} \sim N_C(\boldsymbol{0}, \rho^2 \boldsymbol{I}) \tag{5.1.3}$$

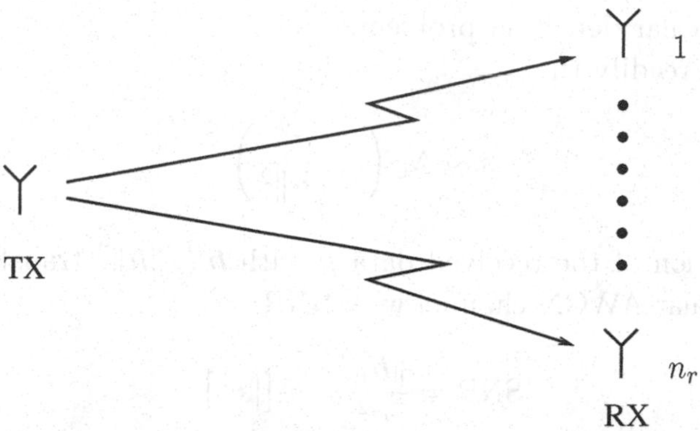

Figure 5.1. A receive diversity system (here $n_t = 1$, $n_r \geq 1$).

then Theorem 4.2 (see page 45) is applicable and it shows that the system can achieve a diversity of order n_r, provided that we use ML decoding to detect the transmitted symbol s. To see this, observe that (5.1.1) can be written as:

$$\boldsymbol{y} = (s\boldsymbol{I})\boldsymbol{h} + \boldsymbol{e} \tag{5.1.4}$$

which is of the form of (4.2.1).

We can study the ML receiver structure associated with the system (5.1.1) in more detail as follows. For a given $\boldsymbol{h}$, the ML detection of s given $\boldsymbol{y}$ amounts to minimizing the following Euclidean metric (cf. the discussion in Section 4.2.2):

$$\|\boldsymbol{y} - \boldsymbol{h}s\|^2 \tag{5.1.5}$$

with respect to $s \in \mathcal{S}$. Since

$$\begin{aligned} \|\boldsymbol{y} - \boldsymbol{h}s\|^2 &= \|\boldsymbol{y}\|^2 + |s|^2\|\boldsymbol{h}\|^2 - 2\mathrm{Re}\{s^*\boldsymbol{h}^H\boldsymbol{y}\} \\ &= \|\boldsymbol{h}\|^2 \cdot \left| s - \frac{\boldsymbol{h}^H\boldsymbol{y}}{\|\boldsymbol{h}\|^2} \right|^2 + \text{const.} \end{aligned} \tag{5.1.6}$$

it follows that minimizing (5.1.5) is equivalent to minimizing

$$|s - \hat{s}|^2 \tag{5.1.7}$$

where

$$\hat{s} \triangleq \frac{\boldsymbol{h}^H\boldsymbol{y}}{\|\boldsymbol{h}\|^2} \tag{5.1.8}$$

This is a simple scalar detection problem.

It also follows readily that

$$\hat{s} \sim N_C\left(s, \frac{\sigma^2}{\|\boldsymbol{h}\|^2}\right) \tag{5.1.9}$$

Hence multiplication of the received data $\boldsymbol{y}$ with $\boldsymbol{h}^H/\|\boldsymbol{h}\|^2$ transforms the SIMO channel into a scalar AWGN channel with SNR

$$\text{SNR} = \frac{\|\boldsymbol{h}\|^2}{\sigma^2} \cdot E\left[|s|^2\right] \tag{5.1.10}$$

Clearly, the error probability of this system depends on $\boldsymbol{h}$ only via the SNR in (5.1.10). Hence Theorem 4.1 (see page 42) can be applied to compute an expression for the error rate. In particular, a direct application of that theorem shows (once again) that the system achieves a diversity of order n_r.

The above simple calculation shows that the ML detection of s given the vector $\boldsymbol{y}$ is equivalent to linear processing of the received data (viz., multiplication of $\boldsymbol{y}$ with $\boldsymbol{h}^H/\|\boldsymbol{h}\|^2$) followed by the solution of a scalar ML detection problem. The vector $\boldsymbol{h}^H/\|\boldsymbol{h}\|^2$ can be interpreted as a spatial matched filter, or a *beamforming* vector, although the latter interpretation might (depending on the antenna configuration) lack the interpretation of standard beamforming in the context of classical array signal processing [Van Trees, 2002].

Example 5.1: Non-optimal linear processing of received data.

The linear processing in (5.1.8) is only one possible way of forming a decision statistic for the detection of s. If we instead let

$$\hat{s}_* = \frac{\boldsymbol{w}^H\boldsymbol{y}}{\boldsymbol{w}^H\boldsymbol{h}} \tag{5.1.11}$$

where $\boldsymbol{w}$ is an arbitrary but *fixed* vector, then

$$\hat{s}_* \sim N_C\left(s, \frac{\|\boldsymbol{w}\|^2}{|\boldsymbol{w}^H\boldsymbol{h}|^2} \cdot \sigma^2\right) \tag{5.1.12}$$

Clearly, $\hat{s}_*$ can be interpreted as the output of a scalar AWGN channel, and therefore it can be used as a decision statistic, in the same way as $\hat{s}$ in (5.1.8). The SNR in $\hat{s}_*$ is found to be

$$\text{SNR}_* = \frac{|\boldsymbol{w}^H\boldsymbol{h}|^2}{\|\boldsymbol{w}\|^2 \cdot \sigma^2} \cdot E\left[|s|^2\right] \tag{5.1.13}$$

Under the assumption (5.1.3),

$$\boldsymbol{w}^H \boldsymbol{h} \sim N_C(0, \|\boldsymbol{w}\|^2 \cdot \rho^2) \tag{5.1.14}$$

Hence, provided that $\boldsymbol{w}$ is fixed and non-random (which implies in particular that $\boldsymbol{w} = \boldsymbol{h}$ is *not* possible here), Theorem 4.1 (see page 42) can be applied (with $\gamma^2 = |\boldsymbol{w}^H \boldsymbol{h}|^2$) to show that *diversity is not achieved for any* $\boldsymbol{w}$.

■

5.2 Frequency-Selective Channels

Next we study a receive diversity system operating on a frequency-selective fading channel. For such a system, the sequence of transmitted vectors $\{\boldsymbol{x}(n)\}$ reduces to a sequence of scalars $\{x(n)\}$. As for the case of flat fading, no transmit encoding is necessary and hence we set $x(n) = s(n)$. Also, in the case under study, the channel "impulse response" $\boldsymbol{H}(z^{-1})$ reduces to a vector-valued FIR filter $\boldsymbol{h}(z^{-1})$ with coefficients $\{\boldsymbol{h}_0, \ldots, \boldsymbol{h}_L\}$ of dimension $n_r \times 1$.

The received data samples $\boldsymbol{y}(n)$ for $n = 0, \ldots, N_0 + N_{\text{post}} - 1$ can be written:

$$\boldsymbol{y}(n) = \sum_{l=0}^{L} \boldsymbol{h}_l \boldsymbol{x}(n-l) + \boldsymbol{e}(n) = \boldsymbol{h}(z^{-1})\boldsymbol{x}(n) + \boldsymbol{e}(n) \tag{5.2.1}$$

where $\boldsymbol{e}(n)$ is noise. Let

$$\boldsymbol{S} \triangleq \begin{bmatrix} s(0) & s(-1) & \cdots & s(-L) \\ s(1) & s(0) & & s(1-L) \\ \vdots & & \ddots & \vdots \\ \vdots & & & \vdots \\ \vdots & & & s(0) \\ \vdots & & & \vdots \\ s(N_0 + N_{\text{post}} - 2) & s(N_0 + N_{\text{post}} - 3) & & s(N_0 + N_{\text{post}} - L - 2) \\ s(N_0 + N_{\text{post}} - 1) & s(N_0 + N_{\text{post}} - 2) & \cdots & s(N_0 + N_{\text{post}} - L - 1) \end{bmatrix} \tag{5.2.2}$$

be a Toeplitz matrix that contains the transmitted symbols. If we arrange the received data in a vector as follows:

$$\boldsymbol{y} = \left[\boldsymbol{y}^T(0) \quad \cdots \quad \boldsymbol{y}^T(N_0 + N_{\text{post}} - 1)\right]^T \tag{5.2.3}$$

we can write

$$y = (S \otimes I_{n_r})h + e \tag{5.2.4}$$

where

$$h = \text{vec}\left(\begin{bmatrix} h_0 & \cdots & h_L \end{bmatrix} \right) \tag{5.2.5}$$

and

$$e = \begin{bmatrix} e^T(0) & \cdots & e^T(N_0 + N_{\text{post}} - 1) \end{bmatrix}^T \tag{5.2.6}$$

is a vector of noise (see (2.2.13) in Section 2.2.2).

Let $s_0(n)$ be the true transmitted sequence of data and let $s(n)$ be any other hypothetical data string. Also, let $S_0 - S$ be a Toeplitz matrix formed from the error signal $s_0(n) - s(n)$ in the same way as S in (5.2.2) is built from $s(n)$. Then we can infer from Theorem 4.2 (see page 45) that a diversity of order $n_r(L+1)$ can be achieved if the matrix $S_0 - S$ has full column rank for all possible nonzero error signals $\{s_0(n) - s(n)\}$, and provided that $E[hh^H]$ is nonsingular. By studying (5.2.2) we can expect that the rank properties of $S_0 - S$ will depend heavily on the choice of preambles and postambles (and as we will see below this is indeed the case). Note that all the above equations are valid also in the special case of one receive antenna ($n_r = 1$), in which case the achievable diversity order becomes $L + 1$.

5.2.1 Transmission with Known Preamble and Postamble

Assume that we use a fixed preamble and postamble of length N_{pre} and N_{post}, respectively, where $N_{\text{pre}} \geq L$ and $N_{\text{post}} \geq L$, and that *at least one of these sequences is known to the receiver.* Data are transmitted for $n = -N_{\text{pre}}, \ldots, N_0 + N_{\text{post}} - 1$ and received for $n = 0, \ldots, N_0 + N_{\text{post}} - 1$. See Figure 5.2 for an illustration of the transmission scheme that we consider. (Figure 5.2 is similar to Figure 2.5, but no assumptions on the preambles and postambles were made in Chapter 2.)

Diversity

Note first that at least one element of $s_0(n) - s(n)$ must be nonzero for $n = 0, \ldots, N_0 - 1$ and hence at least one diagonal of $S_0 - S$ is nonzero. If the preamble is known to the receiver, then the error signal $s_0(n) - s(n)$ must be zero during the preamble:

$$s_0(n) - s(n) = 0, \quad n = -N_{\text{pre}}, \ldots, -1 \tag{5.2.7}$$

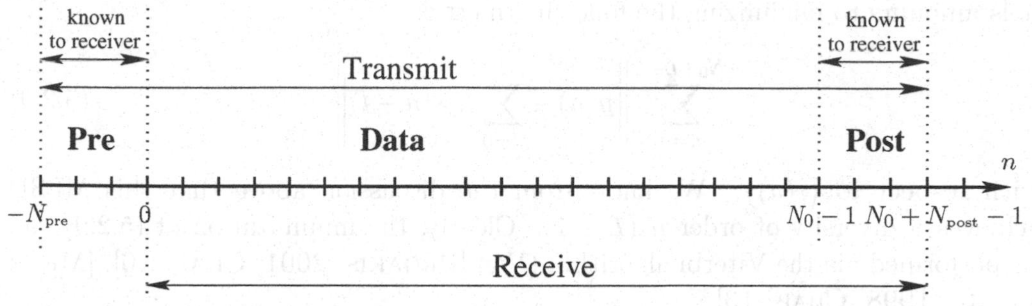

Figure 5.2. Transmission with a known preamble and postamble.

Consequently the matrix $\boldsymbol{S}_0 - \boldsymbol{S}$ will have zeros above its main diagonal; therefore its column rank will be equal to $L + 1$. Conversely, if the postamble is known at the receiving end, the following error signal must be identical to zero:

$$s_0(n) - s(n) = 0, \quad n = N_0, \ldots, N_0 + N_{\text{post}} - 1 \tag{5.2.8}$$

and therefore $\boldsymbol{S}_0 - \boldsymbol{S}$ will have only zeros below its N_0th diagonal; thus its column rank will be equal to $L + 1$ also in this case. Consequently, with ML decoding, a diversity of order $n_r(L+1)$ can be achieved if either the preamble or the postamble is known to the receiver.

If both the preamble and postamble are known, we can expect an increase in the BER performance compared to the case when only one of them is known (although the slope of the BER curve, which determines the diversity order, will be the same in both these cases). This is so since in this case there are fewer unknowns in the detection problem. Transmission schemes where the preambles and postambles are known to the receiver are very common in practice. For instance, such a scheme is used in the GSM system [Mouly and Pautet, 1992].

Equalization via Exact and Approximate MLSD

Let us consider the transmission of N_0 symbols together with an associated preamble and postamble of length N_{pre} and N_{post}, respectively. Without loss of generality, we assume in this subsection, to simplify the notation, that $L = N_{\text{pre}} = N_{\text{post}}$ (the channel can be appropriately padded with zeros). Assume that the preamble and postamble are known to the receiver. Then ML detection of the transmitted symbols amounts to minimizing the following metric:

$$\sum_{n=0}^{N_0+L-1} \left\| \boldsymbol{y}(n) - \sum_{l=0}^{L} \boldsymbol{h}_l s(n-l) \right\|^2 \tag{5.2.9}$$

with respect to $\{s(n)\}$. We know from the discussion above that this MLSD achieves a diversity of order $n_r(L+1)$. Clearly, the minimization of (5.2.9) can be performed via the Viterbi algorithm (VA) [PROAKIS, 2001, CHAP. 10], [MEYR ET AL., 1998, CHAP. 13].

Since the quantities involved in (5.2.9) are n_r-vectors the computational complexity associated with the application of VA is rather high. This computational burden can be reduced as follows. Let us rewrite (5.2.9) as:

$$\begin{aligned} &\sum_{n=0}^{N_0+L-1} \left\| \boldsymbol{y}(n) - \sum_{l=0}^{L} \boldsymbol{h}_l s(n-l) \right\|^2 \\ &= \sum_{n=0}^{N_0+L-1} \|\boldsymbol{y}(n)\|^2 - 2\mathrm{Re}\left\{ \sum_{n=0}^{N_0+L-1} \left(\boldsymbol{h}(z^{-1})s(n)\right)^H \boldsymbol{y}(n) \right\} \\ &+ \sum_{n=0}^{N_0+L-1} \left(\boldsymbol{h}(z^{-1})s(n)\right)^H \left(\boldsymbol{h}(z^{-1})s(n)\right) \end{aligned} \tag{5.2.10}$$

We have that

$$\begin{aligned} \sum_{n=0}^{N_0+L-1} \left(\boldsymbol{h}(z^{-1})s(n)\right)^H \boldsymbol{y}(n) &= \sum_{n=0}^{N_0+L-1} \sum_{l=0}^{L} (\boldsymbol{h}_l s(n-l))^H \boldsymbol{y}(n) \\ \approx \sum_{n=0}^{N_0+L-1} \sum_{l=0}^{L} s^*(n) \boldsymbol{h}_l^H \boldsymbol{y}(n+l) &= \sum_{n=0}^{N_0+L-1} s^*(n) \boldsymbol{h}^H(z) \boldsymbol{y}(n) \end{aligned} \tag{5.2.11}$$

where the approximation denoted by "≈" is due to end-effects and hence the approximation error is $O(1)$ for $N_0 \gg 1$ (while the whole expression in (5.2.11) is $O(N_0)$). Note that (5.2.11) depends on $\boldsymbol{y}(n)$ for values of n for which $\boldsymbol{y}(n)$ is not available, and this problem can be overcome by setting these values to zero. The approximation errors induced in that way are also $O(1)$.

Next, let $\gamma(z^{-1})$ be the z-transform of the autocorrelation sequence associated with the channel impulse response:

$$\gamma(z^{-1}) = \sum_{l=-L}^{L} \gamma_l z^{-l} = \boldsymbol{h}^H(z)\boldsymbol{h}(z^{-1}) \tag{5.2.12}$$

and note that $[\boldsymbol{h}(z^{-1})s(n)]^H[\boldsymbol{h}(z^{-1})s(n)]$ is real. Hence

$$
\begin{aligned}
&\sum_{n=0}^{N_0+L-1} \left(\boldsymbol{h}(z^{-1})s(n)\right)^H \left(\boldsymbol{h}(z^{-1})s(n)\right)\\
=&\mathrm{Re}\left\{\sum_{n=0}^{N_0+L-1}\sum_{l=0}^{L}\left(\boldsymbol{h}(z^{-1})s(n)\right)^H \boldsymbol{h}_l s(n-l)\right\}\\
\approx&\mathrm{Re}\left\{\sum_{n=0}^{N_0+L-1}\sum_{l=0}^{L}\left(\boldsymbol{h}(z^{-1})s(n+l)\right)^H \boldsymbol{h}_l s(n)\right\}\\
=&\mathrm{Re}\left\{\sum_{n=0}^{N_0+L-1}\left(\sum_{l=0}^{L}\boldsymbol{h}_l^H\left(\boldsymbol{h}(z^{-1})s(n+l)\right)\right)^* s(n)\right\}\\
=&\mathrm{Re}\left\{\sum_{n=0}^{N_0+L-1}\left(\boldsymbol{h}^H(z)\boldsymbol{h}(z^{-1})s(n)\right)^* s(n)\right\}\\
=&\mathrm{Re}\left\{\sum_{n=0}^{N_0+L-1}\sum_{l=-L}^{L}\gamma_l^* s^*(n-l)s(n)\right\}\\
=&\mathrm{Re}\left\{\sum_{n=0}^{N_0+L-1}\sum_{l=-L}^{L}\gamma_l s^*(n)s(n-l)\right\}\\
=&\mathrm{Re}\left\{\sum_{n=0}^{N_0+L-1}\left(\sum_{l=-L}^{-1}\gamma_l s^*(n)s(n-l)\right.\right.\\
&\left.\left.+\sum_{l=1}^{L}\gamma_l s^*(n)s(n-l)+\gamma_0 s^*(n)s(n)\right)\right\}\\
\approx&\mathrm{Re}\left\{\sum_{n=0}^{N_0+L-1}\left(\gamma_0 s^*(n)s(n)+2\sum_{l=1}^{L}\gamma_l s^*(n)s(n-l)\right)\right\}
\end{aligned}
\tag{5.2.13}
$$

where in the last step we also used the fact that $\gamma_{-l}^* = \gamma_l$.

Equations (5.2.10), (5.2.11) and (5.2.13) show that maximizing the likelihood function in (5.2.9) is equivalent (to within an $O(1)$ approximation attributable to end-effects) to maximizing the following metric:

$$
\mathrm{Re}\left\{\sum_{n=0}^{N_0+L-1} s^*(n)\left(z(n)-\sum_{l=1}^{L}\gamma_l s(n-l)-\frac{1}{2}\gamma_0 s(n)\right)\right\} \tag{5.2.14}
$$

where $z(n)$ is the following *matched filtered* sequence:

$$z(n) = \boldsymbol{h}^H(z)\boldsymbol{y}(n) \tag{5.2.15}$$

The simplified metric in (5.2.14) can be maximized by a scalar Viterbi equalizer with memory length L, as opposed to (5.2.9) which requires a *vector-valued* Viterbi algorithm with memory length L. Note also that modulo the approximation made, the matched filtered data $z(n)$ is a sufficient statistic for ML detection.

Linear MMSE Equalization

Even with the approximation made above the computational complexity of Viterbi equalization is typically rather high. An alternative to using the Viterbi algorithm is to use a linear equalizer, i.e., to form estimates of $s(n)$ that are linear in the received data. Once such linear equalizer is the *minimum-mean square error* (MMSE) method that we describe next.

Specifically, we seek a vector-valued linear FIR filter $\boldsymbol{\xi}(z^{-1})$ such that the output of the filtered sequence

$$\hat{s}(n) = \boldsymbol{\xi}^T(z^{-1})\boldsymbol{y}(n) = \boldsymbol{\xi}^T(z^{-1})\boldsymbol{h}(z^{-1})s(n) + \boldsymbol{\xi}^T(z^{-1})\boldsymbol{e}(n) \tag{5.2.16}$$

recovers $s(n)$ in a minimum mean-square error (MMSE) sense (to obtain estimates of the transmitted symbols, $\hat{s}(n)$ is projected onto the signal constellation $\mathcal{S}$). Note that the fully optimal MMSE filter $\boldsymbol{\xi}(z^{-1})$ may have an infinite length. We will look for a filter $\boldsymbol{\xi}(z^{-1})$ of a *given* length $2M+1$, which is suboptimal in general. If we increase M, the set of filters $\{\boldsymbol{\xi}(z^{-1})\}$ increases and hence we approach the optimal filter. On the other hand, if M is too large, the approximations due to end-effects may not be negligible unless the block length N is correspondingly large. Moreover, the computational burden increases with increasing M. Hence, finding the optimal value of M in a practical application is a tradeoff between performance and computational complexity.

The MMSE filter design problem amounts to minimizing:

$$\begin{aligned} E\left[|\hat{s}(n) - s(n)|^2\right] =& E\left[|\boldsymbol{\xi}^T(z^{-1})\boldsymbol{y}(n) - s(n)|^2\right] \\ =& E\left[|\boldsymbol{\xi}^T(z^{-1})\boldsymbol{h}(z^{-1})s(n) - s(n)|^2\right] \\ &+ E\left[|\boldsymbol{\xi}^T(z^{-1})\boldsymbol{e}(n)|^2\right] \end{aligned} \tag{5.2.17}$$

with respect to the coefficients of $\boldsymbol{\xi}(z^{-1})$. Before we proceed we note that the MMSE filter output may depend on values of $\boldsymbol{y}(n)$ for time indices n where $\boldsymbol{y}(n)$ is not available. This "end-effect" problem has to be overcome by a suitable approximation, such as setting the unavailable values to zero.

To minimize (5.2.17), assume for simplicity that $s(n)$ is *white*, and that

$$\rho^2 = E\left[|s(n)|^2\right] \tag{5.2.18}$$

is *known*. Note from (5.2.17) that (provided $M \geq L$):

$$E\left[|\hat{s}(n) - s(n)|^2\right] = \rho^2 \cdot \sum_{l=-M}^{M} \left| \left(\sum_{k=-M}^{M} \boldsymbol{\xi}_k^T \boldsymbol{h}_{l-k} \right) - \delta_l \right|^2 + \sigma^2 \cdot \sum_{k=-M}^{M} \|\boldsymbol{\xi}_k\|^2 \tag{5.2.19}$$

where, as before, σ^2 is the variance of the elements of $\boldsymbol{e}(n)$. In (5.2.19), the filter coefficients are assumed by convention to be zero for indices where they are not defined. Let

$$\boldsymbol{\xi} = [\boldsymbol{\xi}_{-M}^T \cdots \boldsymbol{\xi}_M^T]^T \tag{5.2.20}$$

be the coefficient vector of the sought filter, and let $\boldsymbol{\chi}$ be a $(2M+1) \times n_r(2M+1)$ Toeplitz matrix defined according to:

$$\boldsymbol{\chi} = \begin{bmatrix} \boldsymbol{h}_0^T & \boldsymbol{0} & \cdots & & \cdots & \boldsymbol{0} \\ \boldsymbol{h}_1^T & \boldsymbol{h}_0^T & \ddots & & \ddots & \vdots \\ \vdots & \ddots & \ddots & & & \\ \boldsymbol{h}_L^T & & \ddots & & \ddots & \\ \boldsymbol{0} & \ddots & & & \ddots & \vdots \\ \vdots & & & & & \boldsymbol{0} \\ \boldsymbol{0} & \cdots & \boldsymbol{0} & \boldsymbol{h}_L^T & \cdots & \boldsymbol{h}_0^T \end{bmatrix} \tag{5.2.21}$$

Finally, let $\boldsymbol{f}_0$ be a $(2M+1)$-vector whose $(M+1)$th element is equal to one and all other elements are equal to zero. Then we can write (5.2.19) in a matrix form as:

$$\begin{aligned} E\left[|\hat{s}(n) - s(n)|^2\right] &= \rho^2 \cdot (\boldsymbol{\chi}\boldsymbol{\xi} - \boldsymbol{f}_0)^H (\boldsymbol{\chi}\boldsymbol{\xi} - \boldsymbol{f}_0) + \sigma^2 \cdot \|\boldsymbol{\xi}\|^2 \\ &= \left\| \begin{bmatrix} \sqrt{\rho^2}\boldsymbol{f}_0 \\ \boldsymbol{0} \end{bmatrix} - \begin{bmatrix} \sqrt{\rho^2} \cdot \boldsymbol{\chi} \\ \sqrt{\sigma^2} \cdot \boldsymbol{I} \end{bmatrix} \boldsymbol{\xi} \right\|^2 \end{aligned} \tag{5.2.22}$$

Some standard matrix algebra shows that the coefficient vector $\boldsymbol{\xi}$ that minimizes (5.2.22) is given by:

$$\boldsymbol{\xi} = \left(\boldsymbol{\chi}^H \boldsymbol{\chi} + \frac{\sigma^2}{\rho^2} \boldsymbol{I} \right)^{-1} \boldsymbol{\chi}^H \boldsymbol{f}_0 \tag{5.2.23}$$

and the derivation is complete.

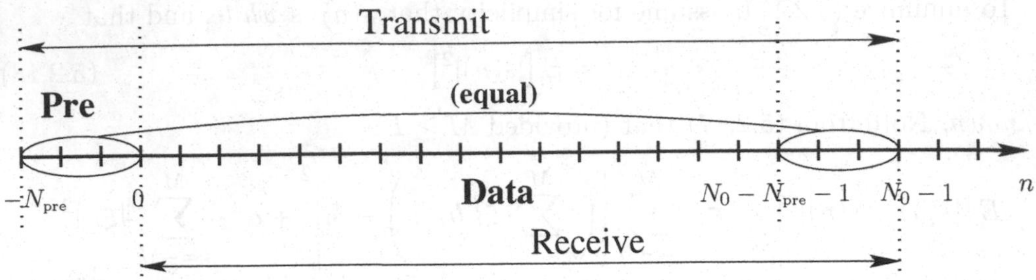

Figure 5.3. Transmission with a cyclic preamble, but no postamble.

5.2.2 Orthogonal Frequency Division Multiplexing

An increasingly popular transmission technique for frequency-selective channels is the so-called orthogonal frequency division multiplexing (OFDM). OFDM has received a considerable amount of attention during the last decade. It is also a transmission technique included in the IEEE 802.11a wireless local area network standard [VAN NEE ET AL., 1999]. An OFDM system uses a preamble that is equal to the last L symbols of the transmitted block (and hence unknown to the receiver). It turns out that the complexity of the signal processing associated with OFDM is low, but that the multipath diversity is lost, i.e., the diversity order of the system is reduced from $n_r(L+1)$ to n_r, as explained in detail in the rest of this section.

Cyclic Prefix

Suppose that there is no postamble ($N_{\text{post}} = 0$), but that we use a preamble of length N_{pre}. Also assume that the symbols in the preamble are equal to the N_{pre} last elements of the transmitted data string:

$$s(n) = s(n + N_0), \quad n = -N_{\text{pre}}, \ldots, -1 \tag{5.2.24}$$

Hence, in the present case, data are transmitted for $n = -N_{\text{pre}}, \ldots, N_0 - 1$ and received during $n = 0, \ldots, N_0 - 1$. The transmission scheme under consideration is illustrated in Figure 5.3. As we will see, the main appealing feature of the cyclic prefix is that the linear convolution induced by the propagation channel $\boldsymbol{h}(z^{-1})$ is transformed into a *circular* convolution, which corresponds exactly to a multiplication in the frequency domain.

With reference to the diversity analysis based on (5.2.2)–(5.2.4), for a system that uses a cyclic prefix we can easily find an error signal $s_0(n) - s(n)$ for which the column rank of $\boldsymbol{S}_0 - \boldsymbol{S}$ is equal to one. Take, for instance, an arbitrary $s(n) \neq 0$ for all n, and assuming a symmetric constellation, let $s_0(n) = -s(n)$. This gives

$$s_0(n) - s(n) = -2s(n) \tag{5.2.25}$$

for all n, and the discussion following (5.2.2)–(5.2.4) indicates that the diversity order can be as low as n_r. Indeed, let's say that $s(n) = \text{const.}$ for $n = 0, \ldots, N_0 - 1$ is a feasible choice: then the matrix $\boldsymbol{S}_0 - \boldsymbol{S}$ corresponding to (5.2.25) has rank one. This should be compared to the case of known preamble or postamble (see Section 5.2.1) where a diversity of order $n_r(L+1)$ was achieved. Note that the above discussion does not really prove that the diversity order cannot be more than n_r for the system under consideration; yet we will see shortly that ML detection gives a BER that behaves as σ^{2n_r} in the high SNR case, and hence that a transmission scheme with cyclic preambles cannot exploit multipath diversity.

As a further remark, note that for this transmission scheme the signal received for $n = 0, \ldots, N_0 - 1$ is *not* a sufficient statistic for ML detection; in particular the samples received during the preamble, as well as some of those for $n \geq N_0$, are discarded even though they contain information about the transmitted data. It is clear that this fact is connected to the loss of multipath diversity.

Transmitted and Received Data

In OFDM, we transmit the following inverse-Fourier-transform encoded data:

$$x(n) = \frac{1}{\sqrt{N_0}} \sum_{k=0}^{N_0-1} s(k) \exp\left(i\frac{2\pi}{N_0}nk\right), \quad n = 0, \ldots, N_0 - 1 \tag{5.2.26}$$

augmented by a cyclic preamble as follows (cf. (5.2.24)):

$$x(n) = x(n + N_0), \quad n = -N_{\text{pre}}, \ldots, -1 \tag{5.2.27}$$

The sinusoids $\{e^{i2\pi nk/N_0}\}$ are often called subcarriers; hence one symbol $s(k)$ is transmitted on each subcarrier $e^{i2\pi nk/N_0}$. For an observation time of N_0, these subcarriers are orthogonal and this fact provides a motivation for the name "OFDM."

The received data vector (of length n_r) can be written:

$$\begin{aligned} \boldsymbol{y}(n) &= \sum_{k=0}^{L} \boldsymbol{h}_k x(n-k) + \boldsymbol{e}(n) \\ &= \frac{1}{\sqrt{N_0}} \sum_{k=0}^{L} \sum_{m=0}^{N_0-1} s(m)\boldsymbol{h}_k \exp\left(i\frac{2\pi}{N_0}(n-k)m\right) + \boldsymbol{e}(n) \end{aligned} \tag{5.2.28}$$

for $n = 0, \ldots, N_0 - 1$. Note that the definition of the cyclic prefix in (5.2.27) is used in (5.2.28) in order to make $x(n)$ well-defined for $n < 0$, but that the data received during the cyclic preamble are discarded. In (5.2.28), $\{\boldsymbol{e}(n)\}$ is a sequence of noise vectors.

At the receiver we take the Fourier transform of the received data, to get:

$$
\begin{aligned}
\boldsymbol{z}(n) &= \frac{1}{\sqrt{N_0}} \sum_{l=0}^{N_0-1} \boldsymbol{y}(l) e^{-i\frac{2\pi}{N_0}ln} \\
&= \frac{1}{N_0} \sum_{l=0}^{N_0-1} \sum_{m=0}^{N_0-1} \sum_{k=0}^{L} s(m) \boldsymbol{h}_k e^{i\frac{2\pi}{N_0}(l-k)m} e^{-i\frac{2\pi}{N_0}ln} + \frac{1}{\sqrt{N_0}} \sum_{l=0}^{N_0-1} \boldsymbol{e}(l) e^{-i\frac{2\pi}{N_0}ln} \\
&= \frac{1}{N_0} \sum_{l=0}^{N_0-1} \sum_{m=0}^{N_0-1} \sum_{k=0}^{L} s(m) \boldsymbol{h}_k e^{-i\frac{2\pi}{N_0}(n-m)l} e^{-i\frac{2\pi}{N_0}km} + \frac{1}{\sqrt{N_0}} \sum_{l=0}^{N_0-1} \boldsymbol{e}(l) e^{-i\frac{2\pi}{N_0}ln} \\
&= s(n) \cdot \sum_{k=0}^{L} \boldsymbol{h}_k e^{-i\frac{2\pi}{N_0}kn} + \frac{1}{\sqrt{N_0}} \sum_{l=0}^{N_0-1} \boldsymbol{e}(l) e^{-i\frac{2\pi}{N_0}ln} \\
&= s(n) \cdot \boldsymbol{h}\left(\frac{2\pi}{N_0}n\right) + \frac{1}{\sqrt{N_0}} \sum_{l=0}^{N_0-1} \boldsymbol{e}(l) e^{-i\frac{2\pi}{N_0}ln}
\end{aligned}
\tag{5.2.29}
$$

where we have made use of (5.2.28) and where we have defined

$$
\boldsymbol{h}\left(\frac{2\pi}{N_0}n\right) = \sum_{l=0}^{L} \boldsymbol{h}_l \exp\left(-i\frac{2\pi}{N_0}ln\right) \tag{5.2.30}
$$

To show the fourth equality in (5.2.29), note that the summation over l yields zero unless $m = n$. Hence in the summation over m, only the term for which $m = n$ remains.

Equation (5.2.29) shows that inverse Fourier transformation at the transmitter together with Fourier transformation of the received data make the frequency selective channel act as a *flat* fading channel with gain $\boldsymbol{h}(2\pi n/N_0)$ for each data symbol, or subcarrier. Moreover, it demonstrates that all signal processing associated with the equalization in an OFDM system can be carried out via fast Fourier transforms. This fact is one of the major advantages of OFDM compared to direct transmission as in Section 5.2.1.

A Matrix-Algebraic Framework

It is instructive to study OFDM in a matrix-algebraic framework. Doing so will also help us establish a compact and convenient expression for the ML metric associated with the detection of $s(n)$ given $\boldsymbol{y}(n)$, as well as analyze the performance of OFDM with receive diversity.

If we let $\boldsymbol{T}$ be an $N_0 \times N_0$ Fourier matrix whose $(k,l)th$ element is equal to:

$$\boldsymbol{T}_{k,l} = \frac{1}{\sqrt{N_0}} \exp\left(-i2\pi \frac{(k-1)(l-1)}{N_0}\right) \tag{5.2.31}$$

we can write the transmitted data in (5.2.26) in a vector form as:

$$\boldsymbol{x} = \boldsymbol{T}^H \boldsymbol{s} \tag{5.2.32}$$

where

$$\boldsymbol{x} = [x(0) \ \cdots \ x(N_0-1)]^T \tag{5.2.33}$$

$$\boldsymbol{s} = [s(0) \cdots s(N_0-1)]^T \tag{5.2.34}$$

Let $\boldsymbol{H}_c$ be the following $n_r N_0 \times N_0$ matrix:

$$\boldsymbol{H}_c \triangleq \begin{bmatrix} \boldsymbol{h}_0 & \mathbf{0} & \cdots & & \mathbf{0} & \boldsymbol{h}_L & & \cdots & \boldsymbol{h}_1 \\ \boldsymbol{h}_1 & \boldsymbol{h}_0 & \ddots & & & \ddots & \boldsymbol{h}_L & \cdots & \boldsymbol{h}_2 \\ \vdots & \ddots & \ddots & & & & & \ddots & \vdots \\ & & \ddots & & \ddots & & & \ddots & \boldsymbol{h}_L \\ & & & & \ddots & & \ddots & \ddots & \mathbf{0} \\ \boldsymbol{h}_L & & & & & & & & \vdots \\ \mathbf{0} & \ddots & & & & & \ddots & \ddots & \\ & \ddots & & & & & & & \mathbf{0} \\ \mathbf{0} & \cdots & \mathbf{0} & \boldsymbol{h}_L & & \cdots & & \boldsymbol{h}_1 & \boldsymbol{h}_0 \end{bmatrix} \tag{5.2.35}$$

Owing to the structure of the cyclic preamble, the received signal

$$\boldsymbol{y} = \left[\boldsymbol{y}^T(0) \ \cdots \ \boldsymbol{y}^T(N_0-1)\right]^T \tag{5.2.36}$$

can be written as:

$$\boldsymbol{y} = \boldsymbol{H}_c \boldsymbol{x} + \boldsymbol{e} = \boldsymbol{H}_c \boldsymbol{T}^H \boldsymbol{s} + \boldsymbol{e} \tag{5.2.37}$$

The matrix $\boldsymbol{H}_c$ is a block-circulant matrix and hence it can be diagonalized as follows (see, e.g., [Moon and Stirling, 2000, Chap. 8] or Appendix A.2):

$$\boldsymbol{H}_c = (\boldsymbol{T}^H \otimes \boldsymbol{I}_{n_r})\boldsymbol{\Delta}\boldsymbol{T} \tag{5.2.38}$$

where

$$\boldsymbol{\Delta} = \begin{bmatrix} \boldsymbol{h}(0) & \mathbf{0} & \cdots & \mathbf{0} \\ \mathbf{0} & \boldsymbol{h}\left(\frac{2\pi}{N_0}\right) & \ddots & \vdots \\ \vdots & \ddots & \ddots & \mathbf{0} \\ \mathbf{0} & \cdots & \mathbf{0} & \boldsymbol{h}\left(\frac{2\pi}{N_0}(N_0-1)\right) \end{bmatrix} \tag{5.2.39}$$

is an $n_r N_0 \times N_0$ block-diagonal matrix.

ML Detection

The ML detector for OFDM with receive diversity is most easily derived in the matrix-algebraic framework of the last section. From (5.2.37) we see that the ML detection of $\boldsymbol{s}$ given $\boldsymbol{y}$ amounts to minimizing the following metric:

$$\left\|\boldsymbol{y} - \boldsymbol{H}_c \boldsymbol{T}^H \boldsymbol{s}\right\|^2 \tag{5.2.40}$$

Since $\boldsymbol{T}^H\boldsymbol{T} = \boldsymbol{I}$, the ML metric can be simplified as follows:

$$\begin{aligned} \left\|\boldsymbol{y} - \boldsymbol{H}_c \boldsymbol{T}^H \boldsymbol{s}\right\|^2 &= \left\|\boldsymbol{y} - (\boldsymbol{T}^H \otimes \boldsymbol{I}_{n_r})\boldsymbol{\Delta}\boldsymbol{T}\boldsymbol{T}^H\boldsymbol{s}\right\|^2 \\ &= \|(\boldsymbol{T} \otimes \boldsymbol{I}_{n_r})\boldsymbol{y} - \boldsymbol{\Delta}\boldsymbol{s}\|^2 \\ &= \left\|(\boldsymbol{\Delta}^H\boldsymbol{\Delta})^{-1/2}\boldsymbol{\Delta}^H(\boldsymbol{T} \otimes \boldsymbol{I}_{n_r})\boldsymbol{y} - (\boldsymbol{\Delta}^H\boldsymbol{\Delta})^{1/2}\boldsymbol{s}\right\|^2 + \text{const.} \\ &= \sum_{n=0}^{N_0-1} \left|\frac{\boldsymbol{h}^H(2\pi n/N_0)}{\|\boldsymbol{h}(2\pi n/N_0)\|}\boldsymbol{z}(n) - \|\boldsymbol{h}(2\pi n/N_0)\| s(n)\right|^2 + \text{const.} \\ &= \sum_{n=0}^{N_0-1} \left(\|\boldsymbol{h}(2\pi n/N_0)\|^2 \cdot \left|s(n) - \hat{s}(n)\right|^2 \right) + \text{const.} \end{aligned} \tag{5.2.41}$$

where

$$\hat{s}(n) \triangleq \frac{\boldsymbol{h}^H(2\pi n/N_0)\boldsymbol{z}(n)}{\|\boldsymbol{h}(2\pi n/N_0)\|^2} \tag{5.2.42}$$

and $\{z(n)\}$ are n_r-vectors defined according to (5.2.29), or equivalently, using matrix notation

$$\begin{bmatrix} z(0) \\ \vdots \\ z(N_0 - 1) \end{bmatrix} \triangleq (T \otimes I_{n_r}) y \tag{5.2.43}$$

Equation (5.2.41) shows that the ML detection of s decouples into N_0 easily solved scalar detection problems for *flat* channels.

Clearly, $\hat{s}(n)$ is linear in y and hence it is a circular Gaussian random variable. We have that

$$\begin{aligned} E\left[\hat{s}(n)\right] &= s(n) \\ E\left[|\hat{s}(n) - s(n)|^2\right] &= \frac{\sigma^2}{\|h(2\pi n/N_0)\|^2} \end{aligned} \tag{5.2.44}$$

It follows that

$$\hat{s}(n) \sim N_C\Big(s(n), \frac{\sigma^2}{\|h(2\pi n/N_0)\|^2}\Big) \tag{5.2.45}$$

Therefore, $\hat{s}(n)$ can be interpreted as the output of a flat fading scalar channel with SNR:

$$\mathrm{SNR} = \frac{\|h(2\pi n/N_0)\|^2}{\sigma^2} \cdot E\left[|s(n)|^2\right] \tag{5.2.46}$$

By applying Theorem 4.1 (see page 42) with $\gamma^2 = \|h(2\pi n/N_0)\|^2$, it follows that a diversity order equal to n_r is achieved.

Note that all computations involved in the above algorithms for the equalization of an OFDM system can easily be carried out via fast Fourier transforms. The price paid for this simple receiver structure (as compared to that using a known preamble and postamble and a Viterbi equalizer) is the loss of multipath diversity, although multipath diversity can be recovered (to some extent) via coding across the subcarriers (see Section 8.2.6).

5.3 Summary and Discussion

Receive diversity is implemented in many practical systems. In this chapter, we have derived optimal (in a maximum-likelihood sense) detection rules for both frequency flat and frequency-selective SIMO channels. For flat fading channels (see

Section 5.1), we deduced the classical maximum-ratio combining formula and concluded that a diversity order equal to the number of receive antennas (viz. n_r) can be achieved in Rayleigh fading. For frequency-selective channels (see Section 5.2), we first provided a maximum-likelihood framework for detection using the symbol-spaced discrete-time model of Chapter 2, and then established conditions under which the maximum diversity order, viz. $n_r(L+1)$ can be achieved. In particular, we found that a diversity order equal to $n_r(L+1)$ can be achieved if all data are transmitted in blocks (cf. Section 2.2) and the preambles or postambles are known to the receiver.

We also studied orthogonal frequency division multiplexing (OFDM) for receive diversity systems in some detail. The principle of OFDM is fairly simple: by means of a cyclic preamble and a Fourier-transformation of the transmitted data, the frequency-selective channel is essentially converted into a set of parallel flat-fading channels called subcarriers, which are affected by uncorrelated noise (yet, the fading is *not* independent on the different subcarriers). Although OFDM is suboptimal in that it sacrifices the multipath diversity (i.e., it achieves a diversity order equal to n_r only), it does have a number of distinct advantages. For example, provided that the preamble is chosen properly (viz., taken equal to the last samples of the Fourier-transformed symbol sequence), ML symbol detection can be accomplished via a single fast Fourier transform; for this reason OFDM is generally considered to be computationally attractive even for channels with very long delay spread. A general discussion of OFDM and its advantages and disadvantages can be found in [May and Rohling, 2001].

The results in this chapter are based on the model in Chapter 2 and the error probability results of Chapter 4, and they can be viewed as background material for the rest of the text, which will focus on transmit diversity. Note that the general focus of this book is on all-digital implementations of antenna diversity systems and therefore our treatment has omitted a discussion on suboptimal diversity combining strategies (such as selection or switched combining). Details about these topics, along with a more general discussion of diversity combining can be found in, for example, [Stüber, 2001, Chap. 6] and [Eng et al., 1996].

5.4 Problems

1. Extend the calculations in Section 5.1 to the case where the receiver noise is spatially colored, i.e.,

$$\boldsymbol{e} \sim N_C(\mathbf{0}, \boldsymbol{\Lambda}) \tag{5.4.1}$$

2. Extend the analysis in Section 5.1 to the case in which the fading is correlated,

that is,

$$h \sim N_C(\mathbf{0}, \boldsymbol{\Upsilon}) \tag{5.4.2}$$

instead of (5.1.3).

3. Prove that $\gamma_{-l}^* = \gamma_l$ in (5.2.12)

4. Explain why (5.2.29) does *not* hold if a fixed preamble is used (as opposed to a cyclic one).

5. Derive the ML detector for OFDM, along with its error performance, starting from (5.2.29) and without going via the matrix-algebraic framework.

Hint: Prove first that the noise term in (5.2.29) is white (i.e., independent for different values of n) and that $z(n)$ is a sufficient statistic for the detection of $s(n)$.

6. A receive-diversity system with one transmit antenna and two receive antennas transmits BPSK symbols over a frequency-flat Rayleigh fading channel. Suppose that the two channel gains are zero-mean complex Gaussian random variables with covariance matrix as follows:

$$E\left[\begin{bmatrix} h_1 \\ h_2 \end{bmatrix} \begin{bmatrix} h_1^* & h_2^* \end{bmatrix}\right] = \begin{bmatrix} 1 & \rho \\ \rho^* & 1 \end{bmatrix} \tag{5.4.3}$$

where ρ is a complex number such that $|\rho| < 1$. Hence, the two channels are "equally good," but they are correlated with one another. Assume that the system uses ML detection.

(a) Determine the Chernoff bound on the average (over the channel) error probability, as a function of the noise variance σ^2 and ρ.

(b) Find the value of ρ that minimizes this error probability bound. Discuss the result.

7. Study a system operating over a frequency flat channel with one transmit antenna and n_r receive antennas. Suppose that a single symbol s is transmitted and let $\boldsymbol{h}$ be a vector of complex channel gains.

(a) Derive the probability of a detection error (conditioned on $\boldsymbol{h}$), assuming ML detection.

(b) Suppose that

$$\boldsymbol{h} \sim N_C(\mathbf{0}, \rho^2 \boldsymbol{I}) \tag{5.4.4}$$

Derive a bound on average error probability.

(c) Let $\rho^2 = 1$. What happens to the average error probability when the number of receive antennas grows without bound ($n_r \to \infty$)?

(d) Suppose that n_r grows without bound, but at the same time the transmit power is scaled down according to $\rho^2 = 1/n_r$. Study the average error probability when $n_r \to \infty$. Compare to the Chernoff bound of the error probability for an AWGN channel with unit channel gain.

Chapter 6

TRANSMIT DIVERSITY AND SPACE-TIME CODING

This chapter will take the first step towards exposing the reader to the concept of transmit diversity. We begin by studying a system where the channel is known to the transmitter. Next we assume that the channel is unknown at the transmitter site and show how transmit diversity can be achieved via repeated transmission of a single symbol, and how such transmit diversity is related to the receive diversity discussed in the previous chapter. Finally, we provide a more systematic discussion of space-time coding and also set the framework for the rest of this book. In most parts of this chapter, we assume a frequency flat channel (frequency selective channels will be discussed in Chapter 8).

6.1 Optimal Beamforming with Channel Known at Transmitter

As a preparation we shall begin by studying a system with $n_r \geq 1$ receive and $n_t > 1$ transmit antennas, *where both the transmitter and receiver know the propagation channel.* This knowledge about the channel can be used to adapt the weights for each transmit antenna in such a way that the SNR at the receiver is maximized. Doing so is sometimes called "beamforming," although it (like the beamforming in Section 5.1) may not have the physical interpretation of forming a beam. All the analysis presented in this section is straightforward; some of it can also be found in [Ganesan and Stoica, 2001b].

Consider the transmission of a certain information symbol s. Effectively, by using beamforming, we transmit the following n_t-vector:

$$\boldsymbol{x} = \boldsymbol{w}s \tag{6.1.1}$$

where $\boldsymbol{w}$ is a weight vector to be chosen by the transmitter. The received signal vector, of length n_r, is equal to:

$$\boldsymbol{y} = \boldsymbol{H}\boldsymbol{w}s + \boldsymbol{e} \tag{6.1.2}$$

where $\boldsymbol{H}$ is the $n_r \times n_t$ channel gain matrix, and $\boldsymbol{e}$ is a vector of noise. Assuming again that the noise is white with variance σ^2, and proceeding in a manner similar to that in Section 5.1, we find that the ML detection of s given $\boldsymbol{y}$ amounts to minimizing the following metric:

$$\begin{aligned} \|\boldsymbol{y} - \boldsymbol{H}\boldsymbol{w}s\|^2 &= \|\boldsymbol{y}\|^2 - 2\mathrm{Re}\left\{\boldsymbol{y}^H\boldsymbol{H}\boldsymbol{w}s\right\} + \|\boldsymbol{H}\boldsymbol{w}\|^2|s|^2 \\ &= \|\boldsymbol{H}\boldsymbol{w}\|^2 \cdot \left|s - \frac{\boldsymbol{w}^H\boldsymbol{H}^H\boldsymbol{y}}{\|\boldsymbol{H}\boldsymbol{w}\|^2}\right|^2 + \text{const.} \\ &= \|\boldsymbol{H}\boldsymbol{w}\|^2 \cdot |s - \hat{s}|^2 + \text{const.} \end{aligned} \tag{6.1.3}$$

where the constant term does not depend on s and where the quantity

$$\hat{s} \triangleq \frac{\boldsymbol{w}^H\boldsymbol{H}^H\boldsymbol{y}}{\|\boldsymbol{H}\boldsymbol{w}\|^2} \sim N_C\left(s, \frac{\sigma^2}{\|\boldsymbol{H}\boldsymbol{w}\|^2}\right) \tag{6.1.4}$$

can be interpreted as the output of an AWGN channel with SNR:

$$\text{SNR} = \frac{\|\boldsymbol{H}\boldsymbol{w}\|^2}{\sigma^2} \cdot E\left[|s|^2\right] \tag{6.1.5}$$

The BER performance of the ML detector depends only on the SNR in (6.1.5). Hence, given that the transmitter knows the channel, it can optimize $\boldsymbol{w}$ so as to maximize the SNR in $\hat{s}$ (or essentially equivalent, minimize the BER). For this optimization to be meaningful we must constrain the transmit power, otherwise the problem would yield an unbounded solution (an arbitrarily high SNR can be achieved by choosing $\boldsymbol{w}$ large enough).

Let us maximize the SNR subject to the constraint that the norm of $\boldsymbol{w}$ is bounded by a constant γ^2, i.e.:

$$\|\boldsymbol{w}\|^2 \leq \gamma^2 \tag{6.1.6}$$

If $E[|s|^2] = 1$, this assumption corresponds to requiring that the total transmit power (for all antennas) is bounded by γ^2. In mathematical terms, the optimization problem takes on the following form:

$$\begin{aligned} \max_{\boldsymbol{w}} \quad & \|\boldsymbol{H}\boldsymbol{w}\|^2 \\ \text{s.t.} \quad & \|\boldsymbol{w}\|^2 \leq \gamma^2 \end{aligned} \tag{6.1.7}$$

To solve this maximization problem note that

$$\frac{\|\boldsymbol{H}\boldsymbol{w}\|^2}{\|\boldsymbol{w}\|^2} = \frac{\boldsymbol{w}^H\boldsymbol{H}^H\boldsymbol{H}\boldsymbol{w}}{\boldsymbol{w}^H\boldsymbol{w}} \leq \lambda_{\max}(\boldsymbol{H}^H\boldsymbol{H}) \tag{6.1.8}$$

with equality if $\boldsymbol{w}$ is proportional to the eigenvector of $\boldsymbol{H}^H\boldsymbol{H}$ that corresponds to the largest eigenvalue (see, e.g., [HORN AND JOHNSON, 1985, TH. 4.2.2]). Hence, the $\boldsymbol{w}_{\text{opt}}$ that solves (6.1.7) is the eigenvector of $\boldsymbol{H}^H\boldsymbol{H}$ that corresponds to the largest eigenvalue and is normalized such that $\|\boldsymbol{w}_{\text{opt}}\|^2 = \gamma^2$. The resulting SNR is

$$\text{SNR}_{\text{opt}} = \frac{\gamma^2}{\sigma^2}\lambda_{\max}(\boldsymbol{H}^H\boldsymbol{H}) \cdot E\left[|s|^2\right] \tag{6.1.9}$$

The special case of $n_r = 1$ is treated in the Example 6.1 below. In the general case of $n_r \geq 1$, we can bound the SNR in (6.1.9) as follows:

$$\text{SNR}_{\text{opt}} = \frac{\gamma^2}{\sigma^2}\lambda_{\max}(\boldsymbol{H}^H\boldsymbol{H}) \cdot E\left[|s|^2\right] \geq \frac{\gamma^2}{n_t\sigma^2}\|\boldsymbol{H}\|^2 \cdot E\left[|s|^2\right] \tag{6.1.10}$$

Suppose that the matrix $\boldsymbol{H}$ has independent zero-mean Gaussian entries (i.e., that the channel is Rayleigh fading), and that the transmitter chooses a new optimal transmit weight vector $\boldsymbol{w}$ for each channel realization. Then application of Theorem 4.1 (see page 42) (with $\gamma^2 = \|\boldsymbol{H}\|^2$) together with (6.1.10) shows that the BER of the system decays as $\sigma^{2n_tn_r}$ when $\sigma^2 \to 0$. Hence, we can say that a diversity of order n_tn_r is achieved (although the term "diversity order" appears to be more rarely used in the presence of channel state information at the transmitter).

Example 6.1: Beamforming for MISO-systems.
In the special case of $n_r = 1$, the optimization of (6.1.7) simplifies somewhat since in that case $\boldsymbol{H}$ becomes a row-vector $\boldsymbol{h}^T$. By the Cauchy-Schwarz inequality, it holds that

$$|\boldsymbol{h}^T\boldsymbol{w}|^2 \leq \|\boldsymbol{h}\|^2\|\boldsymbol{w}\|^2 = \gamma^2\|\boldsymbol{h}\|^2 \tag{6.1.11}$$

with equality if and only if $\boldsymbol{w}$ is proportional to $\boldsymbol{h}^*$; applying the constraint on the transmit power, we find that

$$\boldsymbol{w}_{\text{opt}} = \gamma \cdot \frac{\boldsymbol{h}^*}{\|\boldsymbol{h}\|} \tag{6.1.12}$$

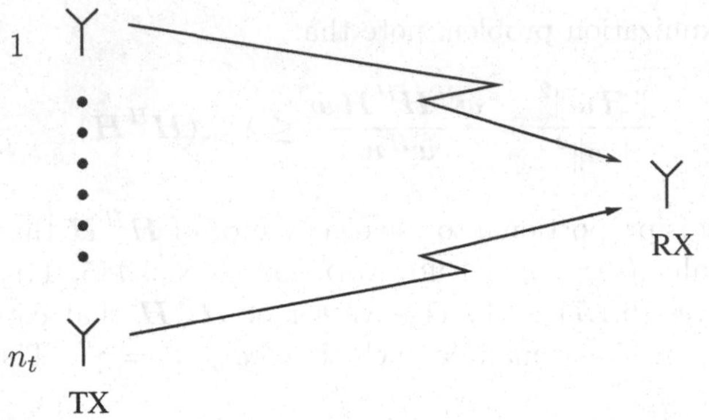

Figure 6.1. A transmit diversity system with $n_t \geq 1$, $n_r = 1$.

The resulting SNR is

$$\text{SNR}_{\text{opt}} = \frac{|\boldsymbol{h}^T \boldsymbol{w}_{\text{opt}}|^2}{\sigma^2} \cdot E\left[|s|^2\right] = \frac{\gamma^2}{\sigma^2}\frac{(\boldsymbol{h}^H \boldsymbol{h})^2}{\|\boldsymbol{h}\|^2} \cdot E\left[|s|^2\right] = \frac{\gamma^2}{\sigma^2}\|\boldsymbol{h}\|^2 \cdot E\left[|s|^2\right] \tag{6.1.13}$$

If the propagation coefficients are i.i.d. Gaussian random variables as in the previous discussion, and the transmitter adapts $\boldsymbol{w}$ for each realization of $\boldsymbol{h}$, it follows immediately from Theorem 4.1 and (6.1.13) that the BER decays as σ^{2n_t} when σ^2 decreases. Moreover, if the total radiated power from all n_t antennas is fixed to unity, i.e., $\gamma^2 = 1$, the SNR in (6.1.13) becomes

$$\text{SNR} = \frac{\|\boldsymbol{h}\|^2}{\sigma^2} \cdot E\left[|s|^2\right] \tag{6.1.14}$$

which is equal to the corresponding SNR for receive diversity (5.1.10). Hence, if the transmitter knows the channel, it is easy to reproduce the performance of a receive diversity system by using transmit diversity. ■

6.2 Achieving Transmit Diversity

Let us now study the case when the channel is unknown to the transmitter but known to the receiver, and investigate whether it is possible to achieve transmit diversity by using beamforming techniques. First suppose that the transmitter simply chooses a constant beamforming vector $\boldsymbol{w}$ (independent of $\boldsymbol{H}$), and transmits $\boldsymbol{w} \cdot s$ as in the preceding Section 6.1. To analyze the performance of the system we observe that for a fixed $\boldsymbol{w}$, and provided that $\boldsymbol{H}$ has i.i.d. Gaussian elements with variance ρ^2, $\boldsymbol{Hw}$ in (6.1.2) will have a Gaussian distribution:

$$\boldsymbol{Hw} \sim N_C(\boldsymbol{0}, \rho^2\|\boldsymbol{w}\|^2\boldsymbol{I}) \tag{6.2.1}$$

Consequently, the SNR in (6.1.5) is distributed in the same way as if we had a rank-one channel with complex gain vector equal to $\boldsymbol{Hw}$; hence the diversity order achieved is equal to n_r, and consequently no transmit diversity is gained (cf. Theorem 4.1).

In the absence of channel state information at the transmitter, it is possible to achieve transmit diversity by transmitting the same symbol s during N time epochs using different transmit weight vectors $\{\boldsymbol{w}_n\}_{n=0}^{N-1}$. Let

$$\boldsymbol{W} = [\boldsymbol{w}_0 \ \cdots \ \boldsymbol{w}_{N-1}] \tag{6.2.2}$$

Then the "code matrix" associated with this transmission technique is of dimension $n_t \times N$ and given by:

$$\boldsymbol{X} = \boldsymbol{W} \cdot s \tag{6.2.3}$$

The received signal is the following matrix of dimension $n_r \times N$:

$$\boldsymbol{Y} = \boldsymbol{HX} + \boldsymbol{E} = \boldsymbol{HW}s + \boldsymbol{E} \tag{6.2.4}$$

where $\boldsymbol{E}$ is an $n_r \times N$ matrix of noise. The model (6.2.4) is of the same form as that studied in Chapter 4 and hence application of Theorem 4.2 shows that transmit diversity is achieved as long as $\boldsymbol{W}$ has rank n_t. Clearly,

$$\text{rank}\{\boldsymbol{W}\} = \text{rank}\{\boldsymbol{W}^H\boldsymbol{W}\} \leq n_t \tag{6.2.5}$$

and equality is only possible if $N \geq n_t$. Therefore we expect that to achieve full transmit diversity, we need to spread the energy from one information symbol over at least as many time intervals as there are transmit antennas. Note that Theorem 4.2 does not prove that $\text{rank}\{\boldsymbol{W}\} = n_t$ is a necessary condition to achieve transmit diversity: owing to the inequality in (4.2.9), Theorem 4.2 only establishes the sufficiency of the previous condition. However, we will see in the next subsection that the exponent of the error rate behaves as $n_r \cdot \text{rank}\{\boldsymbol{W}^H\boldsymbol{W}\}$ and therefore that $N \geq n_t$ is a *necessary* condition to achieve diversity.

6.2.1 The ML Detector

Next we shall address the question of choosing the optimal matrix $\boldsymbol{W}$ (in an error probability sense). However, before doing so, we derive the maximum-likelihood detector of s given $\boldsymbol{Y}$ and analyze some of its properties. The ML detection of s given $\boldsymbol{Y}$ amounts to minimizing

$$\begin{aligned}\|\boldsymbol{Y}-\boldsymbol{HW}s\|^2 &= \|\boldsymbol{Y}\|^2+|s|^2\|\boldsymbol{HW}\|^2-2\mathrm{Re}\{\mathrm{Tr}\{\boldsymbol{HWY}^H s\}\}\\ &= \|\boldsymbol{HW}\|^2\cdot|s-\hat{s}|^2+\text{const.}\end{aligned} \tag{6.2.6}$$

where the constant term does not depend on s and where

$$\hat{s} \triangleq \frac{\mathrm{Tr}\{\boldsymbol{W}^H\boldsymbol{H}^H\boldsymbol{Y}\}}{\|\boldsymbol{HW}\|^2} \sim N_C\left(s, \frac{\sigma^2}{\|\boldsymbol{HW}\|^2}\right) \tag{6.2.7}$$

Clearly, $\hat{s}$ has the same AWGN-interpretation as before, and the SNR in $\hat{s}$ is equal to:

$$\mathrm{SNR} = \frac{\|\boldsymbol{HW}\|^2}{\sigma^2}\cdot E\left[|s|^2\right] \tag{6.2.8}$$

Under the assumption of an i.i.d. Gaussian channel as before, we have that

$$\mathrm{vec}(\boldsymbol{HW}) = (\boldsymbol{W}^T\otimes\boldsymbol{I})\,\mathrm{vec}(\boldsymbol{H}) \sim N_C\big(\boldsymbol{0},\rho^2(\boldsymbol{W}^T\boldsymbol{W}^*\otimes\boldsymbol{I})\big) \tag{6.2.9}$$

and hence by Theorem 4.1 (see page 42) it follows that a diversity of at least order $n_r\cdot\mathrm{rank}\{\boldsymbol{W}^H\boldsymbol{W}\}$ is achieved (cf. the remark following (6.2.5)).

6.2.2 Minimizing the Conditional Error Probability

Suppose that s_0 is the true transmitted symbol. Then the probability that the ML detector takes an incorrect decision in favor of a wrong symbol $s \neq s_0$ is, for a given $\boldsymbol{H}$ (see Theorem 4.2):

$$P(s_0\to s) = Q\left(\sqrt{\frac{\|\boldsymbol{HW}\|^2}{2\sigma^2}\cdot|s-s_0|^2}\right) \tag{6.2.10}$$

The above error probability is a decreasing function of the SNR in (6.2.8). Hence, choosing $\boldsymbol{W}$ to minimize $P(s_0\to s)$ is equivalent to choosing $\boldsymbol{W}$ to maximize the SNR.

Let us consider the maximization of the SNR with respect to $\boldsymbol{W}$, and subject to the following *elemental power constraint:*

$$\lambda_{\max}\{\boldsymbol{WW}^H\} \leq \frac{\gamma^2}{n_t} \tag{6.2.11}$$

In words, the constraint in (6.2.11) states that the transmitted power, in any particular direction, is bounded by γ^2. See, for instance, [STOICA AND GANESAN, 2002A] for more interpretations and details on the eigenvalue constraint in (6.2.11); similar constraints have also been used and are discussed in [SCAGLIONE ET AL., 2002]. In this section, this constraint is chosen for mathematical convenience. It implies, but is *not* equivalent to, the condition that the total transmitted power accumulated over the N time intervals is less than γ^2, i.e., that

$$\|\boldsymbol{W}\|^2 = \mathrm{Tr}\{\boldsymbol{W}\boldsymbol{W}^H\} \leq \gamma^2 \tag{6.2.12}$$

The resulting optimization problem can be formulated as:

$$\begin{aligned} &\max_{\boldsymbol{W}} \quad \|\boldsymbol{H}\boldsymbol{W}\|^2 \\ &\text{s.t.} \quad \lambda_{\max}\{\boldsymbol{W}\boldsymbol{W}^H\} \leq \frac{\gamma^2}{n_t} \end{aligned} \tag{6.2.13}$$

We have that

$$\|\boldsymbol{H}\boldsymbol{W}\|^2 = \mathrm{Tr}\{\boldsymbol{H}\boldsymbol{W}\boldsymbol{W}^H\boldsymbol{H}^H\} \leq \|\boldsymbol{H}\|^2 \cdot \lambda_{\max}(\boldsymbol{W}\boldsymbol{W}^H) \leq \|\boldsymbol{H}\|^2 \cdot \frac{\gamma^2}{n_t} \tag{6.2.14}$$

with equality in the first inequality if $\boldsymbol{W}\boldsymbol{W}^H$ is proportional to an identity matrix. It follows that the optimal $\boldsymbol{W}$ is given by any $\boldsymbol{W}$ that satisfies

$$\boldsymbol{W}\boldsymbol{W}^H = \frac{\gamma^2}{n_t}\boldsymbol{I} \tag{6.2.15}$$

In words, (6.2.15) states that we should distribute the transmitted power *uniformly* in space.

The SNR in (6.2.8) corresponding to the choice of $\boldsymbol{W}$ in (6.2.15) is given by:

$$\mathrm{SNR}_{\mathrm{opt}} = \frac{\gamma^2}{n_t}\frac{\|\boldsymbol{H}\|^2}{\sigma^2} \cdot E\left[|s|^2\right] \tag{6.2.16}$$

When $n_r = 1$, the SNR in (6.2.16) is exactly a factor n_t less than the SNR in (6.1.13) corresponding to the case when the transmitter knows the channel, provided that the total transmit power is the same in both cases. For the case $n_r > 1$, the difference in SNR between an informed and an uninformed transmitter can be less than a factor n_t (cf. (6.1.10)). The difference in the achieved SNR between an informed and an uninformed transmitter is sometimes called *array gain*. The array gain can be understood intuitively as follows: a transmitter that knows the channel can "steer a beam" in the direction of the receiver, but this is not possible if the channel is unknown. In the latter case, some energy will always be wasted in the directions where the receiver is not present.

6.2.3 Minimizing the Average Error Probability

We next minimize the *average* (over $\boldsymbol{H}$) error probability under the constraint (6.2.12), which is more commonly used than (6.2.11).

If we assume as before that

$$\text{vec}(\boldsymbol{H}) \sim N_C(\boldsymbol{0}, \rho^2 \boldsymbol{I}) \tag{6.2.17}$$

then the average error probability can be bounded by using Theorem 4.2:

$$E_{\boldsymbol{H}}[P(s_0 \to s)] \leq |\boldsymbol{W}\boldsymbol{W}^H|^{-n_r} |s - s_0|^{-n_r} \left(\frac{\rho^2}{4\sigma^2}\right)^{-n_r n_t} \tag{6.2.18}$$

Clearly, a diversity order equal to $n_r n_t$ is achieved as long as $\boldsymbol{W}$ has rank n_t. Let us minimize the average error probability with respect to $\boldsymbol{W}$ under the power constraint (6.2.12):

$$\|\boldsymbol{W}\|^2 = \text{Tr}\{\boldsymbol{W}\boldsymbol{W}^H\} \leq \gamma^2 \tag{6.2.19}$$

We have that

$$|\boldsymbol{W}\boldsymbol{W}^H| \leq \left(\frac{1}{n_t}\text{Tr}\{\boldsymbol{W}\boldsymbol{W}^H\}\right)^{n_t} \leq \left(\frac{\gamma^2}{n_t}\right)^{n_t} \tag{6.2.20}$$

with equality if and only if $\boldsymbol{W}$ is unitary and satisfies:

$$\boldsymbol{W}\boldsymbol{W}^H = \frac{\gamma^2}{n_t}\boldsymbol{I} \tag{6.2.21}$$

which is the same as the optimality result (6.2.15) derived by minimizing the conditional error probability (under the elemental power constraint).

6.2.4 Discussion

Under an elemental power constraint, the conditional error probability is minimized by (6.2.15) and so is the average error probability under a total power constraint. One possible choice of $\boldsymbol{W}$ that satisfies (6.2.15) is

$$\boldsymbol{W} = \frac{\gamma}{\sqrt{n_t}}\boldsymbol{I} \tag{6.2.22}$$

With this choice of $\boldsymbol{W}$, the transmission scheme becomes equivalent to spatial cycling, i.e., the symbol s is transmitted during time interval n via antenna number n for $n = 1, \ldots, n_t$ and the remaining antennas are quiet. This transmission scheme does achieve transmit diversity, yet in a crude way. Its main problem is that only one complex symbol is transmitted during $N = n_t$ time intervals and hence the effective data rate is $R = 1/N = 1/n_t$.

To conclude, we have found that the task of achieving transmit diversity, without any requirements on transmission rate, is quite easy. The major question is whether we can find a more sophisticated transmission scheme with a higher rate than $1/n_t$ that achieves transmit diversity and at the same time keeps the decoding complexity low. This question has motivated the last decade's research efforts in MIMO communication theory, and a systematic approach to this problem leads to the concept of *space-time coding.*

6.3 Space-Time Coding

From the discussion in Section 6.2 we know that if the transmitter does not know the channel, it is necessary to code *across both space and time* to achieve transmit diversity. We saw how diversity could be achieved by transmitting one symbol s during a number of time intervals equal to that of transmit antennas; however, the rate of the so-obtained scheme was equal to $R = 1/n_t$. This observation leads inevitably to the question: can we spread symbols over space and time in a spirit similar to that in Section 6.2, but transmitting several symbols at the same time?

The question asked in the previous paragraph has led to the area that is now referred to as *space-time coding.* Many researchers have made important contributions to this area. Although an exhaustive list of all of them would be quite long, in the bibliography of this book (see page 264) we have tried to list a few often cited contributions. In this section, we will give a general overview of space-time coding methods. Of these methods, one particularly interesting technique (namely linear space-time block coding) will become the major topic of all subsequent chapters in this book.

6.3.1 Alamouti's Space-Time Code

One of the first space-time codes is due to Alamouti [ALAMOUTI, 1998], who studied the case of two transmit antennas ($n_t = 2$) and suggested to simultaneously transmit $n_s = 2$ complex symbols s_1 and s_2 during $N = 2$ time intervals by transmitting the following matrix:

$$\boldsymbol{X} = \begin{bmatrix} s_1 & s_2^* \\ s_2 & -s_1^* \end{bmatrix} \tag{6.3.1}$$

The matrix $\boldsymbol{X}$ is transmitted via the two transmit antennas as described in Section 2.1. For the moment, we ignore any scaling factor needed to normalize the transmit power. Note that the rate of the Alamouti code is equal to $R = 2/2 = 1$ (which is double the rate of the technique in Section 6.2).

To understand the benefits of Alamouti's code, we assume for simplicity of the discussion that we have one receive antenna, i.e., $n_r = 1$. In this case, the MIMO channel matrix $\boldsymbol{H}$ in Section 2.1 becomes a row vector, which we denote by

$$\boldsymbol{h}^T = \begin{bmatrix} h_1 & h_2 \end{bmatrix} \tag{6.3.2}$$

Let

$$\check{\boldsymbol{y}} = \begin{bmatrix} \check{y}_1 \\ \check{y}_2 \end{bmatrix} \tag{6.3.3}$$

be the received signal during the two time intervals (the reason for the notation $\check{(\cdot)}$ is that we will later take the complex conjugate of $\check{y}_2$, and denote the so-obtained signal vector by $\boldsymbol{y}$). This received signal can be written (cf. Section 2.1):

$$\check{\boldsymbol{y}}^T = \boldsymbol{h}^T \boldsymbol{X} + \check{\boldsymbol{e}}^T \tag{6.3.4}$$

where $\check{\boldsymbol{e}}$ is a vector of white noise with variance σ^2:

$$\check{\boldsymbol{e}} \sim N_C(\boldsymbol{0}, \sigma^2 \boldsymbol{I}) \tag{6.3.5}$$

Now let

$$\boldsymbol{y} \triangleq \begin{bmatrix} \check{y}_1 \\ \check{y}_2^* \end{bmatrix} \tag{6.3.6}$$

be the received signal but with the second element complex-conjugated and let

$$\boldsymbol{s} = [s_1 \; s_2]^T \tag{6.3.7}$$

Also, define the following matrix:

$$\boldsymbol{F} \triangleq \begin{bmatrix} h_1 & h_2 \\ -h_2^* & h_1^* \end{bmatrix} \tag{6.3.8}$$

Then from the above equations it follows that we can write

$$\boldsymbol{y} = \boldsymbol{F}\boldsymbol{s} + \boldsymbol{e} \tag{6.3.9}$$

where

$$\boldsymbol{e} \triangleq \begin{bmatrix} e_1 \\ e_2 \end{bmatrix} \triangleq \begin{bmatrix} \check{e}_1 \\ \check{e}_2^* \end{bmatrix} \sim N_C(\boldsymbol{0}, \sigma^2 \boldsymbol{I}) \tag{6.3.10}$$

is white noise. The ML detection of s_1 and s_2 amounts to minimizing:

$$\|\boldsymbol{y}-\boldsymbol{F}\boldsymbol{s}\|^2 = \|\boldsymbol{y}\|^2 + \|\boldsymbol{F}\boldsymbol{s}\|^2 - 2\mathrm{Re}\{\boldsymbol{s}^H\boldsymbol{F}^H\boldsymbol{y}\} \tag{6.3.11}$$

It is easy to see that $\boldsymbol{F}$ in (6.3.8) enjoys the following unitary property:

$$\boldsymbol{F}^H\boldsymbol{F} = \|\boldsymbol{h}\|^2 \cdot \boldsymbol{I} \tag{6.3.12}$$

Let us define:

$$\hat{\boldsymbol{s}} \triangleq \begin{bmatrix} \hat{s}_1 \\ \hat{s}_2 \end{bmatrix} \triangleq \frac{1}{\|\boldsymbol{h}\|^2}\boldsymbol{F}^H\boldsymbol{y} = \boldsymbol{s} + \frac{1}{\|\boldsymbol{h}\|^2}\boldsymbol{F}^H\boldsymbol{e} \tag{6.3.13}$$

Then the metric in (6.3.11) can be written:

$$\|\boldsymbol{y}-\boldsymbol{F}\boldsymbol{s}\|^2 = \|\boldsymbol{h}\|^2 \cdot \|\boldsymbol{s}-\hat{\boldsymbol{s}}\|^2 + \text{const.} \tag{6.3.14}$$

where the constant term does not depend on $\boldsymbol{s}$. Equation (6.3.14) shows that the ML detection of s_1 and s_2, given $\boldsymbol{y}$, decouples into two independent *scalar* detection problems. Moreover, since

$$E\left[\frac{1}{\|\boldsymbol{h}\|^4}\boldsymbol{F}^H\boldsymbol{e}\boldsymbol{e}^H\boldsymbol{F}\right] = \frac{\sigma^2}{\|\boldsymbol{h}\|^2}\boldsymbol{I} \tag{6.3.15}$$

the noise term in (6.3.13) is white and consequently

$$\hat{\boldsymbol{s}} \sim N_C\left(\boldsymbol{s}, \frac{\sigma^2}{\|\boldsymbol{h}\|^2}\boldsymbol{I}\right) \tag{6.3.16}$$

Equation (6.3.16) shows not only that the space-time channel decouples into two independent scalar channels, but also that for a given $\boldsymbol{h}$, the linear processing in (6.3.13) transforms the space-time channel into two parallel and independent AWGN channels. The SNR in each sub-channel is equal to

$$\mathrm{SNR} = \frac{\|\boldsymbol{h}\|^2}{\sigma^2} \cdot E\left[|s_n|^2\right] \tag{6.3.17}$$

and hence it follows from Theorem 4.1 (see page 42) that a diversity of order two is achieved in Rayleigh fading. Therefore, for the case of $n_t = 2$ transmit antennas and $n_r = 1$ receive antenna, the Alamouti code obtains the same diversity benefit as if we instead had a system with one transmit and two receive antennas.

By comparing (6.3.17) to the SNR for a two-branch receive diversity system (5.1.10) and to the SNR in (6.1.14) for transmit diversity with channel state information at the transmitter (and with $n_r = 1$), we find that these three SNR expressions are equal to one another. Hence, the performance of a receive diversity system, that of transmit beamforming and that of the Alamouti scheme are equal. However, using the Alamouti code as described in this section we transmit with power $E\left[|s_n|^2\right]$ via each antenna, and hence the total transmission power is effectively doubled compared to the receive diversity system, and compared to the beamforming system. We therefore conclude that *relative to a receive diversity system, as well as relative to a system that uses optimal transmit beamforming, the Alamouti code (using the same total transmit power) provides the same diversity order, but has a 3 dB loss in error performance.*

If we normalize the transmit power so that the total transmitted energy per information symbol is equal to γ^2, the SNR per sub-channel becomes:

$$\text{SNR} = \frac{\gamma^2}{2} \cdot \frac{\|\boldsymbol{h}\|^2}{\sigma^2} \cdot E\left[|s_n|^2\right] \tag{6.3.18}$$

This formula is valid for a general number of receive antennas n_r (we leave the proof to the reader), and it is directly comparable to (5.1.10) (for receive diversity), (6.1.9) and (6.1.14) (for beamforming with perfect channel state information at the transmitter), and (6.2.16) (for transmit diversity without channel state information at the transmitter).

Extending the receiver structure of the Alamouti code to more than one receive antenna is straightforward: this extension will follow as a special case of the discussion in Chapter 7. A more interesting question than this extension is whether it is possible to generalize the Alamouti code to more than two transmit antennas. The answer to this question is orthogonal STBC, which was introduced by [Tarokh et al., 1999a]. In this text, OSTBC will be discussed in Section 7.4.

6.3.2 Space-Time Block Coding (STBC)

We have already defined the concept of space-time block coding (STBC) implicity in Chapter 2. In its most general form, STBC can seen as a way of mapping a set of n_s complex symbols $\{s_1, \ldots, s_{n_s}\}$ onto a matrix $\boldsymbol{X}$ of dimension $n_t \times N$ that is transmitted as described in Chapter 2. The mapping

$$\{s_1, \ldots, s_{n_s}\} \to \boldsymbol{X} \tag{6.3.19}$$

may in principle take on any form.

6.3.3 Linear STBC

Linear space-time block codes choose the mapping (6.3.19) to be linear in the symbols $\{s_n\}$. Specifically, if $\{\boldsymbol{A}_n, \boldsymbol{B}_n\}$ is a set of fixed matrices, $\boldsymbol{X}$ is formed according to:

$$\boldsymbol{X} = \sum_{n=1}^{n_s} (\bar{s}_n \boldsymbol{A}_n + i\tilde{s}_n \boldsymbol{B}_n) \tag{6.3.20}$$

All transmission techniques discussed so far in this chapter can be interpreted as linear STBCs. Linear STBC is an important subclass of space-time coding techniques, which will form the focus of the rest of this book.

6.3.4 Nonlinear STBC

Although we will see that linear STBCs have quite appealing properties from both a performance and implementation point of view, they constitute only a small subclass of the possible mappings $\{s_1, \ldots, s_{n_s}\} \to \boldsymbol{X}$. The current knowledge on nonlinear STBCs appears to be rather limited but there are already a few contributions available. For instance, [SANDHU ET AL., 2002] suggests to first obtain a set of P complex numbers $\{\phi_p\}$ via a nonlinear function

$$\{s_1, \ldots, s_{n_s}\} \to \{\phi_1, \ldots, \phi_P\} \tag{6.3.21}$$

and then form $\boldsymbol{X}$ in a fashion similar to (6.3.20):

$$\boldsymbol{X} = \sum_{n=1}^{P} (\bar{\phi}_n \boldsymbol{A}_n + i\tilde{\phi}_n \boldsymbol{B}_n) \tag{6.3.22}$$

As the topic of non-linear STBC appears to be little understood, it will not be treated further in this text.

6.3.5 Space-Time Trellis Coding

Space-time trellis coding (STTC) constitutes one of the first proposed and studied classes of space-time codes. In a space-time trellis coding scheme, a stream of data $s(n)$ is encoded via n_t convolutional encoders (or via one convolutional encoder with n_t outputs) to obtain n_t streams $x_1(n), \ldots, x_{n_t}(n)$. These n_t streams are then transmitted via the n_t transmit antennas. This is illustrated in Figure 6.2.

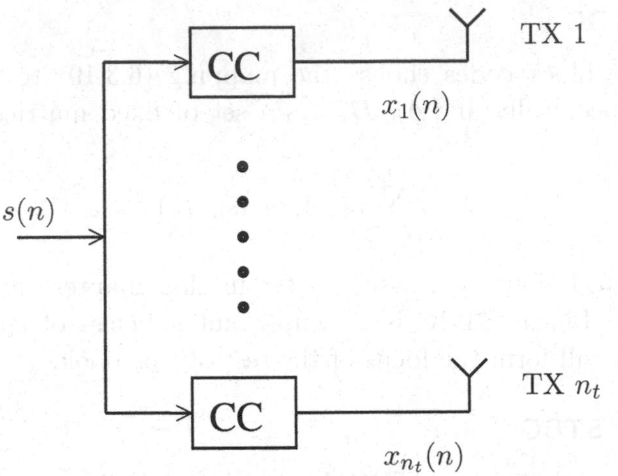

Figure 6.2. Space-time trellis coding. Here CC stands for convolutional coding.

Design Criteria for Space-Time Trellis Codes

Neglecting the preambles and postambles (cf. Section 2.2), the transmitted data matrix in a STTC scheme can be written as:

$$\boldsymbol{X} = \begin{bmatrix} \cdots & x_1(n) & x_1(n+1) & \cdots & x_1(n+l) & \cdots \\ \cdots & x_2(n) & x_2(n+1) & \cdots & x_2(n+l) & \cdots \\ \cdots & \cdots & \cdots & \cdots & \cdots & \cdots \\ \cdots & x_{n_t}(n) & x_{n_t}(n+1) & \cdots & x_{n_t}(n+l) & \cdots \end{bmatrix} \tag{6.3.23}$$

According to the results in Chapter 4, the performance of a space-time trellis code is related to the eigenvalues and the rank properties of the difference between pairs of different code matrices formed as in (6.3.23). The design of trellis codes is a relatively hard problem; some codes were first handcrafted in [TAROKH ET AL., 1998], and more systematic studies can be found in [BLUM, 2002], [GAMAL AND HAMMONS, JR., 2002]. In principle, the detection of a space-time trellis code requires a ML sequence detection algorithm, which is commonly implemented via the Viterbi algorithm. The complexity of such an algorithm grows exponentially with the memory length of the trellis code, and some authors have argued that space-time trellis are computationally rather burdensome.

It is outside the scope of this book to provide a detailed treatment of space-time trellis codes and their design. However, in a certain sense, it can be argued that the matrix $\boldsymbol{X}$ associated with STTC can be seen as a wide STBC matrix and that STTC and STBC therefore are embodiments of the same ideas. This

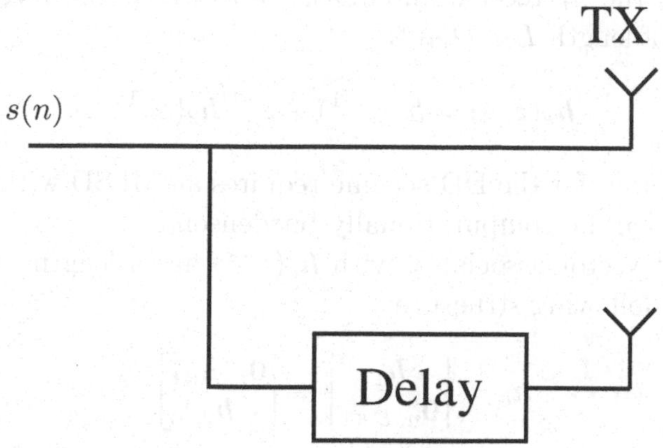

Figure 6.3. Delay diversity.

is certainly true at least to the extent that all analysis techniques (in particular including those in Chapter 4) are applicable to STBC as well as to STTC.

Delay Diversity (DD)

A special case of trellis coding is delay diversity (DD). We describe DD only for $n_t = 2$ transmit antennas, although the idea is in principle applicable to any number of transmit antennas. For DD, the first convolutional encoder is absent, and the second is simply replaced by a time delay. This means that the first antenna transmits the bitstream (or symbol stream) $\{s(n), s(n+1), \ldots\}$ whereas the second antenna transmits the same stream delayed by D symbol intervals: $\{s(n-D), s(n-D+1), \ldots\}$. Therefore the transmitted data are given by:

$$\begin{aligned} x_1(n) &= s(n) \\ x_2(n) &= s(n-D) \end{aligned} \tag{6.3.24}$$

where the delay D is a constant (see Figure 6.3).

Delay diversity as a means of achieving multipath and transmit diversity simultaneously was proposed by [WITTNEBEN, 1991]. Delay diversity is the simplest (and perhaps the first proposed) member of the class of space-time trellis codes, and it is relatively easy to understand intuitively and analyze. It can be used directly on a frequency-selective channel; therefore we assume such a channel in the rest of this section. If $\boldsymbol{h}_1(z^{-1})$ and $\boldsymbol{h}_2(z^{-1})$ denote the two n_r-vector valued FIR filters corresponding to the channel impulse response between the two transmit antennas and the n_r receive antennas, respectively, the receiver will see an effective channel of length $L + D + 1$:

$$\boldsymbol{h}_e(z^{-1}) = \boldsymbol{h}_1(z^{-1}) + z^{-D}\boldsymbol{h}_2(z^{-1}) \tag{6.3.25}$$

Therefore equalization for the DD scheme requires an MLSD with memory length $L + D + 1$, which can be computationally burdensome.

The coefficient vector associated with $\boldsymbol{h}_e(z^{-1})$ has a length of $n_r(L + D + 1)$ and possesses the following structure:

$$\boldsymbol{h}_e = \begin{bmatrix} \boldsymbol{h}_1 \\ \boldsymbol{0}_{n_r D\times 1} \end{bmatrix} + \begin{bmatrix} \boldsymbol{0}_{n_r D\times 1} \\ \boldsymbol{h}_2 \end{bmatrix} \tag{6.3.26}$$

Provided that the covariance matrix corresponding to $\boldsymbol{h}_e$ in (6.3.26) has full rank, the system is equivalent to a SIMO system with one transmit antenna and n_r receive antennas operating on a channel with $L + D + 1$ not fully correlated taps. Hence, assuming that the receiver knows the preamble or postamble, the analysis in Section 5.2.1 shows that a diversity of order $n_r(L + D + 1)$ can be achieved. A necessary condition for the covariance matrix of $\boldsymbol{h}_e$ to have full rank is that $D \leq L + 1$ (indeed, if $D > L+1$ then the vector $\boldsymbol{h}_e$ would contain zero elements); therefore, the maximum achievable diversity order with DD is $2n_r(L+1)$. Provided that this necessary condition holds, a sufficient condition is that $\boldsymbol{h}_1$ and $\boldsymbol{h}_2$ are independent and that the two covariance matrices associated with $\boldsymbol{h}_1$ and $\boldsymbol{h}_2$ both have full rank. The diversity order of DD can also be established in the block-transmission framework of Section 2.2, but we defer doing so to Chapter 8 (see Example 8.1 on page 131) where we discuss transmit diversity for frequency selective channels in a more systematic manner.

6.4 Summary and Discussion

While receive diversity (see Chapter 5) is a fairly mature technique, transmit diversity is an emerging topic. This chapter has introduced the concept of transmit diversity (for flat fading channels) from first principles. In Section 6.1 we studied optimal strategies for a transmitter that knows the channel perfectly; we found that the best scheme is to weigh the signal associated with each transmit antennas in a fashion similar to matched filtering, or maximum-ratio combining (cf. Section 5.1). Next, in Section 6.2 we went on to analyze the best transmission scheme for a transmitter with no knowledge of the propagation channel, given that there are no constraints on the information rate. We found that the error probability is minimized when the transmitted matrix $\boldsymbol{X}$ is proportional to a semi-unitary matrix. This is intuitively appealing since such a transmission scheme effectively sends uncorrelated streams with the same power via all antennas, which roughly speaking does not favor any given direction in space.

The problem of simultaneously achieving optimum error performance (as considered in Section 6.2) and maximal information rate leads to the concept of space-time coding, which was introduced in Section 6.3. In that section, we explained the general concept of space-time block codes and space-time trellis codes. Two special schemes that have received some attention due to their simplicity, namely delay diversity and Alamouti's code, were explained in some detail. A large and general class of space-time codes briefly introduced here, namely linear space-time block codes, will be the main topic of Chapter 7.

Most results in this chapter are based on the model in Section 2.1 and the error probability formulas in Chapter 4, and they are of a fundamental nature. These results, which will be implicitly used in the rest of the text, also relate to the discussion in Chapter 5 in that, to a large extent, the performance of receive diversity can be reproduced by using transmit diversity.

6.5 Problems

1. In Section 6.3.5, suppose that $\boldsymbol{h}_1$ and $\boldsymbol{h}_2$ are independent and that the two covariance matrices associated with $\boldsymbol{h}_1$ and $\boldsymbol{h}_2$ both have full rank. Also assume that $D \leq L + 1$. Show that $E[\boldsymbol{h}_e \boldsymbol{h}_e^H]$ (with $\boldsymbol{h}_e$ defined as in (6.3.26)) has full rank.

2. Prove Equations (6.1.10) and (6.2.14).

3. Consider a system with n_r receive antennas and two transmit antennas, that transmits BPSK symbols with rate one. Assume that the channel matrix $\boldsymbol{H}$ has i.i.d. Gaussian elements. Use Monte-Carlo simulation to plot the BER for $n_r = 1, 2, 3$ using

   (a) optimal beamforming with channel knowledge at the transmitter (as in Section 6.1), and

   (b) the Alamouti code.

   Discuss the result.

4. Assume that the Alamouti code is implemented in a system, but that during the implementation, the complex conjugates are "forgotten" and the code

matrix

$$\boldsymbol{X} = \begin{bmatrix} s_1 & s_2 \\ s_2 & -s_1 \end{bmatrix} \tag{6.5.1}$$

is transmitted instead. Suppose that s_1 and s_2 are complex symbols in general.

(a) Give the ML rule for detection. Does the symbol detection decouple?

(b) Can the code be used to achieve transmit diversity?

5. Consider an OFDM system with one transmit and one receive antenna. Assume that the propagation channel has at least two independently fading taps. Suppose that a reduction of the transmission rate with 50% would be acceptable, provided that we could find a simple modification of the transmission scheme that achieves a multipath diversity of order two. Suggest a method to accomplish this goal.

6. Consider the minimization of the error probability for a fixed $\boldsymbol{H}$ in Section 6.2.2. Explain why it is unfeasible to carry out the minimization subject to the power constraint (6.2.12) instead of the eigenvalue constraint (6.2.11) that we used in the text.

7. Consider a system where neither the transmitter nor the receiver knows the channel $\boldsymbol{H}$, and suppose that we want to transmit one complex symbol s.

(a) Suppose that a fixed matrix $\boldsymbol{W}$ is chosen, and that the matrix $\boldsymbol{X} = s \cdot \boldsymbol{W}$ is transmitted (as in Section 6.2). Is it possible to choose $\boldsymbol{W}$ such that the error probability goes to zero as SNR$\to \infty$? If yes, find such a matrix $\boldsymbol{W}$. If no, explain why it is impossible.

(b) Suppose that instead we let the transmitted matrix have the following structure:

$$\boldsymbol{X} = \begin{bmatrix} \boldsymbol{I} & s \cdot \boldsymbol{W} \end{bmatrix} \tag{6.5.2}$$

Find a condition on the matrix $\boldsymbol{W}$ for which the error rate (corresponding to ML detection) goes to zero as SNR$\to \infty$. Can you formulate a criterion on $\boldsymbol{W}$ that attempts to minimize the error probability?

Hint: The discussion in Section 4.2.3 may be useful.

Chapter 7

LINEAR STBC FOR FLAT FADING CHANNELS

The class of linear space-time block codes is the major category of space-time codes that we will study in this book. Linear STBC have a relatively simple structure: the transmitted code matrix is linear in the real and imaginary parts of the data symbols, or equivalently, in the symbols and their complex conjugates. The present chapter offers a general treatment of linear STBC for frequency flat channels.

7.1 A General Framework for Linear STBC

An STBC code matrix takes on the following form:

$$\boldsymbol{X} = \sum_{n=1}^{n_s} (\bar{s}_n \boldsymbol{A}_n + i\tilde{s}_n \boldsymbol{B}_n) \tag{7.1.1}$$

where $\{s_1, \ldots, s_{n_s}\}$ is a set of symbols to be transmitted, and $\{\boldsymbol{A}_n, \boldsymbol{B}_n\}$ are fixed (in general complex-valued) code matrices of dimension $n_t \times N$. Using an encoding as in (7.1.1), we transmit n_s complex symbols over N time intervals, and hence the transmission rate is equal to:

$$R = \frac{n_s}{N} \tag{7.1.2}$$

Note that $\boldsymbol{X}$ is linear in $\{\bar{s}_n, \tilde{s}_n\}$, but in general it cannot be written as a linear function of only the complex symbols $\{s_n\}$. Also, note that the set $\{\boldsymbol{X}\}$ of matrices that have the structure (7.1.1) is identical to the set of matrices that are formed according to

$$\boldsymbol{X} = \sum_{n=1}^{n_s} (s_n \check{\boldsymbol{A}}_n + s_n^* \check{\boldsymbol{B}}_n) \tag{7.1.3}$$

for code matrices $\{\breve{\boldsymbol{A}}_n, \breve{\boldsymbol{B}}_n\}$ that satisfy:

$$\begin{aligned}\breve{\boldsymbol{A}}_n &= \frac{\boldsymbol{A}_n + \boldsymbol{B}_n}{2}\\ \breve{\boldsymbol{B}}_n &= \frac{\boldsymbol{A}_n - \boldsymbol{B}_n}{2}\end{aligned} \tag{7.1.4}$$

To see this, note that for any complex number s_n, we have that

$$\begin{aligned}\bar{s}_n\boldsymbol{A}_n + i\tilde{s}_n\boldsymbol{B}_n &= \frac{s_n + s_n^*}{2}\boldsymbol{A}_n + \frac{s_n - s_n^*}{2}\boldsymbol{B}_n\\ &= s_n \cdot \left(\frac{\boldsymbol{A}_n + \boldsymbol{B}_n}{2}\right) + s_n^* \cdot \left(\frac{\boldsymbol{A}_n - \boldsymbol{B}_n}{2}\right)\end{aligned} \tag{7.1.5}$$

In this book, we will use the definition in (7.1.1) instead of that in (7.1.3), although both formulations have been used in the literature.

Because of its simple appearance the structure in (7.1.1) is intuitively appealing, and the class of such linear codes is quite large. It includes as a special instance the Alamouti code in Section 6.3.1. Of course, the main problem connected to linear STBC is the design of matrices $\{\boldsymbol{A}_n, \boldsymbol{B}_n\}$ that possess certain desired properties.

Let

$$\begin{aligned}\bar{\boldsymbol{s}} &\triangleq \begin{bmatrix}\bar{s}_1 & \cdots & \bar{s}_{n_s}\end{bmatrix}^T\\ \tilde{\boldsymbol{s}} &\triangleq \begin{bmatrix}\tilde{s}_1 & \cdots & \tilde{s}_{n_s}\end{bmatrix}^T\\ \boldsymbol{s}' &= \begin{bmatrix}\bar{\boldsymbol{s}}\\ \tilde{\boldsymbol{s}}\end{bmatrix}\end{aligned} \tag{7.1.6}$$

and

$$\begin{aligned}\boldsymbol{F}_a &\triangleq \begin{bmatrix}\operatorname{vec}(\boldsymbol{H}\boldsymbol{A}_1) & \cdots & \operatorname{vec}(\boldsymbol{H}\boldsymbol{A}_{n_s})\end{bmatrix}\\ \boldsymbol{F}_b &\triangleq \begin{bmatrix}i\operatorname{vec}(\boldsymbol{H}\boldsymbol{B}_1) & \cdots & i\operatorname{vec}(\boldsymbol{H}\boldsymbol{B}_{n_s})\end{bmatrix}\\ \boldsymbol{F} &= \begin{bmatrix}\boldsymbol{F}_a & \boldsymbol{F}_b\end{bmatrix}\end{aligned} \tag{7.1.7}$$

Then the received space-time signal corresponding to $\boldsymbol{X}$ for a flat fading channel can be written as:

$$\boldsymbol{y} = \operatorname{vec}(\boldsymbol{Y}) = \operatorname{vec}(\boldsymbol{H}\boldsymbol{X} + \boldsymbol{E}) = \boldsymbol{F}_a\bar{\boldsymbol{s}} + \boldsymbol{F}_b\tilde{\boldsymbol{s}} + \boldsymbol{e} = \boldsymbol{F}\boldsymbol{s}' + \boldsymbol{e} \tag{7.1.8}$$

where $\boldsymbol{e} = \operatorname{vec}(\boldsymbol{E})$. Expressed differently,

$$\boldsymbol{y}' = \boldsymbol{F}'\boldsymbol{s}' + \boldsymbol{e}' \tag{7.1.9}$$

where

$$\boldsymbol{y}' = \begin{bmatrix} \bar{\boldsymbol{y}} \\ \tilde{\boldsymbol{y}} \end{bmatrix} \tag{7.1.10}$$

and

$$\boldsymbol{F}' = \begin{bmatrix} \bar{\boldsymbol{F}} \\ \tilde{\boldsymbol{F}} \end{bmatrix} = \begin{bmatrix} \bar{\boldsymbol{F}}_a & \bar{\boldsymbol{F}}_b \\ \tilde{\boldsymbol{F}}_a & \tilde{\boldsymbol{F}}_b \end{bmatrix} \tag{7.1.11}$$

Also,

$$\boldsymbol{e}' = \begin{bmatrix} \bar{\boldsymbol{e}} \\ \tilde{\boldsymbol{e}} \end{bmatrix} \tag{7.1.12}$$

is a vector of real-valued white Gaussian noise whose elements have variance $\sigma^2/2$.

Therefore the problem of detecting the transmitted data $\boldsymbol{s}$ given $\boldsymbol{y}$ amounts to minimizing the metric

$$\|\boldsymbol{Y} - \boldsymbol{H}\boldsymbol{X}\|^2 = \|\boldsymbol{y} - \boldsymbol{F}\boldsymbol{s}'\|^2 = \|\boldsymbol{y}' - \boldsymbol{F}'\boldsymbol{s}'\|^2 \tag{7.1.13}$$

This is an important minimization problem to which we will return several times in this text.

Example 7.1: Transmission of a single symbol.
The simplest example of a linear STBC is perhaps the transmission technique considered in Section 6.2. In this case, we simply have

$$\boldsymbol{X} = \boldsymbol{W} \cdot s \tag{7.1.14}$$

for some constant matrix $\boldsymbol{W}$ of size $n_t \times N$. Hence $n_s = 1$ and

$$\begin{aligned} \boldsymbol{A}_1 &= \boldsymbol{W} \\ \boldsymbol{B}_1 &= \boldsymbol{W} \end{aligned} \tag{7.1.15}$$

in (7.1.1). ■

7.2 Spatial Multiplexing

A more practical simple instance of linear STBC is what we will refer to as *spatial multiplexing*. In a strict sense, spatial multiplexing is not a space-time coding technique. Even so, this transmission technique can be interpreted in the framework of linear STBC.

Different spatial multiplexing schemes have been developed by, for example, [FOSCHINI, JR., 1996]. Here we describe only a simple variant that can be interpreted as a special case of the linear STBC in (7.1.1) by setting $N = 1$, $n_s = n_t$ and choosing $\{\boldsymbol{A}_n, \boldsymbol{B}_n\}$ equal to the nth column of the $n_t \times n_t$ identity matrix. Thereby, n_t symbols are transmitted simultaneously via the n_t transmit antennas. The transmitted matrix at a given time instant reduces to the following column vector:

$$\boldsymbol{X} = \begin{bmatrix} s_1 \\ \vdots \\ s_{n_s} \end{bmatrix} \tag{7.2.1}$$

Since no coding over time is performed, spatial multiplexing does not achieve transmit diversity (cf. the discussion in Section 6.2). Clearly, spatial multiplexing achieves a data rate of $R = n_s = n_t$; however, like for general linear STBC, the associated symbol detection is rather complex (see Section 9.1.3).

7.3 Linear Dispersion Codes

Space-time codes with the general linear structure (7.1.1) have been studied by many authors and they have been given other names than "linear STBC." In [HASSIBI AND HOCHWALD, 2002B], linear STBC was studied in the context of maximizing (with respect to the matrices $\{\boldsymbol{A}_n, \boldsymbol{B}_n\}$) the mutual information between the transmitter and receiver and the resulting codes were given the name "linear dispersion" codes.

From (7.1.9) along with the discussion in Section 3.1 it follows that the average (over the channel $\boldsymbol{H}$) mutual information between $\boldsymbol{s}$ and the received data $\boldsymbol{y}$ (assuming that the receiver knows the channel $\boldsymbol{H}$ perfectly) is equal to

$$\begin{aligned} \frac{1}{2}E_{\boldsymbol{H}}\left[\log_2\left|\boldsymbol{I} + \frac{2}{\sigma^2}\boldsymbol{F}'^T\boldsymbol{F}'\right|\right] &= \frac{1}{2}E_{\boldsymbol{H}}\left[\log_2\left|\boldsymbol{I} + \frac{2}{\sigma^2}(\bar{\boldsymbol{F}}^T\bar{\boldsymbol{F}} + \tilde{\boldsymbol{F}}^T\tilde{\boldsymbol{F}})\right|\right] \\ &= \frac{1}{2}E_{\boldsymbol{H}}\left[\log_2\left|\boldsymbol{I} + \frac{2}{\sigma^2}\mathrm{Re}\left\{\boldsymbol{F}^H\boldsymbol{F}\right\}\right|\right] \end{aligned} \tag{7.3.1}$$

assuming that the elements of $\boldsymbol{s}'$ are i.i.d. zero-mean Gaussian with variance one, and that the elements of $\boldsymbol{e}'$ are i.i.d. Gaussian with variance $\sigma^2/2$. In [HASSIBI AND HOCHWALD, 2002B] it was suggested to optimize (7.3.1) over the set of $\{\boldsymbol{A}_n, \boldsymbol{B}_n\}$ subject to a power constraint; this optimization problem was solved numerically in the cited article. The so-obtained code matrices satisfy an information theoretic optimality criterion that should make these codes useful for high-rate systems. However, the design criterion (7.3.1) does not provide any explicit guarantee that the code has full diversity.

Instead of optimizing $\{\boldsymbol{A}_n, \boldsymbol{B}_n\}$ via an information-theoretic criterion (like in [Hassibi and Hochwald, 2002b]), it is possible to optimize $\{\boldsymbol{A}_n, \boldsymbol{B}_n\}$ using an approach that attempts to minimize the error probability. Results along these lines can be found in, for instance, [Heath and Paulraj, 2002], [Heath et al., 2001a].

7.4 Orthogonal STBC

Orthogonal STBC is an important subclass of linear STBC, that will be the main topic for most of the remaining part of this book. As we will see in this section, OSTBC guarantees that the (coherent) ML detection of different symbols $\{s_n\}$ is *decoupled*, and at the same time achieves a diversity order equal to $n_r n_t$. For two transmit antennas, OSTBC along with its various extensions to ISI channels (see Chapter 8) is currently being considered as a means for improving the performance of wireless local area networks [Liu et al., 2001b], GSM [Lindskog and Paulraj, 2000], and enhanced data rates for GSM evolution (EDGE) [Coupechoux and Braun, 2000]. Also, the simplest form of OSTBC, namely the Alamouti code, has been adopted in the third generation cellular standard W-CDMA [Derryberry et al., 2002].

OSTBC is a linear space-time block code that has the following unitary property:

$$\boldsymbol{X}\boldsymbol{X}^H = \sum_{n=1}^{n_s} |s_n|^2 \cdot \boldsymbol{I} \tag{7.4.1}$$

This property of OSTBC is the main reason for the name it bears. Note that the identity matrix on the right hand side of (7.4.1) could be scaled by an arbitrary constant factor.

The design criterion (7.4.1) can be motivated and derived in a number of different ways. The original derivation of OSTBC is due to [Tarokh et al., 1999a], who studied the error performance associated with unitary matrices $\boldsymbol{X}$. More recently, [Ganesan and Stoica, 2001a], [Wang and Xia, 2002] streamlined the derivations of many of the results associated with OSTBC and established an important link to the theory of orthogonal and amicable orthogonal designs. In this text we will first prove that (7.4.1) guarantees that the symbol detection is decoupled. Next, we will establish a necessary and sufficient condition on $\{\boldsymbol{A}_n, \boldsymbol{B}_n\}$ for (7.4.1) to hold. In Section 7.4.1 we will provide a link to the general framework for ML detection of STBC discussed above in Section 7.1. Section 7.4.2 gives an error performance analysis of OSTBC, and the following subsections discuss various optimality properties of OSTBC.

To understand why the code matrices $\{\boldsymbol{X}\}$ that are proportional to unitary matrices guarantee that the detection of $\{s_n\}$ is decoupled, let us study the ML metric for symbol detection. If $\boldsymbol{X}$ satisfies (7.4.1), then:

$$
\begin{aligned}
&\|\boldsymbol{Y}-\boldsymbol{H}\boldsymbol{X}\|^2 \\
=&\|\boldsymbol{Y}\|^2 - 2\mathrm{Re}\mathrm{Tr}\{\boldsymbol{Y}^H\boldsymbol{H}\boldsymbol{X}\} + \|\boldsymbol{H}\boldsymbol{X}\|^2 \\
=&\|\boldsymbol{Y}\|^2 - 2\sum_{n=1}^{n_s}\mathrm{Re}\mathrm{Tr}\{\boldsymbol{Y}^H\boldsymbol{H}\boldsymbol{A}_n\}\bar{s}_n + 2\sum_{n=1}^{n_s}\mathrm{Im}\,\mathrm{Tr}\{\boldsymbol{Y}^H\boldsymbol{H}\boldsymbol{B}_n\}\tilde{s}_n + \|\boldsymbol{H}\|^2\cdot\|\boldsymbol{s}\|^2 \\
=&\sum_{n=1}^{n_s}\Big(-2\mathrm{Re}\mathrm{Tr}\{\boldsymbol{Y}^H\boldsymbol{H}\boldsymbol{A}_n\}\bar{s}_n + 2\,\mathrm{Im}\,\mathrm{Tr}\{\boldsymbol{Y}^H\boldsymbol{H}\boldsymbol{B}_n\}\tilde{s}_n + |s_n|^2\|\boldsymbol{H}\|^2\Big) \\
&+\text{const.} \\
=&\|\boldsymbol{H}\|^2\cdot\sum_{n=1}^{n_s}\left|s_n - \frac{\mathrm{Re}\mathrm{Tr}\{\boldsymbol{Y}^H\boldsymbol{H}\boldsymbol{A}_n\} - i\,\mathrm{Im}\,\mathrm{Tr}\{\boldsymbol{Y}^H\boldsymbol{H}\boldsymbol{B}_n\}}{\|\boldsymbol{H}\|^2}\right|^2 + \text{const.}
\end{aligned}
\tag{7.4.2}
$$

Equation (7.4.2) demonstrates that the ML metric decouples into a sum of n_s terms, where each term depends on exactly one complex symbol. Consequently, the detection of s_n is decoupled from the detection of s_p for $n \neq p$.

The following result establishes conditions on $\{\boldsymbol{A}_n, \boldsymbol{B}_n\}$ for $\boldsymbol{X}$ to be proportional to a unitary matrix.

Theorem 7.1: Relation between OSTBC and amicable orthogonal designs.
Let $\boldsymbol{X}$ *be a matrix with the structure (7.1.1). Then*

$$
\boldsymbol{X}\boldsymbol{X}^H = \sum_{n=1}^{n_s}|s_n|^2\cdot\boldsymbol{I} \tag{7.4.3}
$$

holds for all complex $\{s_n\}$ if and only if $\{\boldsymbol{A}_n, \boldsymbol{B}_n\}$ *is an amicable orthogonal design, i.e.:*

$$
\begin{aligned}
&\boldsymbol{A}_n\boldsymbol{A}_n^H = \boldsymbol{I}, \boldsymbol{B}_n\boldsymbol{B}_n^H = \boldsymbol{I} \\
&\boldsymbol{A}_n\boldsymbol{A}_p^H = -\boldsymbol{A}_p\boldsymbol{A}_n^H, \boldsymbol{B}_n\boldsymbol{B}_p^H = -\boldsymbol{B}_p\boldsymbol{B}_n^H, \quad n \neq p \\
&\boldsymbol{A}_n\boldsymbol{B}_p^H = \boldsymbol{B}_p\boldsymbol{A}_n^H
\end{aligned}
\tag{7.4.4}
$$

for $n = 1, \ldots, n_s, p = 1, \ldots, n_s$.

Proof: We have that

$$\begin{aligned}
XX^H &= \sum_{n=1}^{n_s}\sum_{p=1}^{n_s}(\bar{s}_n A_n + i\tilde{s}_n B_n)(\bar{s}_p A_p + i\tilde{s}_p B_p)^H \\
&= \sum_{n=1}^{n_s}(\bar{s}_n^2 A_n A_n^H + \tilde{s}_n^2 B_n B_n^H) \\
&+ \sum_{n=1}^{n_s}\sum_{p=1,p>n}^{n_s}\left(\bar{s}_n\bar{s}_p(A_n A_p^H + A_p A_n^H) + \tilde{s}_n\tilde{s}_p(B_n B_p^H + B_p B_n^H)\right) \\
&+ i\sum_{n=1}^{n_s}\sum_{p=1}^{n_s}\tilde{s}_n\bar{s}_p(B_n A_p^H - A_p B_n^H)
\end{aligned} \tag{7.4.5}$$

which shows that (7.4.3) is satisfied whenever (7.4.4) holds. The converse is shown in [Geramita and Geramita, 1978] (using a more abstract notation), but for completeness we give a proof in Section 7.7. ■

A set of matrices $\{A_n, B_n\}$ for which (7.4.4) holds is called an amicable orthogonal design. Therefore, Theorem 7.1 establishes an important link between the theory of amicable orthogonal designs and OSTBC. While the condition (7.4.3) (or (7.4.4)) that guarantees decoupled ML detection was easy to derive, the problem of constructing a set of matrices $\{A_n, B_n\}$ that satisfy (7.4.4) is difficult. In this text we will defer a systematic treatment of this construction problem to Appendix B, but it is worth mentioning that although in general $\{A_n, B_n\}$ are complex-valued throughout this book, for OSTBC these matrices can be chosen to be real-valued (see Appendix B).

In the next four examples we show the best known orthogonal STBC matrices for $n_t = 2$, 3, 4 and 8. Note that taking $\hat{n}_t$ arbitrary rows out of an OSTBC matrix for n_t transmit antennas yields a (non-square) OSTBC matrix for $\hat{n}_t$ antennas; hence OSTBC matrices for $n_t = 5, 6$ and 7 can easily be constructed by using the results of the following examples.

Example 7.2: The Alamouti Code is an OSTBC.
Consider the Alamouti code in Section 6.3.1:

$$X = \begin{bmatrix} s_1 & s_2^* \\ s_2 & -s_1^* \end{bmatrix} \tag{7.4.6}$$

Identification of $\boldsymbol{A}_n$ and $\boldsymbol{B}_n$ in (7.1.1) shows that

$$\begin{aligned} \boldsymbol{A}_1 &= \begin{bmatrix} 1 & 0 \\ 0 & -1 \end{bmatrix} & \boldsymbol{A}_2 &= \begin{bmatrix} 0 & 1 \\ 1 & 0 \end{bmatrix} \\ \boldsymbol{B}_1 &= \begin{bmatrix} 1 & 0 \\ 0 & 1 \end{bmatrix} & \boldsymbol{B}_2 &= \begin{bmatrix} 0 & -1 \\ 1 & 0 \end{bmatrix} \end{aligned} \tag{7.4.7}$$

Here $N = n_s = 2$ and hence the code rate is equal to $R = 1$. It is easy to verify both that $\boldsymbol{X}$ in (7.4.6) satisfies (7.4.3) and that (7.4.7) satisfies (7.4.4). Hence the Alamouti code is an orthogonal space-time block code for $n_t = 2$. A consequence of this observation is that the code achieves full diversity as well as decoupled ML decoding. We have already seen this in the direct analysis in Section 6.3.1. ■

Example 7.3: OSTBC for $n_t = 3$.
For $n_t = 3$, $N = 4$, $n_s = 3$ the following code is an orthogonal STBC:

$$\boldsymbol{X} = \begin{bmatrix} s_1 & 0 & s_2 & -s_3 \\ 0 & s_1 & s_3^* & s_2^* \\ -s_2^* & -s_3 & s_1^* & 0 \end{bmatrix} \tag{7.4.8}$$

This code has a rate of $R = 3/4$. An alternative OSTBC for $n_t = 3$, that has rate $R = 3/4$ as well, is:

$$\boldsymbol{X} = \begin{bmatrix} s_1 & -s_2^* & s_3^* & 0 \\ s_2 & s_1^* & 0 & -s_3^* \\ s_3 & 0 & -s_1^* & s_2^* \end{bmatrix} \tag{7.4.9}$$

■

Example 7.4: OSTBC for $n_t = 4$.
For $n_t = 4$, $N = 4$, $n_s = 3$ the following code is an orthogonal STBC:

$$\boldsymbol{X} = \begin{bmatrix} s_1 & 0 & s_2 & -s_3 \\ 0 & s_1 & s_3^* & s_2^* \\ -s_2^* & -s_3 & s_1^* & 0 \\ s_3^* & -s_2 & 0 & s_1^* \end{bmatrix} \tag{7.4.10}$$

This code has also a rate of $R = 3/4$. ■

Example 7.5: OSTBC for $n_t = 8$.
For $n_t = 8$, $N = 8$, $n_s = 4$ the following code is an orthogonal STBC (with rate

$R = 1/2$):

$$
\boldsymbol{X} = \begin{bmatrix} s_1 & 0 & 0 & 0 & s_4 & 0 & s_2 & -s_3 \\ 0 & s_1 & 0 & 0 & 0 & s_4 & s_3^* & s_2^* \\ 0 & 0 & s_1 & 0 & -s_2^* & -s_3 & s_4^* & 0 \\ 0 & 0 & 0 & s_1 & s_3^* & -s_2 & 0 & s_4^* \\ -s_4^* & 0 & s_2 & -s_3 & s_1^* & 0 & 0 & 0 \\ 0 & -s_4^* & s_3^* & s_2^* & 0 & s_1^* & 0 & 0 \\ -s_2^* & -s_3 & -s_4 & 0 & 0 & 0 & s_1^* & 0 \\ s_3^* & -s_2 & 0 & -s_4 & 0 & 0 & 0 & s_1^* \end{bmatrix} \tag{7.4.11}
$$

■

The next example shows that $\boldsymbol{X}$ can be unitary for all $\{s_n\}$ that belong to a certain constellation, without (7.4.4) being satisfied. Hence the assumption in Theorem 7.1 that $\boldsymbol{X}$ is unitary for *all* complex $\{s_n\}$ is necessary.

Example 7.6: A Diagonal Code.
Consider the following *diagonal code*:

$$
\boldsymbol{X} = \begin{bmatrix} s_1 & 0 \\ 0 & s_2 \end{bmatrix} \tag{7.4.12}
$$

where we assume that s_1 and s_2 belong to a unitary (scalar) constellation such that $|s_n| = 1$. The corresponding matrices $\boldsymbol{A}_n$ and $\boldsymbol{B}_n$ are easily identified:

$$
\begin{aligned}
\boldsymbol{A}_1 &= \begin{bmatrix} 1 & 0 \\ 0 & 0 \end{bmatrix} & \boldsymbol{A}_2 &= \begin{bmatrix} 0 & 0 \\ 0 & 1 \end{bmatrix} \\
\boldsymbol{B}_1 &= \begin{bmatrix} 1 & 0 \\ 0 & 0 \end{bmatrix} & \boldsymbol{B}_2 &= \begin{bmatrix} 0 & 0 \\ 0 & 1 \end{bmatrix}
\end{aligned} \tag{7.4.13}
$$

Clearly, $\boldsymbol{X}\boldsymbol{X}^H = \boldsymbol{I}$ for all s_n in the constellation, yet (7.4.13) does not satisfy (7.4.4). Hence, even though the code matrix $\boldsymbol{X}$ satisfies (7.4.1) for the values of s_n in the constellation, it is not an OSTBC; it can also be shown that the code does not provide transmit diversity (see Exercise 7.4 on page 128). ■

7.4.1 ML Detection of OSTBC in a General Framework

The already derived Equation (7.4.2) gives an explicit expression for the ML metric for detection of the symbols $\{s_n\}$. It is useful to establish a link between this result and the more general framework for linear STBC in Section 7.1. We have the following theorem.

Theorem 7.2: Channel decoupling property of OSTBC.
The code defined by $\{A_n, B_n\}$ is an orthogonal STBC if and only if the matrix F in (7.1.7) satisfies:

$$Re\{F^H F\} = \|H\|^2 \cdot I \tag{7.4.14}$$

for all H.

Proof: Since $\text{vec}(HX) = Fs'$ by construction (see Section 7.1), we have that

$$\|Fs'\|^2 = s'^T \text{Re}\{F^H F\} s' = \|HX\|^2 \tag{7.4.15}$$

for all s and H. If X is an OSTBC, then by (7.4.3):

$$\|HX\|^2 = \|H\|^2 \cdot \|s'\|^2 \tag{7.4.16}$$

for all H and s'. Therefore, by (7.4.15) and (7.4.16):

$$\|H\|^2 \cdot \|s'\|^2 = s'^T \text{Re}\{F^H F\} s' \tag{7.4.17}$$

for all s'. Hence all eigenvalues of $\text{Re}\{F^H F\}$ must be equal to $\|H\|^2$ and consequently (7.4.14) must hold for all H.

Conversely, suppose that (7.4.14) holds for all H. Then (7.4.16) holds for all H and s' (cf. (7.4.15)) and hence

$$\text{Tr}\left\{H(XX^H - \|s'\|^2 \cdot I)H^H\right\} = 0 \tag{7.4.18}$$

for all H. Suppose that $XX^H - \|s\|^2 \cdot I$ is nonzero and let h be an eigenvector of $XX^H - \|s\|^2 \cdot I$ corresponding to a nonzero eigenvalue. Taking H to be the matrix with h^H as its first row and with all other elements equal to zero, we find that (7.4.18) is contradicted. Therefore $XX^H = \|s\|^2 \cdot I$ for all s, so X must be an OSTBC. ■

For ML detection, we do not need the above-shown fact that X is an orthogonal STBC *only* if (7.4.14) holds (the "if" part would be enough). However, this stronger result makes the picture of OSTBC more complete and it will also be used later in Section 7.4.4.

From Theorem 7.2 it follows that for OSTBC the ML metric in (7.1.13) can be written

$$\begin{aligned}\|y - Fs'\|^2 &= \|y\|^2 - 2\text{Re}\{y^H Fs'\} + \text{Re}\{s'^T FF^H s'\} \\ &= \|y\|^2 - 2\text{Re}\{y^H Fs'\} + \|H\|^2 \cdot \|s'\|^2 \\ &= \|H\|^2 \cdot \|s' - \hat{s}'\|^2 + \text{const.}\end{aligned} \tag{7.4.19}$$

where

$$\hat{s}' \triangleq \frac{\text{Re}\{\boldsymbol{F}^H \boldsymbol{y}\}}{\|\boldsymbol{H}\|^2} \tag{7.4.20}$$

Equation (7.4.19) reinforces the result (7.4.2) that the ML detection of $\{s_n\}$ is equivalent to solving n_s scalar detection problems, one for each s_n. It also shows that since the likelihood function for symbol detection is a function of $\hat{\boldsymbol{s}}'$ only, $\hat{\boldsymbol{s}}'$ is a sufficient statistic for detection *and any subsequent decoding.* Note that the matrix $\boldsymbol{F}$ in (7.4.20) can be interpreted as a space-time *matched filter.*

7.4.2 Error Performance of OSTBC

Let $\hat{\boldsymbol{s}}$ be defined via

$$\begin{bmatrix} \bar{\hat{\boldsymbol{s}}} \\ \tilde{\hat{\boldsymbol{s}}} \end{bmatrix} = \hat{\boldsymbol{s}}' \tag{7.4.21}$$

Since the real and imaginary parts of $\hat{\boldsymbol{s}}$ are linear in the received data (see (7.4.20)), $\hat{\boldsymbol{s}}$ has a Gaussian distribution. We can show that $\hat{\boldsymbol{s}}$ is a circularly symmetric Gaussian vector with independent elements and a variance that is proportional to $\sigma^2/\|\boldsymbol{H}\|^2$. This important result, which quantifies the performance of orthogonal STBC, is formally stated in the following theorem.

Theorem 7.3: Error performance of OSTBC.
Assume that $\boldsymbol{X}$ *is a normalized version of (7.1.1):*

$$\boldsymbol{X} = \sqrt{\frac{\rho^2}{n_t}} \cdot \sum_{n=1}^{n_s} (\bar{s}_n \boldsymbol{A}_n + i\tilde{s}_n \boldsymbol{B}_n) \tag{7.4.22}$$

where ρ^2 *is a constant used to control the transmit power, and that (7.4.4) holds:*

$$\begin{aligned} &\boldsymbol{A}_n \boldsymbol{A}_n^H = \boldsymbol{I}, \boldsymbol{B}_n \boldsymbol{B}_n^H = \boldsymbol{I} \\ &\boldsymbol{A}_n \boldsymbol{A}_p^H = -\boldsymbol{A}_p \boldsymbol{A}_n^H, \boldsymbol{B}_n \boldsymbol{B}_p^H = -\boldsymbol{B}_p \boldsymbol{B}_n^H, n \neq p \\ &\boldsymbol{A}_n \boldsymbol{B}_p^H = \boldsymbol{B}_p \boldsymbol{A}_n^H \end{aligned} \tag{7.4.23}$$

Let

$$\hat{\boldsymbol{s}} = \sqrt{\frac{n_t}{\rho^2}} \cdot \frac{Re\{\boldsymbol{F}_a^H \boldsymbol{y}\} + iRe\{\boldsymbol{F}_b^H \boldsymbol{y}\}}{\|\boldsymbol{H}\|^2} \tag{7.4.24}$$

where $\boldsymbol{F}_a$ *and* $\boldsymbol{F}_b$ *are defined by (7.1.7). Then*

(i) The average transmitted energy during the N time intervals is equal to

$$E\left[Tr\{\boldsymbol{X}\boldsymbol{X}^H\}\right] = \rho^2 \cdot E\left[\|\boldsymbol{s}\|^2\right] \tag{7.4.25}$$

(ii) The vector $\hat{\mathbf{s}}$ is a sufficient statistic for symbol detection, and it has the following distribution:

$$\hat{\boldsymbol{s}} \sim N_C\Big(\boldsymbol{s}, \frac{n_t}{\rho^2}\frac{\sigma^2}{\|\boldsymbol{H}\|^2} \cdot \boldsymbol{I}\Big) \tag{7.4.26}$$

or equivalently,

$$\hat{\boldsymbol{s}}' = \begin{bmatrix} \bar{\hat{\boldsymbol{s}}} \\ \tilde{\hat{\boldsymbol{s}}} \end{bmatrix} \sim N\Big(\boldsymbol{s}', \frac{n_t}{\rho^2}\frac{\sigma^2}{2\|\boldsymbol{H}\|^2}\boldsymbol{I}\Big) \tag{7.4.27}$$

(iii) The SNR for each element in $\hat{\mathbf{s}}$ is equal to:

$$SNR = \frac{\rho^2}{n_t} \cdot \frac{\|\boldsymbol{H}\|^2}{\sigma^2} \cdot E[|s_n|^2] \tag{7.4.28}$$

Proof: Part (i) is immediate and part (iii) follows directly from part (ii). It remains to derive the first and second-order moments of $\hat{\boldsymbol{s}}$. We have that:

$$\begin{aligned} E\left[\hat{\boldsymbol{s}}'\right] &= \frac{1}{\text{Tr}\{\boldsymbol{H}\boldsymbol{H}^H\}} E\left[\text{Re}\{\boldsymbol{F}^H\boldsymbol{y}\}\right] \\ &= \frac{1}{\text{Tr}\{\boldsymbol{H}\boldsymbol{H}^H\}} E\left[\text{Re}\{\boldsymbol{F}^H(\boldsymbol{F}\boldsymbol{s}' + \boldsymbol{e})\}\right] \\ &= \boldsymbol{s}' \end{aligned} \tag{7.4.29}$$

and

$$\begin{aligned} & E\left[(\hat{\boldsymbol{s}}' - \boldsymbol{s}')(\hat{\boldsymbol{s}}' - \boldsymbol{s}')^T\right] \\ =& \frac{n_t}{\rho^2} \frac{1}{\left(\text{Tr}\{\boldsymbol{H}^H\boldsymbol{H}\}\right)^2} E\left[\text{Re}\{\boldsymbol{F}^H\boldsymbol{e}\}\left(\text{Re}\{\boldsymbol{F}^H\boldsymbol{e}\}\right)^T\right] \\ =& \frac{n_t}{\rho^2}\frac{1}{2\|\boldsymbol{H}\|^4} E\left[\text{Re}\{\boldsymbol{F}^H\boldsymbol{e}\boldsymbol{e}^H\boldsymbol{F} + \boldsymbol{F}^H\boldsymbol{e}\boldsymbol{e}^T\boldsymbol{F}^*\}\right] \\ =& \frac{n_t}{\rho^2}\frac{\sigma^2}{2\|\boldsymbol{H}\|^2}\boldsymbol{I} \end{aligned} \tag{7.4.30}$$

where we used the fact that

$$\mathrm{Re}\{\boldsymbol{X}\} \cdot \mathrm{Re}\{\boldsymbol{Y}\} = \frac{1}{2}\mathrm{Re}\{\boldsymbol{X}\boldsymbol{Y} + \boldsymbol{X}\boldsymbol{Y}^*\} \tag{7.4.31}$$

for any two matrices $\boldsymbol{X}$ and $\boldsymbol{Y}$. The stated result (7.4.27) follows.

■

In essence, Theorem 7.3 shows that (for a fixed $\boldsymbol{H}$), the arrangement of the received data in $\boldsymbol{y} = \mathrm{vec}(\boldsymbol{Y})$ and pre-multiplication of $\boldsymbol{y}$ with the matrix $\boldsymbol{F}^H$ in (7.1.7) decouples the space-time channel corresponding to an OSTBC into $2n_s$ independent and real scalar AWGN channels, or equivalently, n_s complex AWGN channels (see Figure 7.1). This decoupling into AWGN channels is indeed one of the main virtues of OSTBC and has two important consequences in the context of concatenation of OSTBC with outer codes. First, any outer coding techniques (including relatively recent developments such as trellis-coded modulation [BIGLIERI ET AL., 1991], [PROAKIS, 2001, SEC. 8.3] and turbo coding [SKLAR, 1997]) designed for scalar channels can be used on each sub-channel. Second, since the likelihood function for the coherent OSTBC detection problem depends on the received data only via the output of these AWGN channels, the output of these AWGN channels is a sufficient statistic not only for symbol detection, but for all subsequent decoding as well. This means that soft information to a channel decoder can be transferred in the same way as for any other AWGN channel. Finally, we note already here that a similar decoupling result holds for the extension of OSTBC to frequency selective channels (see Chapter 8).

Diversity Order

From Theorem 7.3 we know that the SNR on the nth complex sub-channel is equal to

$$\mathrm{SNR} = \frac{\rho^2}{n_t} \cdot \frac{\|\boldsymbol{H}\|^2}{\sigma^2} \cdot E\left[|s_n|^2\right] \tag{7.4.32}$$

Suppose that the propagation channel obeys the model in Section 2.1.2:

$$\boldsymbol{h} \triangleq \mathrm{vec}(\boldsymbol{H}) \sim N_C\left(\boldsymbol{0}, \boldsymbol{R}_t^T \otimes \boldsymbol{R}_r\right) \tag{7.4.33}$$

where $\boldsymbol{R}_t$ and $\boldsymbol{R}_r$ are matrices that determine the transmit and receive correlation, respectively. A straightforward application of Theorem 4.1 (see page 42) implies that the diversity order is equal to the rank of the covariance matrix of $\boldsymbol{h}$:

$$G_d = \mathrm{rank}\{\boldsymbol{R}_t^T \otimes \boldsymbol{R}_r\} = \mathrm{rank}\{\boldsymbol{R}_t\} \cdot \mathrm{rank}\{\boldsymbol{R}_r\} \tag{7.4.34}$$

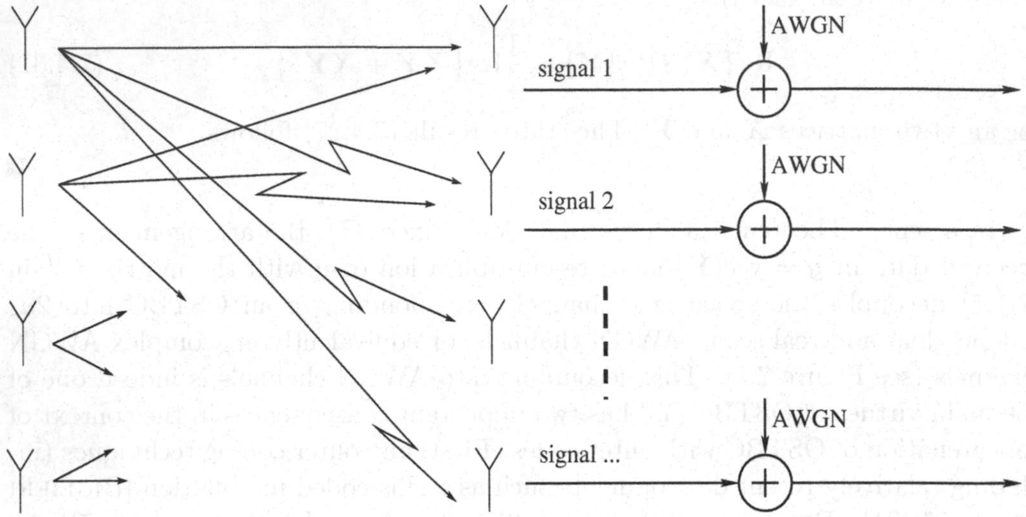

(a) $n_t \times n_r$ space-time channel (b) n_s independent AWGN channels

Figure 7.1. Decoupling of the space-time channel into n_s parallel and independent complex AWGN channels.

If the fading is uncorrelated, then both $\boldsymbol{R}_t$ and $\boldsymbol{R}_r$ have full rank and consequently OSTBC achieves full diversity (i.e., $G_d = n_t n_r$). On the other hand, in the case of completely correlated fading, $\boldsymbol{R}_t$ and $\boldsymbol{R}_r$ are rank deficient and in this case the diversity order will be $G_d < n_r n_t$. See also [Bölcskei and Paulraj, 2000a] for a discussion on the performance of OSTBC in correlated Rayleigh fading.

We can also compute the diversity order of OSTBC, without using the expression for the SNR, as follows. Consider the received data in (7.1.8) and let $\boldsymbol{y} = \text{vec}(\boldsymbol{Y})$, $\boldsymbol{h} = \text{vec}(\boldsymbol{H})$, $\boldsymbol{e} = \text{vec}(\boldsymbol{E})$. It follows that (compare (2.1.8)):

$$\boldsymbol{y} = \text{vec}(\boldsymbol{H}\boldsymbol{X} + \boldsymbol{E}) = (\boldsymbol{X}^T \otimes \boldsymbol{I})\boldsymbol{h} + \boldsymbol{e} \tag{7.4.35}$$

Suppose that the symbols $\{s_n^0\}_{n=1}^{n_s}$ are transmitted, and let $\{s_n\}$ be any hypothetical symbols that are different from $\{s_n^0\}$. Also, let $\boldsymbol{X}$ be the OSTBC matrix associated with $\{s_n\}$ and let $\boldsymbol{X}_0$ be the matrix corresponding to $\{s_n^0\}$. Clearly,

$$\boldsymbol{X} - \boldsymbol{X}_0 = \sum_{n=1}^{n_s} \Big((\bar{s}_n - \bar{s}_n^0)\boldsymbol{A}_n + i(\tilde{s}_n - \tilde{s}_n^0)\boldsymbol{B}_n \Big) \tag{7.4.36}$$

is also an OSTBC matrix. From Theorem 7.1 (see page 102) we conclude that:

$$(\boldsymbol{X} - \boldsymbol{X}_0)(\boldsymbol{X} - \boldsymbol{X}_0)^H = \sum_{n=1}^{n_s} |s_n - s_n^0|^2 \cdot \boldsymbol{I} \tag{7.4.37}$$

Since the matrix in (7.4.37) has always full rank (equal to n_t), application of Theorem 4.2 (see page 45) shows once again that the system achieves a diversity order equal to the rank of the covariance matrix of $\boldsymbol{h}$.

BER Performance Examples

Since OSTBC decouples the space-time channel into parallel scalar channels each with

$$\text{SNR} = \frac{\rho^2}{n_t} \cdot \frac{\|\boldsymbol{H}\|^2}{\sigma^2} \cdot E\left[|s_n|^2\right] \tag{7.4.38}$$

we can use Theorem 4.1 (see page 42) to compute the error probability. If an OSTBC with rate $R = n_s/N < 1$ is used, an increase in the constellation size is needed to preserve the same data rate (in terms of bits/second) as we would achieve with a conventional system using one transmit antenna and no OSTBC. The next example illustrates that provided that the knowledge of the propagation channel is sufficiently accurate, the loss in performance due to the increase in the constellation size is often compensated for by the higher diversity order of OSTBC, as long as the SNR is large enough.

Example 7.7: BER performance of OSTBC.
Assume that the elements of $\boldsymbol{H}$ are independent and Gaussian with unit variance. In Figures 7.2 and 7.3 we plot the average BER versus the SNR for OSTBC using one, two, three and four transmit antennas, and one receive antenna. Monte-Carlo simulation on a standard personal computer was used to obtain the plots. For the case of one transmit antenna, no OSTBC is used. In the case of two antennas we use the Alamouti code (see Section 6.3.1), and for three and four antennas we use the rate-3/4 codes (7.4.8) and (7.4.10). In Figure 7.2, the transmit power is equal to one for each antenna, whereas in Figure 7.3 we have normalized the transmit power so that the total radiated power (accumulated over all antennas) is equal to one. For the case of one and two antennas, we use 8-PSK modulation. To keep the bit rate constant, we use a uniform and rectangular 16-QAM constellation for the case of three and four transmit antennas. Gray coding is used to map the information bits onto the complex constellation. The shift of the curves between Figures 7.2 and 7.3 is a consequence of the difference in the normalization of the

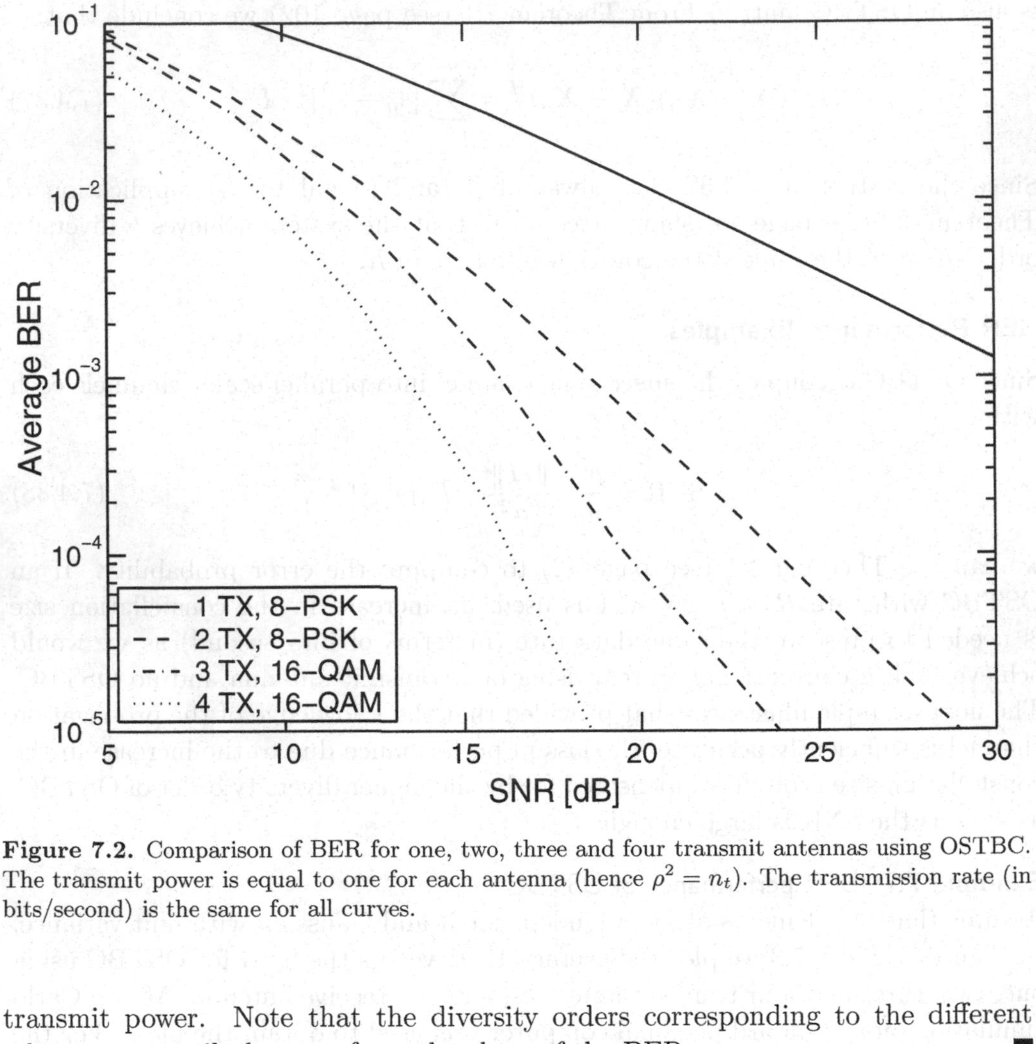

Figure 7.2. Comparison of BER for one, two, three and four transmit antennas using OSTBC. The transmit power is equal to one for each antenna (hence $\rho^2 = n_t$). The transmission rate (in bits/second) is the same for all curves.

transmit power. Note that the diversity orders corresponding to the different schemes can easily be seen from the slope of the BER curves. ■

For receive diversity with $n_t = 1$ transmit and $n_r \geq 1$ receive antennas, the received SNR after optimal combining is (see (5.1.10)):

$$\text{SNR} = \frac{\|\boldsymbol{h}\|^2}{\sigma^2} \tag{7.4.39}$$

provided that the symbols satisfy $E[|s_n|^2] = 1$. On the other hand, the SNR for OSTBC, assuming a total transmitted energy equal to one during each time

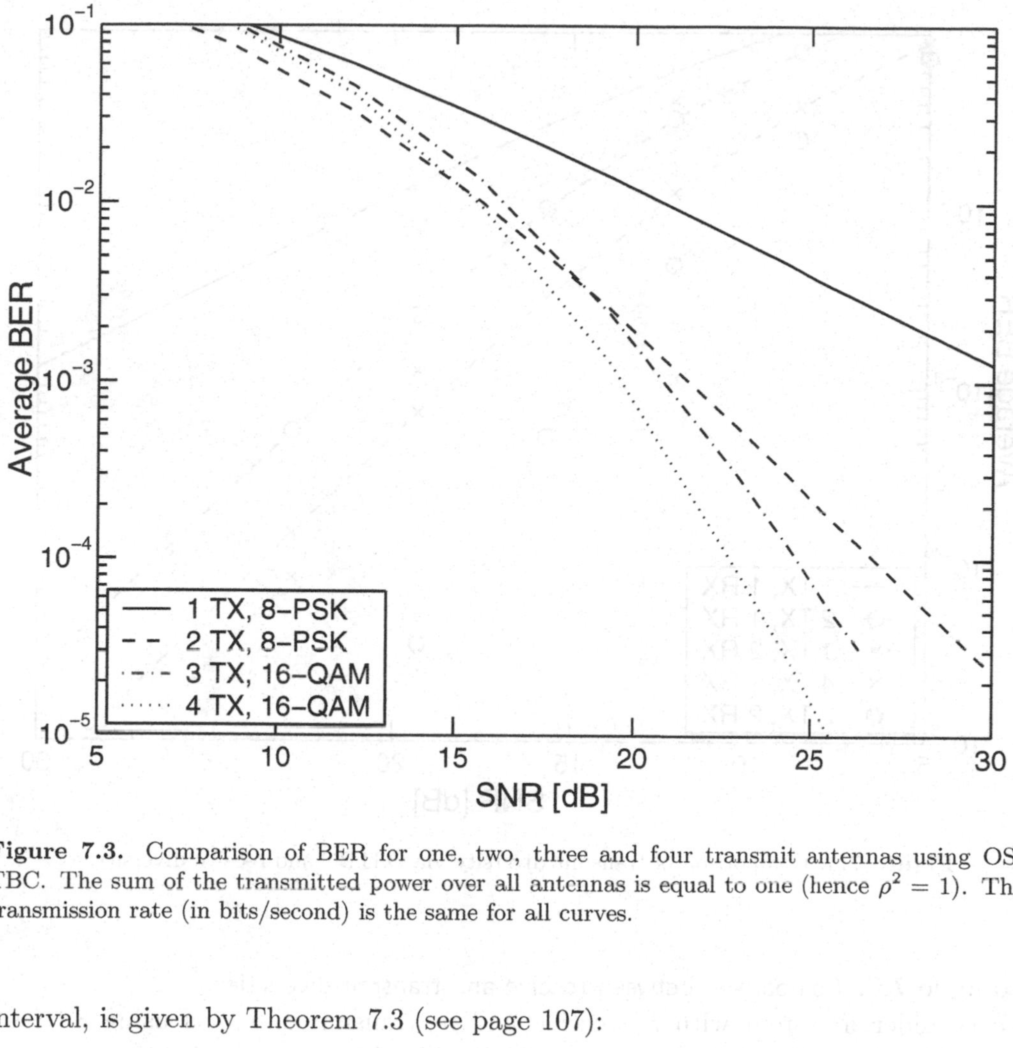

Figure 7.3. Comparison of BER for one, two, three and four transmit antennas using OSTBC. The sum of the transmitted power over all antennas is equal to one (hence $\rho^2 = 1$). The transmission rate (in bits/second) is the same for all curves.

interval, is given by Theorem 7.3 (see page 107):

$$\text{SNR} = \frac{N}{n_s} \cdot \frac{1}{n_t} \frac{\|\boldsymbol{H}\|^2}{\sigma^2} \tag{7.4.40}$$

Hence, the use of transmit diversity via OSTBC incurs a loss in SNR compared to receive diversity (unless $N > n_s n_t$, which is typically not the case). This loss was illustrated already in Section 6.2, and is related to the fact that an OSTBC transmitter does not steer energy in any particular direction. On the other hand, a transmitter that knows the channel can apply transmit weights as in Section 6.1 and steer a beam in the direction of the receiver. The following example provides an illustration of these facts.

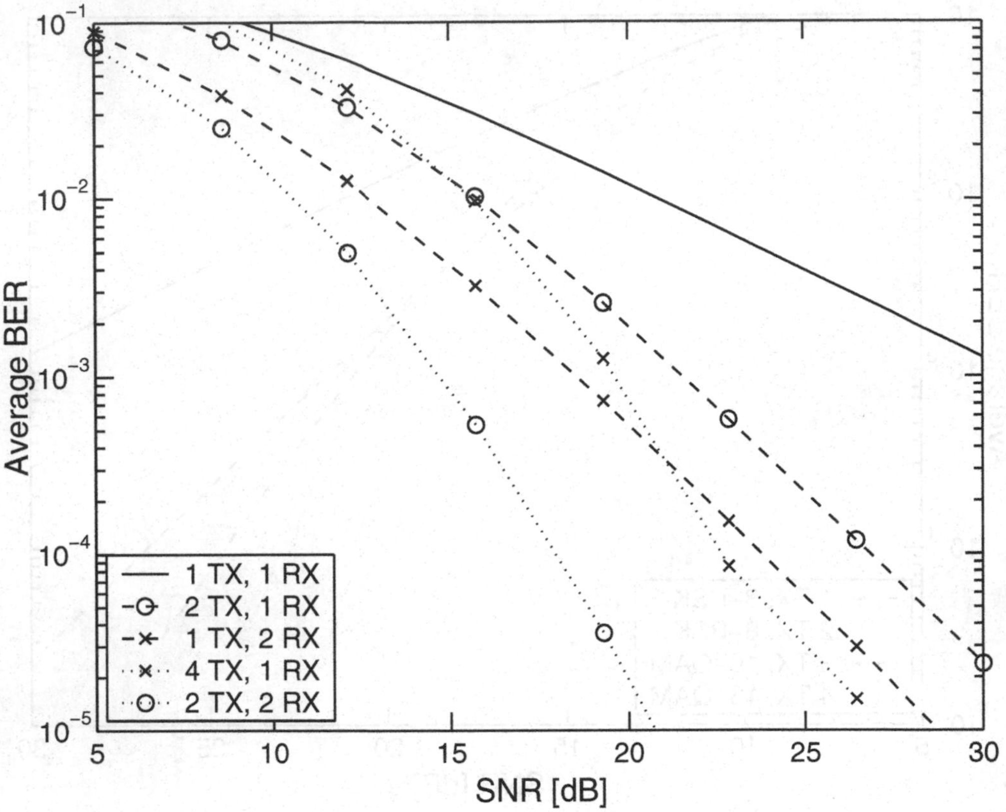

Figure 7.4. Comparison of transmit diversity via OSTBC and receive diversity.

Example 7.8: Comparison between receive and transmit diversities.
We consider a system with n_r receive and n_t transmit antennas using 8-PSK. Figure 7.4 shows the BER performance using ML detection, obtained via Monte-Carlo simulation and assuming that the channel matrix $\boldsymbol{H}$ had i.i.d. zero-mean Gaussian entries. For the case of $n_t = 1$, no space-time coding is used, whereas for $n_t = 2$ the Alamouti code is used to achieve transmit diversity. The total transmitted power (accumulated over all antennas) is the same in both cases. Clearly, the case of $n_t = 1$, $n_r = 2$ is 3 dB superior to the case of $n_t = 2$, $n_r = 1$, as predicted by the theory. A similar conclusion holds for the comparison between the cases $n_t = 2, n_r = 2$ and $n_t = 4, n_r = 1$. ■

7.4.3 Mutual Information Properties of OSTBC

Assuming that the signals transmitted by different antennas are uncorrelated, and that the average transmitted energy per antenna and time interval is $1/n_t$, the capacity of the MIMO channel is, for a given $\boldsymbol{H}$ and unity bandwidth; see (3.2.1):

$$C_{\text{MIMO}}(\boldsymbol{H}) = \log_2 \left| \boldsymbol{I} + \frac{1}{n_t} \frac{\boldsymbol{H}\boldsymbol{H}^H}{\sigma^2} \right| \tag{7.4.41}$$

If OSTBC is used, and under the same power normalization, we know from Theorem 7.3 (see page 107) that the MIMO channel transforms into n_s independent scalar complex AWGN channels with SNR

$$\text{SNR} = \frac{N}{n_s} \cdot \frac{\|\boldsymbol{H}\|^2}{n_t \sigma^2} \tag{7.4.42}$$

The mutual information offered by n_s such channels is n_s times the corresponding mutual information for a SISO system (cf. (3.2.2)), divided by the number of time instants used for the transmission:

$$C_{\text{OSTBC}}(\boldsymbol{H}) = \frac{n_s}{N} \log_2 \left| 1 + \frac{N}{n_s} \frac{\|\boldsymbol{H}\|^2}{n_t \sigma^2} \right| \tag{7.4.43}$$

(This equation can also be derived from (7.3.1) and Theorem 7.2.)

We can expect that the effective channel structure enforced by employing OSTBC will limit the mutual information, and this is indeed the case. The following result, introduced in [Sandhu and Paulraj, 2000], quantifies this limitation.

Theorem 7.4: Mutual information of OSTBC.
Let the average transmitted energy (accumulated over all n_t transmit antennas) per time interval be equal to one. Then it holds that

$$C_{MIMO}(\boldsymbol{H}) \geq C_{OSTBC}(\boldsymbol{H}) \tag{7.4.44}$$

with equality if and only if $\boldsymbol{H}$ has rank one and $n_s = N$.

Proof: Let $\{\lambda_k\}_{k=1}^{n_t}$ be the eigenvalues of $\boldsymbol{H}^H\boldsymbol{H}/(n_t\sigma^2)$. Then, since $n_s \leq N$,

$$\begin{aligned}
C_{\text{MIMO}}(\boldsymbol{H}) - C_{\text{OSTBC}}(\boldsymbol{H}) &= \log_2 \prod_{k=1}^{n_t}(1+\lambda_k) - \frac{n_s}{N}\log_2\left(1+\frac{N}{n_s}\sum_{k=1}^{n_t}\lambda_k\right) \\
&= \log_2\left(1+\sum_{k=1}^{n_t}\lambda_k + \cdots + \prod_{k=1}^{n_t}\lambda_k\right) - \frac{n_s}{N}\log_2\left(1+\frac{N}{n_s}\sum_{k=1}^{n_t}\lambda_k\right) \\
&\geq \log_2\left(1+\sum_{k=1}^{n_t}\lambda_k\right) - \frac{n_s}{N}\log_2\left(1+\frac{N}{n_s}\sum_{k=1}^{n_t}\lambda_k\right) \\
&= \frac{n_s}{N}\left(\log_2\left(1+\sum_{k=1}^{n_t}\lambda_k\right)^{N/n_s} - \log_2\left(1+\frac{N}{n_s}\sum_{k=1}^{n_t}\lambda_k\right)\right) \\
&\geq 0
\end{aligned} \tag{7.4.45}$$

If the rank of $\boldsymbol{H}$ is one, all but one of the $\{\lambda_k\}$ are equal to zero. If also $n_s = N$, it follows immediately from (7.4.45) that $C_{\text{OSTBC}}(\boldsymbol{H}) = C_{\text{MIMO}}(\boldsymbol{H})$. On the other hand, equality in (7.4.45) can hold only when exactly one λ_k is nonzero and $n_s = N$. ■

Despite the pleasing property that OSTBC decouples the space-time channel into parallel and independent AWGN channels, Theorem 7.4 shows that the structure imposed by OSTBC generally limits the maximal error-free throughput that can be achieved, regardless of the amount of outer channel coding that is employed. We show in Appendix B that $n_s = N$ is only possible for $n_t = 2$ (of course, the degenerate case $n_t = 1$ does not require any OSTBC). Hence, Theorem 7.4 states that OSTBC can only achieve the channel capacity in the case of $n_t = 2$ and a rank one channel matrix $\boldsymbol{H}$. This includes the special case of $n_t = 2$, $n_r = 1$, which may be practically quite important. In fact, we can expect OSTBC to be particularly suitable for improving the performance (in terms of BER or outage capacity) of existing SISO systems by employing an extra transmit antenna. However, as we illustrate in the following example, the usefulness of OSTBC may not be limited to that case only.

Example 7.9: Average and 99%-outage capacity of a system using OSTBC.
Figure 7.5 shows a plot of the average and 99%-outage capacity of a MIMO system (see (7.4.41)) with $n_t = 2$ transmit and $n_r = 2$ receive antennas. The figure also shows the corresponding average and outage OSTBC capacity (see (7.4.43)),

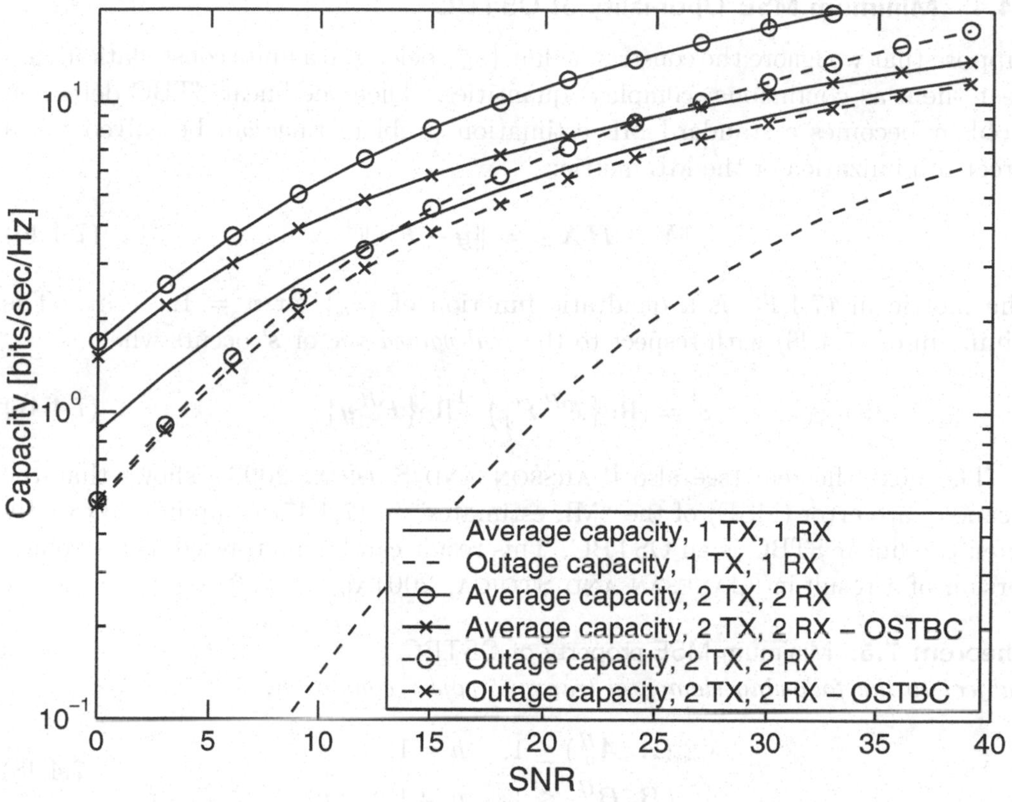

Figure 7.5. Average and 99%-outage capacity for a system with $n_r = 2$ receive and $n_t = 2$ transmit antennas, compared to the channel capacity of the two scalar channels provided by OSTBC and to the corresponding quantities for a SISO system.

compared to the quantities for a SISO system (with $n_t = n_r = 1$). Clearly, $C_{\text{MIMO}}(\boldsymbol{H}) \geq C_{\text{OSTBC}}(\boldsymbol{H})$ as predicted by Theorem 7.4, but the loss associated with OSTBC is not as large for the outage capacity as for the average capacity. Considering the average capacity, the increase in capacity offered by going from $n_t = n_r = 1$ to $n_r = n_t = 2$ is relatively modest, and moreover, only a small part of this increase can be exploited by OSTBC. On the other hand, considering the perhaps more important quantity of outage capacity, we observe that the increase in the channel capacity from $n_t = n_r = 1$ to $n_t = n_r = 2$ is significant, and that OSTBC can exploit a large fraction of this increase. In other words, the capacity limitation of OSTBC indicated in Theorem 7.4 is not severe in this case. ■

7.4.4 Minimum MSE Optimality of OSTBC

Suppose that we ignore the constraint that $\{s_n\}$ belong to a finite constellation, and treat them as continuous (complex) quantities. Then the linear STBC detection problem becomes a standard ML estimation problem that can be solved via a direct minimization of the ML metric:

$$\|\boldsymbol{Y} - \boldsymbol{H}\boldsymbol{X}\|^2 = \|\boldsymbol{y} - \boldsymbol{F}\boldsymbol{s}'\|^2 \tag{7.4.46}$$

The metric in (7.4.46) is a quadratic function of $\{s_n\}$ for $n = 1, \ldots, n_s$. The minimum of (7.4.46) with respect to the *real-valued* vector $\boldsymbol{s}'$ occurs when

$$\hat{\boldsymbol{s}}' = \left(\mathrm{Re}\{\boldsymbol{F}^H\boldsymbol{F}\}\right)^{-1}\mathrm{Re}\{\boldsymbol{F}^H\boldsymbol{y}\} \tag{7.4.47}$$

The next theorem (see also [Larsson and Stoica, 2003]) shows that the mean-square error (MSE) of the "ML estimates" in (7.4.47) is minimized exactly when the linear STBC is an OSTBC. This result can be interpreted as a stronger version of a result in [Ganesan and Stoica, 2001a].

Theorem 7.5: Minimum MSE property of OSTBC.
Subject to the following elemental transmit power constraint:

$$\begin{aligned} \lambda_{max}(\boldsymbol{A}_n\boldsymbol{A}_n^H) &\le 1, \quad n = 1, \ldots, n_s \\ \lambda_{max}(\boldsymbol{B}_n\boldsymbol{B}_n^H) &\le 1, \quad n = 1, \ldots, n_s \end{aligned} \tag{7.4.48}$$

the MSE of $\hat{\boldsymbol{s}}'$ is minimized exactly when $\boldsymbol{X}$ is an OSTBC.

Proof: Using the circular Gaussian distribution of $\boldsymbol{E}$ along with the fact that

$$\mathrm{Re}\{\boldsymbol{X}\}\mathrm{Re}\{\boldsymbol{Y}\} = \frac{1}{2}(\mathrm{Re}\{\boldsymbol{X}\boldsymbol{Y}\} + \mathrm{Re}\{\boldsymbol{X}\boldsymbol{Y}^*\}) \tag{7.4.49}$$

for any two matrices $\boldsymbol{X}$ and $\boldsymbol{Y}$, we find that the MSE of $\hat{\boldsymbol{s}}'$ is

$$\begin{aligned} E\left[\|\hat{\boldsymbol{s}}' - \boldsymbol{s}'\|^2\right] &= E\left[\left\|\left(\mathrm{Re}\{\boldsymbol{F}^H\boldsymbol{F}\}\right)^{-1}\mathrm{Re}\{\boldsymbol{F}^H\boldsymbol{y}\} - \boldsymbol{s}'\right\|^2\right] \\ =& E\left[\left\|\left(\mathrm{Re}\{\boldsymbol{F}^H\boldsymbol{F}\}\right)^{-1}\mathrm{Re}\{\boldsymbol{F}^H\boldsymbol{e}\}\right\|^2\right] \\ =& \frac{1}{2}E\left[\mathrm{Tr}\left\{\left(\mathrm{Re}\{\boldsymbol{F}^H\boldsymbol{F}\}\right)^{-2}\mathrm{Re}\left\{\boldsymbol{F}^H\boldsymbol{e}\boldsymbol{e}^H\boldsymbol{F} + \boldsymbol{F}^H\boldsymbol{e}\boldsymbol{e}^T\boldsymbol{F}^*\right\}\right\}\right] \\ =& \frac{\sigma^2}{2}\mathrm{Tr}\left\{\left(\mathrm{Re}\{\boldsymbol{F}^H\boldsymbol{F}\}\right)^{-1}\right\} \end{aligned} \tag{7.4.50}$$

Under (7.4.48) we have that

$$\begin{aligned}\operatorname{Tr}\{\boldsymbol{F}^H\boldsymbol{F}\} =&\sum_{n=1}^{n_s}\Big(\operatorname{Tr}\{\boldsymbol{A}_n^H\boldsymbol{H}^H\boldsymbol{H}\boldsymbol{A}_n\}\\ &+\operatorname{Tr}\{\boldsymbol{B}_n^H\boldsymbol{H}^H\boldsymbol{H}\boldsymbol{B}_n\}\Big) \leq 2n_s\cdot\|\boldsymbol{H}\|^2\end{aligned} \tag{7.4.51}$$

By the matrix version of the Cauchy-Schwarz inequality [Horn and Johnson, 1985, Th. 5.1.4]:

$$\begin{aligned}&\sqrt{\operatorname{Tr}\{\operatorname{Re}\{\boldsymbol{F}^H\boldsymbol{F}\}\}\cdot\operatorname{Tr}\left\{(\operatorname{Re}\{\boldsymbol{F}^H\boldsymbol{F}\})^{-1}\right\}}\\ \geq&\operatorname{Tr}\left\{(\operatorname{Re}\{\boldsymbol{F}^H\boldsymbol{F}\})^{1/2}\cdot(\operatorname{Re}\{\boldsymbol{F}^H\boldsymbol{F}\})^{-1/2}\right\}\\ =&\operatorname{Tr}\{\boldsymbol{I}_{2n_s}\} = 2n_s\end{aligned} \tag{7.4.52}$$

where $\boldsymbol{I}_{2n_s}$ is the identity matrix of dimension $2n_s \times 2n_s$, and $(\cdot)^{1/2}$ denotes the Hermitian square root of a positive definite matrix. Note that equality in (7.4.52) holds if $\operatorname{Re}\{\boldsymbol{F}^H\boldsymbol{F}\}$ is proportional to an identity matrix.

From (7.4.51) and (7.4.52) we have that (note that $\operatorname{Tr}\{\boldsymbol{F}^H\boldsymbol{F}\}$ is always real):

$$\operatorname{Tr}\left\{(\operatorname{Re}\{\boldsymbol{F}^H\boldsymbol{F}\})^{-1}\right\} \geq \frac{4n_s^2}{\operatorname{Tr}\{\operatorname{Re}\{\boldsymbol{F}^H\boldsymbol{F}\}\}} \geq \frac{4n_s^2}{2n_s\cdot\|\boldsymbol{H}\|^2} = \frac{2n_s}{\|\boldsymbol{H}\|^2} \tag{7.4.53}$$

The above lower bound is achieved when

$$\operatorname{Re}\{\boldsymbol{F}^H\boldsymbol{F}\} = \boldsymbol{I}\cdot\|\boldsymbol{H}\|^2 \tag{7.4.54}$$

Hence

$$E\big[\|\hat{\boldsymbol{s}}' - \boldsymbol{s}'\|^2\big] \geq n_s\cdot\frac{\sigma^2}{\|\boldsymbol{H}\|^2} \tag{7.4.55}$$

with equality if $\boldsymbol{X}$ is an OSTBC (cf. Theorem 7.2 on page 106). ■

Note that (7.4.48) and (7.4.51) imply that

$$\operatorname{Tr}\{\boldsymbol{X}\boldsymbol{X}^H\} = \boldsymbol{s}'^T\check{\boldsymbol{F}}^H\check{\boldsymbol{F}}\boldsymbol{s}' \leq 2n_s n_t\|\boldsymbol{s}'\|^2 \tag{7.4.56}$$

where $\check{\boldsymbol{F}}$ stands for the matrix $\boldsymbol{F}$ in (7.1.7) with $\boldsymbol{H} = \boldsymbol{I}$. Therefore, an STBC that satisfies the elemental power constraint (7.4.48) also satisfies a total power constraint. A more detailed discussion on the relationship and interplay between elemental and total power constraints was given in [Stoica and Ganesan, 2002a] (see also Chapter 6 and [Scaglione et al., 2002]).

7.4.5 Geometric Properties of OSTBC

Let us consider a transmitted matrix $\boldsymbol{X}$ (that may not be an OSTBC matrix). From Theorem 4.2 (see page 45), we find that the probability that the ML detector decides for a wrong codeword $\boldsymbol{X} \neq \boldsymbol{X}_0$ is given by

$$P(\boldsymbol{X}_0 \to \boldsymbol{X}|\boldsymbol{H}) = Q\left(\sqrt{\frac{1}{2\sigma^2}\|\boldsymbol{H}(\boldsymbol{X}_0 - \boldsymbol{X})\|^2}\right) \tag{7.4.57}$$

where $\boldsymbol{H}$ is the propagation channel, and σ^2 is the noise variance. This error probability depends in a monotonically decreasing way on the following Euclidean distance between the noise-free *received* codewords $\boldsymbol{HX}$ and $\boldsymbol{HX}_0$:

$$d^2 = \|\boldsymbol{H}(\boldsymbol{X}_0 - \boldsymbol{X})\|^2 \tag{7.4.58}$$

In a SISO system, the distance between the received (noise-free) codewords

$$d^2 = |h|^2|x_0 - x|^2 \tag{7.4.59}$$

is proportional to the distance between the transmitted codewords $|x_0 - x|^2$. This is not the case in general for a MIMO system, unless $\boldsymbol{H}$ is unitary. However, if $\boldsymbol{X}$ is an OSTBC matrix, we have that (see Theorem 7.3 on page 107):

$$(\boldsymbol{X} - \boldsymbol{X}_0)(\boldsymbol{X} - \boldsymbol{X}_0)^H = \sum_{n=1}^{n_s} |s_n^0 - s_n|^2\boldsymbol{I} \tag{7.4.60}$$

and hence

$$d^2 = \|\boldsymbol{H}(\boldsymbol{X}_0 - \boldsymbol{X})\|^2 = \|\boldsymbol{H}\|^2 \cdot \sum_{n=1}^{n_s} |s_n^0 - s_n|^2 \tag{7.4.61}$$

Consequently, for OSTBC matrices, the distance between the received (noise-free) codewords $d^2 = \|\boldsymbol{H}(\boldsymbol{X}_0 - \boldsymbol{X})\|^2$ is proportional to the distance between the transmitted codewords:

$$\|\boldsymbol{X} - \boldsymbol{X}_0\|^2 = n_t \sum_{n=1}^{n_s} |s_n^0 - s_n|^2 \tag{7.4.62}$$

For the set of transmitted codewords we can define a *distance profile.* The distance profile $\mathcal{D}$ for a given code matrix $\boldsymbol{X}_0$ is defined as the set of the squared distances between that code matrix $\boldsymbol{X}_0$ and all other code matrices $\boldsymbol{X}_k \neq \boldsymbol{X}_0$. In mathematical terms:

$$\mathcal{D}(\boldsymbol{X}_0) = \left\{\|\boldsymbol{X}_k - \boldsymbol{X}_0\|^2 : \boldsymbol{X}_k \neq \boldsymbol{X}_0\right\} \tag{7.4.63}$$

A set of code matrices is said to constitute a *uniform constellation* if its members have the same distance profile [FORNEY, 1991]. This means, in particular, that the minimum Euclidean distance

$$d_{\min}^2 = \min_{\boldsymbol{X}_k \neq \boldsymbol{X}_0} \|\boldsymbol{X}_k - \boldsymbol{X}_0\|^2 \tag{7.4.64}$$

as well as the number of code matrices achieving this minimum distance is the same for all codewords; hence the average error probability is approximately the same for all codewords.

An example of a uniform constellation in the context of OSTBC is the set of OSTBC matrices where the scalar symbols $\{s_k\}$ are taken from an MPSK constellation. This is so because in the case of MPSK, all the constellation points are uniformly distributed over the unit circle and therefore the constellation looks the same from each point. (An example of a constellation that does not have this property is QAM; therefore OSTBC using QAM symbols does not form a uniform constellation.)

7.4.6 Union Bound Optimality of OSTBC

We prove in this section that OSTBC matrices are optimal in the sense that they minimize the union bound on the probability of error, subject to a power constraint. The proof of this fact was first given in [SANDHU, 2002]. Our proof is inspired by that in the cited reference, even though it is slightly different and somewhat more general.

Let $\mathcal{U} = \{\boldsymbol{s}\}$ be the set of all possible signal vectors (of length n_s). Assume that the signal $\{s_1^0, \ldots, s_{n_s}^0\}$ is transmitted, and let $\boldsymbol{s}_0$ denote the corresponding signal vector of length n_s. Let $\boldsymbol{X}_0$ be the transmitted code matrix, and let $\boldsymbol{Y} = \boldsymbol{H}\boldsymbol{X}_0 + \boldsymbol{E}$ be the received data matrix in the usual manner. Then the probability that a certain wrong signal vector $\boldsymbol{s}_k \neq \boldsymbol{s}_0$ is decided in favor of $\boldsymbol{s}_0$ is equal to (see Theorem 4.2 on page 45):

$$\begin{aligned} P(\boldsymbol{s}_0 \to \boldsymbol{s}|\boldsymbol{H}) =& P(\|\boldsymbol{Y} - \boldsymbol{H}\boldsymbol{X}_k\|^2 < \|\boldsymbol{Y} - \boldsymbol{H}\boldsymbol{X}_0\|^2) \\ =& Q\left(\sqrt{\frac{1}{2\sigma^2}\|\boldsymbol{H}(\boldsymbol{X}_k - \boldsymbol{X}_0)\|^2}\right) \end{aligned} \tag{7.4.65}$$

where $\boldsymbol{X}_k$ is the OSTBC matrix corresponding to $\boldsymbol{s}_k$.

Let $K = |\mathcal{U}|$ be the number of elements in $\mathcal{U}$, i.e., the number of possible signal vectors of length n_s. Furthermore, let

$$\breve{\mathcal{U}} = \{\breve{\boldsymbol{s}} = \boldsymbol{s} - \hat{\boldsymbol{s}} : \boldsymbol{s} \in \mathcal{U}, \hat{\boldsymbol{s}} \in \mathcal{U}, \boldsymbol{s} \neq \hat{\boldsymbol{s}}\} \tag{7.4.66}$$

be the set of all possible error signals, and let $\breve{K} = |\breve{\mathcal{U}}|$ be the number of elements in $\breve{\mathcal{U}}$. Note that typically, $\breve{K} \gg K$. If all transmitted signal vectors are equally likely, the probability that the transmitted signal vector is incorrectly detected can be upper bounded by the union bound (see Section 4.2.1):

$$P_{\text{union}} = \frac{1}{K} \sum_{k=1}^{\breve{K}} Q\left(\sqrt{\frac{1}{2\sigma^2} \left\| \boldsymbol{H} \breve{\boldsymbol{X}}_k \right\|^2}\right) \tag{7.4.67}$$

where $\breve{\boldsymbol{X}}_k$ is the STBC matrix associated with the error signal $\breve{\boldsymbol{s}}_k$.

We have the following result.

Theorem 7.6: Union bound optimality of OSTBC.
Let $\boldsymbol{H}$ be fixed. Consider the class of $\{\boldsymbol{A}_n, \boldsymbol{B}_n\}$ that satisfy the same elemental power constraint as in Section 7.4.4:

$$\begin{aligned} \lambda_{max}\{\boldsymbol{A}_n \boldsymbol{A}_n^H\} &\leq 1, \quad n = 1, \ldots, n_s \\ \lambda_{max}\{\boldsymbol{B}_n \boldsymbol{B}_n^H\} &\leq 1, \quad n = 1, \ldots, n_s \end{aligned} \tag{7.4.68}$$

Among the codes that satisfy (7.4.68), it holds that

$$\frac{1}{K} \sum_{k=1}^{\breve{K}} Q\left(\sqrt{\frac{1}{2\sigma^2} \left\| \boldsymbol{H} \breve{\boldsymbol{X}}_k \right\|^2}\right) \geq \frac{1}{K} \sum_{k=1}^{\breve{K}} Q\left(\sqrt{\frac{\|\breve{\boldsymbol{s}}_k\|^2}{2\sigma^2} \cdot \|\boldsymbol{H}\|^2}\right) \tag{7.4.69}$$

Equality holds if $\{\boldsymbol{A}_n, \boldsymbol{B}_n\}$ satisfy (7.4.23).

Proof: See Section 7.7. ■

Note that if $\{\boldsymbol{A}_n, \boldsymbol{B}_n\}$ are constrained to be proportional to unitary matrices, the optimality result in Theorem 7.6 is true also under a power constraint such as (6.2.12). See [Stoica and Ganesan, 2002a] (and Chapter 6) for a discussion on these constraints and the relation between them.

7.5 STBC Based on Linear Constellation Precoding

The technique of constellation precoding was introduced in the context of SISO systems [Boutros and Viterbo, 1998], [Boutros et al., 1996], and later used for space-time diversity, for example, in [Xin et al., 2001]. Related ideas also appear in [Gamal and Damen, 2002]. In principle, the constellation precoding techniques encode a set of symbols s_n by forming a vector of n_s symbols,

$\boldsymbol{s} = [s_1 \ \cdots \ s_{n_s}]^T$, and pre-multiplying this vector by an $n_t \times n_s$ precoding matrix $\boldsymbol{\Phi}$ as follows:

$$\boldsymbol{x} = \boldsymbol{\Phi}\boldsymbol{s} \tag{7.5.1}$$

Next, the transmitted $n_t \times n_t$ code matrix is computed as:

$$\boldsymbol{X} = \boldsymbol{U} \cdot \begin{bmatrix} x_1 & 0 & \cdots & 0 \\ 0 & x_2 & \ddots & \vdots \\ \vdots & \ddots & \ddots & 0 \\ 0 & \cdots & 0 & x_{n_t} \end{bmatrix} \tag{7.5.2}$$

where $\boldsymbol{U}$ is a unitary matrix, and transmitted in the usual way (cf. Section 2.1).

The received data can be written as:

$$\boldsymbol{Y} = \boldsymbol{H}\boldsymbol{X} + \boldsymbol{E} = \boldsymbol{H}\boldsymbol{U} \begin{bmatrix} x_1 & 0 & \cdots & 0 \\ 0 & x_2 & \ddots & \vdots \\ \vdots & \ddots & \ddots & 0 \\ 0 & \cdots & 0 & x_{n_t} \end{bmatrix} + \boldsymbol{E} \tag{7.5.3}$$

Clearly, the unitary matrix $\boldsymbol{U}$ can be "absorbed" into the channel: $\boldsymbol{H}\boldsymbol{U}$; if $\boldsymbol{H}$ has independent and Gaussian elements then so will $\boldsymbol{H}\boldsymbol{U}$ and hence the presence of the matrix $\boldsymbol{U}$ does not affect the error probability (but it can be chosen, for example, to reduce the dynamic range of the transmitted signal). For simplicity, we shall simply set $\boldsymbol{U} = \boldsymbol{I}$.

The probability that a transmitted vector $\boldsymbol{s}_0$ is incorrectly detected as a different vector $\boldsymbol{s}$ is (see Theorem 4.2 on page 45):

$$P(\boldsymbol{s}_0 \to \boldsymbol{s}|\boldsymbol{H}) = Q\left(\sqrt{\frac{1}{2\sigma^2}\|\boldsymbol{H}(\boldsymbol{X} - \boldsymbol{X}_0)\|^2}\right) \tag{7.5.4}$$

where $\boldsymbol{X}$ and $\boldsymbol{X}_0$ are the matrices in (7.5.2) associated with $\boldsymbol{s}$ and $\boldsymbol{s}_0$. If $\boldsymbol{H}$ has independent and zero-mean Gaussian elements with power ρ^2, then the average error probability is bounded by (see Theorem 4.2):

$$E_{\boldsymbol{H}}[P(\boldsymbol{s}_0 \to \boldsymbol{s})] \leq \left| \begin{bmatrix} x_1 - x_1^0 & 0 & \cdots & 0 \\ 0 & x_2 - x_2^0 & \ddots & \vdots \\ \vdots & \ddots & \ddots & 0 \\ 0 & \cdots & 0 & x_{n_t} - x_{n_t}^0 \end{bmatrix} \right|^{-n_r} \cdot \left(\frac{\rho^2}{4\sigma^2}\right)^{-n_r n_t}$$
$$= \left(\prod_{n=1}^{n_t} |\boldsymbol{\phi}_n^T (\boldsymbol{s} - \boldsymbol{s}_0)| \right)^{-n_r} \cdot \left(\frac{\rho^2}{4\sigma^2}\right)^{-n_r n_t} \tag{7.5.5}$$

where $\boldsymbol{\phi}_n^T$ is the nth row of $\boldsymbol{\Phi}$.

Like OSTBC, codes based on constellation precoding are constructed starting from an error performance criterion, and in general they incur (like OSTBC) a loss in the mutual information between the transmitter and the receiver. The detection of codes based on constellation precoding amounts to minimizing a metric similar to (7.1.13), which is a computationally complex problem. A deeper treatment of constellation precoding codes falls outside the scope of this text.

7.6 Summary and Discussion

A linear space-time block code is a matrix $\boldsymbol{X}$ that is formed as a linear combination of the real and imaginary parts of a set of symbols $\{s_1, \ldots, s_{n_s}\}$ according to $\boldsymbol{X} = \sum_{n=1}^{n_s} (\bar{s}_n \boldsymbol{A}_n + i\tilde{s}_n \boldsymbol{B}_n)$, where $\{\boldsymbol{A}_n, \boldsymbol{B}_n\}$ are fixed matrices. Clearly, linear codes constitute only a special case of all possible space-time codes; nevertheless the class of linear STBC is quite rich.

Linear STBC can be constructed either by maximizing the throughput via information-theoretic criteria, or by adopting a more performance-oriented point of view (for example, by attempting to minimize the error probability). In this chapter, we have provided a summary of linear STBC along with their associated maximum-likelihood criteria for coherent detection (see Section 7.1). The intention of the chapter has not been to present an exhaustive review of all available results on linear STBC; instead we have tried to focus on general principles illustrated by a few examples. The linear STBC discussed in the chapter include spatial multiplexing and linear dispersion codes (see Sections 7.2 and 7.3), orthogonal STBC (see Section 7.4), and codes based on constellation precoding (see Section 7.5).

A substantial part of this chapter (viz., Section 7.4) was dedicated to orthogonal STBC, which is an important subclass of linear STBC for which $\boldsymbol{X}$ is proportional to a unitary matrix for all symbols. There are at least three reasons behind the interest that OSTBC has generated. First, although the theory behind their existence (which is presented in Appendix B) is fairly complicated, they are conceptually quite simple. Second, in addition to providing diversity of maximal order (i.e., $n_r n_t$), they possess a number of optimality properties; for example, they minimize the union bound on the error probability for a given channel. Last but not least, OSTBC leads to decoupled ML detection of the symbols used to form $\boldsymbol{X}$.

The developments in this chapter are based on the channel models of Chapter 2, as well as results from Chapters 3–4. Most remaining parts of the book will deal with linear STBC and therefore build on results from the present chapter. For instance, in Chapter 8 we will discuss linear STBC for frequency-selective channels, and in Chapter 9 we will present detection algorithms for linear STBC. The material in Chapters 10 and 11 will also be primarily concerned with linear STBC.

7.7 Proofs

Proof of Theorem 7.1

For all $\boldsymbol{s}$, we have that

$$
\begin{aligned}
\|\boldsymbol{s}\|^2 \boldsymbol{I} = & \sum_{n=1}^{n_s} (\bar{s}_n^2 \boldsymbol{A}_n \boldsymbol{A}_n^H + \tilde{s}_n^2 \boldsymbol{B}_n \boldsymbol{B}_n^H) \\
& + \sum_{n=1}^{n_s} \sum_{p=1, p>n}^{n_s} \Big(\bar{s}_n \bar{s}_p (\boldsymbol{A}_n \boldsymbol{A}_p^H + \boldsymbol{A}_p \boldsymbol{A}_n^H) + \tilde{s}_n \tilde{s}_p (\boldsymbol{B}_n \boldsymbol{B}_p^H + \boldsymbol{B}_p \boldsymbol{B}_n^H) \Big) \\
& + i \sum_{n=1}^{n_s} \sum_{p=1}^{n_s} \tilde{s}_n \bar{s}_p (\boldsymbol{B}_n \boldsymbol{A}_p^H - \boldsymbol{A}_p \boldsymbol{B}_n^H)
\end{aligned}
\tag{7.7.1}
$$

- Let k be an arbitrary integer and set $s_n = 1$ for $n = k$ and $s_n = 0$ for $n \neq k$. Then insertion in (7.7.1) shows that we must have $\boldsymbol{A}_k \boldsymbol{A}_k^H = \boldsymbol{I}$.
- In a similar way, let k be arbitrary and set $s_n = i$ for $n = k$ and $s_n = 0$ for $n \neq k$. Insertion in (7.7.1) shows that $\boldsymbol{B}_k \boldsymbol{B}_k^H = \boldsymbol{I}$.
- Let k, p be two arbitrary integers such that $k \neq p$. Set $s_n = 1$ for $n = k$ and $n = p$, and $s_n = 0$ for $n \neq k$, $n \neq p$. Insertion in (7.7.1) shows that we must have $\boldsymbol{A}_k \boldsymbol{A}_p^H + \boldsymbol{A}_p \boldsymbol{A}_k^H = 0$.
- In a similar manner, let k, p be two arbitrary integers such that $k \neq p$. Set

$s_n = i$ for $n = k$ and $n = p$, and $s_n = 0$ for $n \neq k$, $n \neq p$. Insertion in (7.7.1) shows that $\boldsymbol{B}_k \boldsymbol{B}_p^H + \boldsymbol{B}_p \boldsymbol{B}_k^H = 0$.

- Let k be an arbitrary integer and choose $s_n = 1 + i$ for $n = k$ and $s_n = 0$ for $n \neq k$. It follows that $\boldsymbol{B}_k \boldsymbol{A}_k^H = \boldsymbol{A}_k \boldsymbol{B}_k^H$.

- Finally, let k and p, $k \neq p$, be arbitrary and set $s_n = 1$ for $n = k$, $s_n = i$ for $n = p$, and $s_n = 0$ for $n \neq k$, $n \neq p$. Insertion in (7.7.1) shows that $\boldsymbol{A}_k \boldsymbol{B}_p^H = \boldsymbol{B}_k \boldsymbol{A}_p^H$.

Proof of Theorem 7.6

Clearly, if $\boldsymbol{X}$ is an OSTBC, we have that:

$$\left\| \boldsymbol{H} \breve{\boldsymbol{X}}_k \right\|^2 = \|\breve{\boldsymbol{s}}_k\|^2 \|\boldsymbol{H}\|^2 \tag{7.7.2}$$

Hence, it follows immediately that equality holds in (7.4.69) if $\{\boldsymbol{A}_n, \boldsymbol{B}_n\}$ satisfy (7.4.23). The difficult part is to show the inequality.

For notational convenience, let

$$\begin{aligned} \phi_{n,p} &= 2\mathrm{ReTr}\{\boldsymbol{H}^H \boldsymbol{H} \boldsymbol{A}_n \boldsymbol{A}_p^H\} \\ \psi_{n,p} &= 2\mathrm{ReTr}\{\boldsymbol{H}^H \boldsymbol{H} \boldsymbol{B}_n \boldsymbol{B}_p^H\} \\ \varphi_{n,p} &= 2\,\mathrm{Im}\,\mathrm{Tr}\{\boldsymbol{H}^H \boldsymbol{H} \boldsymbol{A}_p \boldsymbol{B}_n^H\} \end{aligned} \tag{7.7.3}$$

for any $n = 1, \ldots, n_s$ and $p = 1, \ldots, n_s$. Also, for any integer $r = 0, \ldots, n_s$, let us denote by $\mathcal{P}_r$ the set of $\{\boldsymbol{A}_k, \boldsymbol{B}_k\}$ for which

$$\begin{aligned} &\phi_{n,n} \leq 2\|\boldsymbol{H}\|^2, \quad n = 1, \ldots, n_s \\ &\psi_{n,n} \leq 2\|\boldsymbol{H}\|^2, \quad n = 1, \ldots, n_s \\ &\phi_{n,p} = 0, \quad n \neq p, n = r+1, \ldots, n_s, p = r+1, \ldots, n_s \\ &\psi_{n,p} = 0, \quad n \neq p, n = r+1, \ldots, n_s, p = r+1, \ldots, n_s \\ &\varphi_{n,p} = 0, \quad n = r+1, \ldots, n_s, p = r+1, \ldots, n_s \end{aligned} \tag{7.7.4}$$

Note that $\mathcal{P}_{n_s}$ is equal to the set of all permissible matrices $\{\boldsymbol{A}_n, \boldsymbol{B}_n\}$ (see (7.4.68)); in this case none of $\phi_{n,p}$, $\psi_{n,p}$ or $\varphi_{n,p}$ are forced to be zero by (7.7.4).

We will prove the inequality in (7.4.69) by induction over r. Clearly, for $r = 0$, we have that

$$\begin{aligned} \frac{1}{2\sigma^2} \left\| \boldsymbol{H} \breve{\boldsymbol{X}}_k \right\|^2 &= \frac{1}{4\sigma^2} \sum_{n=1}^{n_s} \left(\breve{s}_n^{(k)2} \phi_{n,n} + \breve{s}_n^{(k)2} \psi_{n,n} \right) \\ &\leq \frac{\|\breve{\boldsymbol{s}}_k\|^2}{2\sigma^2} \|\boldsymbol{H}\|^2 \end{aligned} \tag{7.7.5}$$

Consequently, because the Q-function is monotonically decreasing, (7.4.69) holds for all $\{\boldsymbol{A}_n, \boldsymbol{B}_n\} \in \mathcal{P}_0$.

To perform the induction step, assume now that (7.4.69) holds for all $\{\boldsymbol{A}_n, \boldsymbol{B}_n\} \in \mathcal{P}_{r_0}$ where $0 \leq r_0 \leq n_s$. Then for all $\{\boldsymbol{A}_n, \boldsymbol{B}_n\} \in \mathcal{P}_{r_0+1}$, we can write

$$\frac{1}{K}\sum_{k=1}^{\breve{K}} Q\left(\sqrt{\frac{1}{2\sigma^2}\left\|\boldsymbol{H}\breve{\boldsymbol{X}}_k\right\|^2}\right) = \frac{1}{K}\sum_{k=1}^{\breve{K}} Q\left(\sqrt{\frac{1}{2\sigma^2}\left(z_{k,r_0} + z_{k,r_0}^{+}\right)}\right) \tag{7.7.6}$$

where

$$\begin{aligned} z_{k,r_0} = & \frac{1}{2}\sum_{n=1}^{n_s}\left(\breve{\bar{s}}_n^{(k)2}\phi_{n,n} + \breve{\tilde{s}}_n^{(k)2}\psi_{n,n}\right) \\ & + \sum_{n=1}^{r_0}\sum_{p=1,p>n}^{r_0}\left(\breve{\bar{s}}_n^{(k)}\breve{\bar{s}}_p^{(k)}\phi_{n,p} + \breve{\tilde{s}}_n^{(k)}\breve{\tilde{s}}_p^{(k)}\psi_{n,p}\right) \\ & + \sum_{n=1}^{r_0}\sum_{p=1}^{r_0}\breve{\tilde{s}}_n^{(k)}\breve{\bar{s}}_p^{(k)}\varphi_{n,p} \end{aligned} \tag{7.7.7}$$

and

$$z_{k,r_0}^{+} = \sum_{n=1}^{r_0}\left(\breve{\bar{s}}_n^{(k)}\breve{\bar{s}}_{r_0+1}^{(k)}\phi_{n,r_0+1} + \breve{\tilde{s}}_n^{(k)}\breve{\tilde{s}}_{r_0+1}^{(k)}\psi_{n,r_0+1}\right) + \sum_{n=1}^{r_0+1}\breve{\tilde{s}}_n^{(k)}\breve{\bar{s}}_{r_0+1}^{(k)}\varphi_{n,r_0+1} \tag{7.7.8}$$

Due to symmetry reasons (see [Sandhu, 2002] for a more detailed discussion) we must have

$$\begin{aligned} & \frac{1}{K}\sum_{k=1}^{\breve{K}} Q\left(\sqrt{\frac{1}{2\sigma^2}\left\|\boldsymbol{H}\breve{\boldsymbol{X}}_k\right\|^2}\right) \\ & = \frac{1}{2K}\sum_{k=1}^{\breve{K}} Q\left(\sqrt{\frac{1}{2\sigma^2}\left(z_{k,r_0} + z_{k,r_0}^{+}\right)}\right) + Q\left(\sqrt{\frac{1}{2\sigma^2}\left(z_{k,r_0} - z_{k,r_0}^{+}\right)}\right) \\ & \geq \frac{1}{K}\sum_{k=1}^{\breve{K}} Q\left(\sqrt{\frac{1}{2\sigma^2}z_{k,r_0}}\right) \\ & \geq \frac{1}{K}\sum_{k=1}^{\breve{K}} Q\left(\sqrt{\frac{\|\breve{\boldsymbol{s}}_k\|^2}{2\sigma^2}\|\boldsymbol{H}\|^2}\right) \end{aligned} \tag{7.7.9}$$

where the first inequality is due to the easily verified fact that

$$Q\left(\sqrt{x+y}\right) + Q\left(\sqrt{x-y}\right) \geq 2Q\left(\sqrt{x}\right) \tag{7.7.10}$$

for any x and y, and in the second inequality we used the induction assumption.

7.8 Problems

1. Give an example of a linear STBC (according to the definition in (7.1.1)) that is *not* a linear function of only the complex symbol vector $\boldsymbol{s}$.

2. Prove (7.4.47).

3. Study the following code matrix (in the framework of Section 7.1):

$$\boldsymbol{X} = \begin{bmatrix} s_1 & s_1 \\ s_2 & -s_2 \end{bmatrix} \tag{7.8.1}$$

   Suppose that s_1 and s_2 are complex symbols in general.

   (a) Find the matrices $\{\boldsymbol{A}_n, \boldsymbol{B}_n\}$ and $\boldsymbol{F}$.

   (b) Give the ML rule for detection. Is the symbol detection decoupled?

   (c) Can the code be used to achieve transmit diversity?

4. Consider the code in Example 7.6 (see page 105) and verify that it does not give transmit diversity.

5. Consider again the diagonal code in Example 7.6. The reason for the loss of diversity of this code is easy to understand intuitively: energy from a certain information bit is transmitted via only one antenna and hence the resulting SNR will be a function of the path gains corresponding to that particular antenna only. In an attempt to correct this situation, consider the following "ad-hoc" code matrix:

$$\boldsymbol{X} = \frac{1}{2}\begin{bmatrix} s_1 + s_2 & s_1 - s_2 \\ s_1 - s_2 & s_1 + s_2 \end{bmatrix} \tag{7.8.2}$$

   Assume that the symbol constellation $\mathcal{S}$ is unitary; hence $|s_1| = |s_2| = 1$.

   (a) Show that $\boldsymbol{X}\boldsymbol{X}^H = \boldsymbol{I}$ for all $\{s_n\}$ in the constellation, and demonstrate that the ML detection of s_1 and s_2 is decoupled.

(b) Verify that the matrices $\{\boldsymbol{A}_n, \boldsymbol{B}_n\}$ associated with (7.8.2) do not satisfy (7.4.4).

(c) Does the code achieve transmit diversity?

6. Derive (7.4.43) by using (7.3.1) along with Theorem 7.2 (see page 106).

7. Obtain a bound on the error rate for spatial multiplexing using Theorem 4.2 (see page 45).

8. Show that the matrices $\{\boldsymbol{A}_n, \boldsymbol{B}_n\}$ associated with constellation precoding are diagonal (provided that $\boldsymbol{U} = \boldsymbol{I}$), and derive explicit expressions for them.

9. Prove the last inequality in (7.4.45).

10. Suppose that (7.4.4) holds. Prove that $\mathrm{Re}\{\boldsymbol{F}^H \boldsymbol{F}\} = \|\boldsymbol{H}\|^2 \cdot \boldsymbol{I}$ via a direct algebraic approach.

Hint: The trace of a Hermitian matrix is real-valued, and the trace of an anti-Hermitian matrix is imaginary.

Chapter 8

LINEAR STBC FOR FREQUENCY-SELECTIVE CHANNELS

In this chapter we proceed to analyze transmit diversity for frequency-selective channels and, in particular, extensions of linear space-time block coding to such channels. Surprisingly, for a channel with $L+1$ Rayleigh fading taps, it is relatively easy to construct a transmission scheme that achieves a diversity order equal to $n_r n_t(L+1)$; however, it is harder to construct a scheme that achieves this diversity order and at the same time has a tractable receiver complexity.

8.1 Achieving Diversity for a Frequency-Selective Channel

Recall from Section 2.2 that if a scalar sequence $\{s(-N_{\text{pre}}), \ldots, s(N_0 + N_{\text{post}} - 1)\}$, is transmitted via the block-transmission technique in Section 2.2, the received data can be expressed as:

$$y = (X^T \otimes I_{n_r})h + e \tag{8.1.1}$$

where

$$\begin{aligned} h &= \operatorname{vec}\left(\begin{bmatrix} H_0 & \cdots & H_L \end{bmatrix}\right) \\ y &= \begin{bmatrix} y^T(0) & \cdots & y^T(N_0 + L - 1) \end{bmatrix}^T \\ e &= \begin{bmatrix} e^T(0) & \cdots & e^T(N_0 + L - 1) \end{bmatrix}^T \end{aligned} \tag{8.1.2}$$

and X is a Toeplitz matrix built from the transmitted data according to:

$$X = \begin{bmatrix} x(0) & & \cdots & \cdots & \cdots & x(N_0 + L - 1) \\ x(-1) & \ddots & & & & x(N_0 + L - 2) \\ \vdots & & \ddots & & & \vdots \\ x(-L) & \cdots & \cdots & x(0) & \cdots & x(N_0 - 1) \end{bmatrix} \tag{8.1.3}$$

As long as $N_{\text{post}} \geq L$, and the preamble is known to the receiver, the data received during $n = 0, \ldots, N_0 + L - 1$ are sufficient statistics for detection of the transmitted signal. Hence, provided that the receiver noise $\boldsymbol{e}(n)$ is spatially and temporally white and that the propagation channel $\boldsymbol{H}(z^{-1})$ is known, the ML detection of $\{s(0), \ldots, s(N_0 - 1)\}$ amounts to minimizing the metric

$$\sum_{n=0}^{N_0+L-1} \left\| \boldsymbol{y}(n) - \sum_{l=0}^{L} \boldsymbol{H}_l \boldsymbol{x}(n-l) \right\|^2 = \|\boldsymbol{y} - (\boldsymbol{X}^T \otimes \boldsymbol{I}_{n_r})\boldsymbol{h}\|^2 \tag{8.1.4}$$

Equation (8.1.4) is central to both the design of transmission schemes and receiver structures, and will be elaborated on later in this chapter.

An important question is under what conditions the ML decoding of $\{s(n)\}$ can give diversity, or more precisely, what diversity order is achievable. To answer this question, we will capitalize on the formulation (8.1.1) of the received signal. Let $\boldsymbol{X}_0$ denote the Toeplitz matrix corresponding to the true transmitted data sequence $s_0(n)$, and let $\boldsymbol{X}$ denote the Toeplitz matrix corresponding to any hypothetical signal $s(n)$ different from the true transmitted one. Also, let $\boldsymbol{X}_0 - \boldsymbol{X}$ be the Toeplitz matrix associated with the error signal $s_0(n) - s(n)$. Assume that the $n_r n_t(L+1)$ elements of $\boldsymbol{h}$ are complex Gaussian with a positive definite covariance matrix $E[\boldsymbol{h}\boldsymbol{h}^H] > 0$. Then, provided that the matrix $\boldsymbol{X}_0 - \boldsymbol{X}$ has full rank for all possible error sequences $\{s_0(n) - s(n)\}$, we know from Theorem 4.2 (see page 45) that ML detection of $s(n)$ achieves a diversity of order $n_r n_t(L+1)$. The factor n_r in the diversity order will be called receive diversity order, the factor n_t corresponds to the order of transmit diversity, and the last factor $L+1$ is associated with the diversity gained from the multipath propagation. We will see below, as expected, that not all possible transmission schemes (i.e., mappings between $s(n)$ and $\boldsymbol{x}(n)$) can achieve both transmit and multipath diversity simultaneously. In particular, not all choices of preambles and postambles in the block transmission guarantee that the matrix $\boldsymbol{X}_0 - \boldsymbol{X}$ has full rank.

Example 8.1: Delay diversity, revisited.
The "delay diversity" transmission scheme of Section 6.3.5 can easily be analyzed in the present framework. Consider the code matrix (6.3.23) associated with delay diversity (for $n_t = 2$):

$$\begin{bmatrix} \cdots & s(n) & s(n+1) & \cdots & s(n+l) & \cdots \\ \cdots & s(n-D) & s(n+1-D) & \cdots & s(n+l-D) & \cdots \end{bmatrix} \tag{8.1.5}$$

The matrix $\boldsymbol{X}$ in (8.1.3) has $2(L+1)$ rows and takes on the form:

$$\boldsymbol{X} = \begin{bmatrix} \cdots & s(n) & s(n+1) & \cdots & s(n+l) & \cdots \\ \cdots & s(n-D) & s(n+1-D) & \cdots & s(n+l-D) & \cdots \\ \cdots & \vdots & \vdots & \vdots & \vdots & \cdots \\ \cdots & s(n-L) & s(n+1-L) & \cdots & s(n+l-L) & \cdots \\ \cdots & s(n-L-D) & s(n+1-L-D) & \cdots & s(n+l-L-D) & \cdots \end{bmatrix} \tag{8.1.6}$$

If we assume that the symbol stream under consideration is preceded and followed by a known preamble and postamble, similarly to Section 5.2.1, then it is not hard to see that the matrix difference $\boldsymbol{X}_0 - \boldsymbol{X}$ formed according to (8.1.6) associated with DD has rank $L+D+1$ for any nonzero error signal $\{s_0(n) - s(n)\}$. Hence, provided that the covariance matrix of $[\boldsymbol{h}_1^T \ \boldsymbol{h}_2^T]^T$ has full rank, it follows from Theorem 4.2 (see page 45) that a diversity of order $n_r(L+D+1)$ is achieved. (See also [GORE ET AL., 2001] for a direct proof of this fact.) Since $\boldsymbol{X}$ has $2(L+1)$ rows, its rank cannot be more than $2(L+1)$. It follows that the maximum achievable diversity order is equal to $2n_r(L+1)$, which is achieved for $D = L+1$. This reinforces the conclusion in Section 6.3.5. ■

8.2 Space-Time OFDM (ST-OFDM)

For systems with one transmit antenna, we have seen in Chapter 5 that a frequency-selective channel can be converted into parallel and independent flat channels by using orthogonal-frequency division multiplexing (OFDM). In this section we will show how the OFDM concept can be extended to $n_t > 1$ transmit antennas. The so-obtained scheme will be called space-time OFDM (ST-OFDM), but it has also been referred to as space-*frequency* coding.

8.2.1 Transmit Encoding for ST-OFDM

The principal idea behind ST-OFDM is analogous to that of conventional OFDM, but instead of operating on a single symbol stream of length N_0 we assume that now we are given n_s streams (of length N_0) in parallel. In words, the procedure associated with ST-OFDM encoding can be described as follows. First, the symbol streams are encoded via a space-time block code. Next, the set of code matrices is inverse Fourier transformed. This way N_0 space-time block matrices are obtained. These are transmitted, after the addition of a cyclic prefix, as shown in Figure 8.1. Note that like for conventional OFDM, the transmitted bursts do not have any postambles.

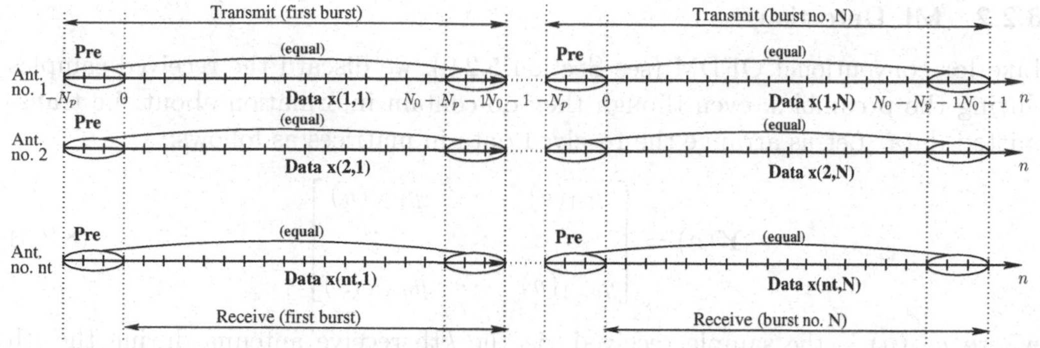

Figure 8.1. The ST-OFDM transmission scheme.

Consider a symbol sequence $s(n)$ of length $n_s N_0$, and partition it into n_s sequences of length N_0 as follows:

$$\begin{aligned} s_1(n) &= s(n), \quad n = 0, \ldots, N_0 - 1 \\ s_2(n) &= s(n + N_0), \quad n = 0, \ldots, N_0 - 1 \\ s_{n_s}(n) &= s(n + (n_s - 1)N_0), \quad n = 0, \ldots, N_0 - 1 \end{aligned} \tag{8.2.1}$$

These symbol streams are "space-frequency" encoded by mapping them onto a sequence of $n_t \times N$-matrices $\{\boldsymbol{\Phi}(n)\}$:

$$\{s_1(n), \ldots, s_{n_s}(n)\} \to \{\boldsymbol{\Phi}(n)\} \tag{8.2.2}$$

where

$$\boldsymbol{\Phi}(n) = \begin{bmatrix} \boldsymbol{\phi}_1(n) & \cdots & \boldsymbol{\phi}_N(n) \end{bmatrix} \tag{8.2.3}$$

for $n = 0, \ldots, N_0 - 1$. For each n, the mapping in (8.2.2) is equivalent to the mapping implicit to space-time block coding for frequency-flat channels in Chapter 7.

The set of matrices $\{\boldsymbol{\Phi}(n)\}$ is now inverse Fourier-transformed according to:

$$\boldsymbol{X}(n) = \frac{1}{\sqrt{N_0}} \sum_{k=0}^{N_0 - 1} \boldsymbol{\Phi}(k) \exp\left(i \frac{2\pi}{N_0} nk \right) \tag{8.2.4}$$

and the matrices $\{\boldsymbol{X}(n)\}$, $n = 0, \ldots, N_0 - 1$, are transmitted according to Figure 8.1.

8.2.2 ML Detection

Like for conventional OFDM (see Section 5.2.2), we discard the received samples during the preambles, even though they do contain information about the transmitted data. Let us arrange the received data in matrices as follows:

$$\boldsymbol{Y}(n) = \begin{bmatrix} y_{1,1}(n) & \cdots & y_{1,N}(n) \\ \vdots & \ddots & \vdots \\ y_{n_r,1}(n) & \cdots & y_{n_r,N}(n) \end{bmatrix} \tag{8.2.5}$$

where $y_{k,l}(n)$ is the sample received via the kth receive antenna during the lth burst and the nth time interval. Then

$$\boldsymbol{Y}(n) = \sum_{l=0}^{L} \boldsymbol{H}_l \boldsymbol{X}(n-l) + \boldsymbol{E}(n) \tag{8.2.6}$$

for $n = 0, \ldots, N_0 - 1$, where $\boldsymbol{E}(n)$ is a noise matrix defined analogously to $\boldsymbol{Y}(n)$. Note that the presence of the cyclic prefix implies that

$$\boldsymbol{X}(n) = \boldsymbol{X}(n + N_0), \quad n = -N_{\text{pre}}, \ldots, -1 \tag{8.2.7}$$

The ML metric for detection of the data is equal to:

$$\sum_{n=0}^{N_0-1} \left\| \boldsymbol{Y}(n) - \sum_{l=0}^{L} \boldsymbol{H}_l \boldsymbol{X}(n-l) \right\|^2 \tag{8.2.8}$$

Similarly to conventional OFDM, this ML metric is equal to a sum of metrics, one for each subcarrier. We state this result next.

Theorem 8.1: ML metric for ST-OFDM.
The ML metric for ST-OFDM is given by:

$$\sum_{n=0}^{N_0-1} \left\| \boldsymbol{Y}(n) - \sum_{l=0}^{L} \boldsymbol{H}_l \boldsymbol{X}(n-l) \right\|^2 = \sum_{n=0}^{N_0-1} \left\| \boldsymbol{Z}(n) - \boldsymbol{H}\left(\frac{2\pi}{N_0} n\right) \boldsymbol{\Phi}(n) \right\|^2 \tag{8.2.9}$$

where

$$\begin{aligned} \boldsymbol{Z}(n) &= \frac{1}{\sqrt{N_0}} \sum_{k=0}^{N_0-1} \boldsymbol{Y}(k) \cdot \exp\left(-i\frac{2\pi}{N_0} nk\right) \\ \boldsymbol{H}(\omega) &= \sum_{l=0}^{L} \boldsymbol{H}_l \exp\left(-i\omega l\right) \end{aligned} \tag{8.2.10}$$

Proof: Follows by a calculation similar to that in Section 5.2.2. ■

To proceed, we need to specify the STBC matrix $\mathbf{\Phi}(n)$ that is used in the transmit encoding.

8.2.3 ST-OFDM with Linear STBC

If a linear space-time block code is used, then (8.2.2) takes on the following form:

$$\mathbf{\Phi}(n) = \sum_{k=1}^{n_s} \left(\bar{s}_k(n)\boldsymbol{A}_k + i\tilde{s}_k(n)\boldsymbol{B}_k\right) \tag{8.2.11}$$

where $\{\boldsymbol{A}_k, \boldsymbol{B}_k\}$ is a set of matrices as discussed in Chapter 7. The ML detection problem for a system using ST-OFDM with linear STBC decouples into N_0 independent ML detection problems equivalent to minimization of N_0 metrics for frequency flat channels (see Section 7.1). A consequence of this important result, which we state below, is that with ST-OFDM a diversity of order no more than $n_t n_r$ can be achieved.

Theorem 8.2: ML metric for ST-OFDM with linear STBC.
The ML metric for ST-OFDM with linear STBC is given by:

$$\sum_{n=0}^{N_0-1} \left\|\boldsymbol{z}(n) - \boldsymbol{F}(n)\boldsymbol{s}'(n)\right\|^2 \tag{8.2.12}$$

where $\boldsymbol{H}(\omega)$ is defined in (8.2.10),

$$\begin{aligned}
\boldsymbol{s}'(n) &\triangleq \begin{bmatrix} \bar{\boldsymbol{s}}(n) \\ \tilde{\boldsymbol{s}}(n) \end{bmatrix} \\
\boldsymbol{F}(n) &\triangleq \left[\boldsymbol{F}_a(n) \quad \boldsymbol{F}_b(n)\right] \\
\boldsymbol{F}_a(n) &\triangleq \left[\operatorname{vec}\left(\boldsymbol{H}\left(2\pi\frac{n}{N_0}\right)\boldsymbol{A}_1\right) \quad \cdots \quad \operatorname{vec}\left(\boldsymbol{H}\left(2\pi\frac{n}{N_0}\right)\boldsymbol{A}_{n_s}\right)\right] \\
\boldsymbol{F}_b(n) &\triangleq \left[i\operatorname{vec}\left(\boldsymbol{H}\left(2\pi\frac{n}{N_0}\right)\boldsymbol{B}_1\right) \quad \cdots \quad i\operatorname{vec}\left(\boldsymbol{H}\left(2\pi\frac{n}{N_0}\right)\boldsymbol{B}_{n_s}\right)\right]
\end{aligned} \tag{8.2.13}$$

and

$$\boldsymbol{z}(n) = \operatorname{vec}(\boldsymbol{Z}(n)) \tag{8.2.14}$$

Proof: Follows by using Theorem 8.1 and paralleling the calculations in Section 7.1. ■

8.2.4 ST-OFDM with OSTBC

A particularly interesting special case of ST-OFDM occurs when *orthogonal* STBC is used as an underlying space-time code. In this case (cf. Theorem 7.2 (see page 106)):

$$\operatorname{Re}\left\{\boldsymbol{F}^H(n)\boldsymbol{F}(n)\right\} = \left\|\boldsymbol{H}\left(\frac{2\pi}{N_0}n\right)\right\|^2 \cdot \boldsymbol{I} \tag{8.2.15}$$

and the minimization of the ML metric for ST-OFDM reduces to that of minimizing

$$\sum_{n=0}^{N_0-1} \left\|\hat{\boldsymbol{s}}'(n) - \boldsymbol{s}'(n)\right\|^2 \tag{8.2.16}$$

where

$$\hat{\boldsymbol{s}}'(n) = \frac{\operatorname{Re}\{\boldsymbol{F}^H(n)\boldsymbol{z}(n)\}}{\left\|\boldsymbol{H}\left(\frac{2\pi}{N_0}n\right)\right\|^2} \tag{8.2.17}$$

which is equivalent to solving N_0 independent OSTBC detection problems (cf. Section 7.4). From Theorem 7.3 (see page 107) we conclude that

$$\hat{\boldsymbol{s}}'(n) \sim N\left(\boldsymbol{s}'(n), \frac{\sigma^2}{2\left\|\boldsymbol{H}\left(\frac{2\pi}{N_0}n\right)\right\|^2}\boldsymbol{I}\right) \tag{8.2.18}$$

and that the SNR on each complex subcarrier is proportional to

$$\text{SNR} = \frac{\left\|\boldsymbol{H}\left(\frac{2\pi}{N_0}n\right)\right\|^2}{\sigma^2} \tag{8.2.19}$$

Equation (8.2.19) reconfirms the fact that a diversity of order $n_r n_t$ is achieved in Rayleigh fading; i.e., that the multipath diversity is lost.

8.2.5 ST-OFDM with the Alamouti Code

For ST-OFDM based on the Alamouti code, the encoding in (8.2.11) becomes:

$$\boldsymbol{\Phi}(n) = \begin{bmatrix} s_1(n) & s_2^*(n) \\ s_2(n) & -s_1^*(n) \end{bmatrix} \tag{8.2.20}$$

During the second burst, we should take the complex conjugate of the data before inverse Fourier-transforming it. Since complex conjugation in the frequency

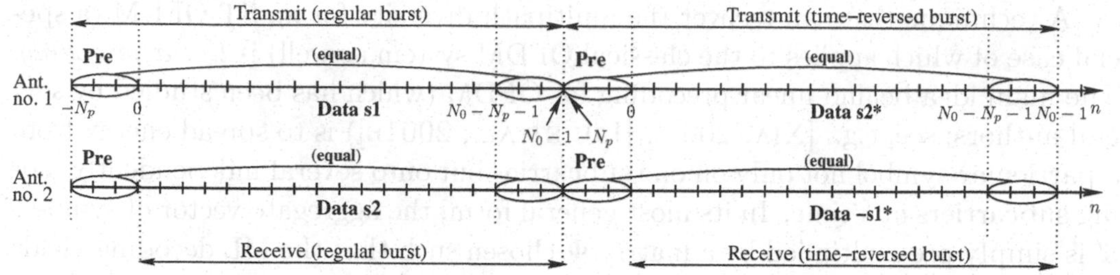

Figure 8.2. The ST-OFDM transmission scheme with the Alamouti code (OSTBC with $n_t = 2$).

domain corresponds to complex conjugation and time-reversal (to within a cyclic shift of one sample period) in the time domain, the data transmitted during the second burst is a time-reversed version of the Fourier transform of $s_k^*(n)$.

The ST-OFDM transmission scheme using the Alamouti code is depicted in Figure 8.2. We call the first block of data the "regular burst" and the second block of data the "time-reversed burst." Both these bursts are transmitted with their associated cyclic preambles.

8.2.6 ST-OFDM with Linear Precoding

It is clear from the above discussion (see Theorem 8.2) that the relation between the transmitted symbols and the received data for an ST-OFDM system using linear STBC can be written:

$$\boldsymbol{z}(n) = \boldsymbol{F}(n)\boldsymbol{s}'(n) + \boldsymbol{e}(n) \tag{8.2.21}$$

where $\boldsymbol{e}(n)$ is white noise (that is also independent for different n). Since the system is represented by N_0 *independent* equations of the form (8.2.21) (one for each n), which can be written concisely as

$$\begin{bmatrix} \boldsymbol{z}(0) \\ \vdots \\ \boldsymbol{z}(N_0-1) \end{bmatrix} = \begin{bmatrix} \boldsymbol{F}(0) & \mathbf{0} & \cdots & \mathbf{0} \\ \mathbf{0} & \boldsymbol{F}(1) & \ddots & \vdots \\ \vdots & \ddots & \ddots & \mathbf{0} \\ \mathbf{0} & \cdots & \mathbf{0} & \boldsymbol{F}(N_0-1) \end{bmatrix} \cdot \begin{bmatrix} \boldsymbol{s}'(0) \\ \vdots \\ \boldsymbol{s}'(N_0-1) \end{bmatrix} + \begin{bmatrix} \boldsymbol{e}(0) \\ \vdots \\ \boldsymbol{e}(N_0-1) \end{bmatrix} \tag{8.2.22}$$

it is clear that the error performance associated with the nth symbol $\boldsymbol{s}'(n)$ is determined only by $\boldsymbol{F}(n)$, and a manifestation of this fact is that multipath diversity is lost.

A technique that can recover the multipath diversity for an ST-OFDM (a special case of which applies to the classical OFDM system as well) is *linear precoding.* The main idea behind linear precoding for OFDM (which has been studied by several authors; see, e.g., [XIA, 2001], [LIU ET AL., 2001B]) is to spread energy from a particular symbol not only on one subcarrier but onto several independently fading subcarriers at a time. In its most general form, the aggregate vector of symbols $\boldsymbol{s}'$ is simply pre-multiplied by a matrix $\boldsymbol{\Psi}$ chosen such that the ML decoding yields full diversity.

In general, although it can yield a significant increase in error performance, linear precoding destroys the decoupling that is one of the most important benefits of OFDM. This means in particular that the detection problem becomes an integer-constrained least-squares problem (see Section 9.1.3), which is in general computationally much more demanding to solve than carrying out the fast Fourier transforms required by OFDM. Hence a linearly precoded OFDM system "competes" on different grounds and must be compared to other block-transmission methods that (unlike ST-OFDM) are not designed with low receiver complexity as a major goal. A more detailed discussion of linearly precoded ST-OFDM, and how it compares to other transmission techniques, is outside the scope of this text.

8.2.7 Discussion

Using the ST-OFDM scheme, and assuming that there are no extra guard intervals between the bursts, we transmit $n_s N_0$ symbols during $NN_0 + NN_{\text{pre}}$ time intervals. The rate of ST-OFDM is therefore:

$$R_{\text{ST-OFDM}} = \frac{n_s N_0}{NN_0 + NN_{\text{pre}}} \leq \frac{n_s N_0}{NN_0 + NL} \tag{8.2.23}$$

In the same way as for conventional OFDM (see Section 5.2.2), all signal processing associated with ST-OFDM can be carried out with N_0-point FFTs. The price paid for this simple receiver structure is the loss of multipath diversity. In analogy to the case of transmission with a single antenna (see Section 5.2), the reason for this diversity loss is the cyclic preamble.

A transmission scheme that is somewhat related to ST-OFDM with OSTBC, but which uses a fixed (and known to the receiver) preamble, and that can be used to achieve multipath diversity as well, is the TR-OSTBC scheme that is discussed in Section 8.3. Although the burst structure of TR-OSTBC is similar to that of ST-OFDM with the Alamouti code, the receiver structure associated with TR-OSTBC is very different from that of ST-OFDM.

8.3 Time-Reversal OSTBC (TR-OSTBC)

The space-time OFDM technique discussed in Section 8.2 is an easy way to achieve transmit diversity, but the simplicity of the receiver comes at the cost of a lost multipath diversity. An alternative to ST-OFDM that has shown some promise both in terms of performance and computational complexity is the so-called time-reversal OSTBC (TR-OSTBC) scheme. TR-OSTBC was originally proposed in [Lindskog and Paulraj, 2000] and later extended and analyzed in [Flore and Lindskog, 2000], [Stoica and Lindskog, 2002], [Larsson et al., 2002c]. This section is devoted to this transmission technique. We shall mostly treat the case of $n_t = 2$ (the extension to $n_t > 2$ is discussed in Section 8.3.7).

For $n_t = 2$, TR-OSTBC is similar to ST-OFDM based on the Alamouti code in what concerns the transmission scheme, but TR-OSTBC and ST-OFDM use different preambles and postambles and the corresponding receiver structures are quite different. Both methods partition the signal block into separate bursts which are related to each other in a similar manner to the elements of an Alamouti matrix (see Section 8.2.5). However, while ST-OFDM relies on encoding via the Fourier transform together with a cyclic prefix, TR-OSTBC uses a known preamble and postamble. We will show that TR-OSTBC approximately decouples the frequency selective MIMO channel into parallel and independent frequency selective SISO channels, to which standard SISO equalization methods can be applied. Finally, while ST-OFDM achieves a diversity of order $n_r n_t$, we will show that TR-OSTBC can achieve a diversity of order $n_r n_t(L+1)$.

8.3.1 Transmission Scheme

Similarly to ST-OFDM, consider the transmission of $2N_0$ symbols $\{s(0), \ldots, s(2N_0 - 1)\}$ and partition the data into two sub-streams:

$$\begin{aligned} s_1(n) &= s(n), \quad n = 0, \ldots, N_0 - 1 \\ s_2(n) &= s(n + N_0), \quad n = 0, \ldots, N_0 - 1 \end{aligned} \tag{8.3.1}$$

Also, let

$$\boldsymbol{s}(n) = \begin{bmatrix} s_1(n) \\ s_2(n) \end{bmatrix} \tag{8.3.2}$$

The TR-OSTBC transmission scheme is illustrated in Figure 8.3. We transmit a regular burst and a time-reversed burst in the same way as for ST-OFDM with the Alamouti code in Section 8.2.5. However, in the present case, these bursts will in general have preambles and postambles which are fixed and known to the receiver. When forming the time-reversed burst, the regular burst *including its*

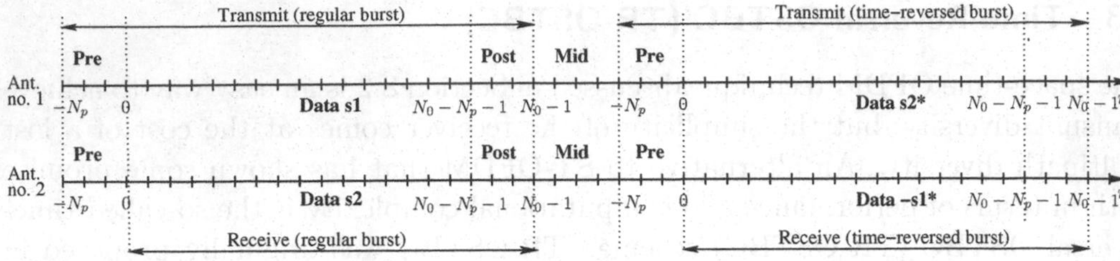

Figure 8.3. The TR-OSTBC transmission scheme.

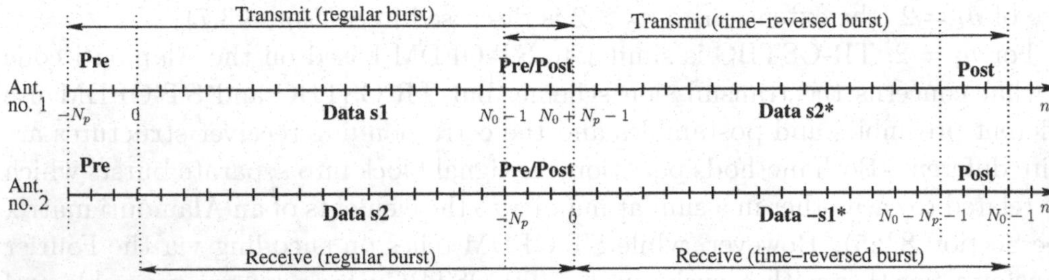

Figure 8.4. The TR-OSTBC transmission scheme without midamble. In the upper part of the figure, the time indices associated with the regular burst are shown, and the lower part of the figure shows the time indices for the time-reversed burst.

preamble and postamble is time-reversed; hence the preamble of the time-reversed burst contains samples from the postamble of the regular burst, and vice versa. For notational simplicity, we will assume that $N_{\text{pre}} = N_{\text{post}} \triangleq N_p$ in the rest of this section.

Between the postamble of the regular burst and the preamble of the time-reversed burst, we can optionally transmit a midamble of length N_m that can be used for synchronization or channel estimation purposes; the structure of the so-obtained transmission scheme is reminiscent of the burst structure in GSM [Mouly and Pautet, 1992]. In both cases, the postamble corresponding to the regular burst and the preamble corresponding to the time-reversed burst can be considered as a part of an $N_m + 2N_p$ symbols long midamble whose head and tail symbols exhibit a special conjugate symmetry property. Also, in the special case that no midamble is desired and all postambles and preambles contain zeros, or whenever the postamble of the regular burst and the preamble of the time-reversed burst possess certain conjugate symmetry properties, these postambles and preambles can be merged into one sequence of length N_p (see Figure 8.4).

If we assume that $s_k(n)$ is defined also for $n = -N_p, \ldots, -1$ (corresponding to the preamble) and for $n = N_0, \ldots, N_0 + N_p - 1$ (corresponding to the postamble), the transmitted vectors can be written as follows.

Regular burst: During the time intervals $n = -N_p, \ldots, N_0 + N_p - 1$, we transmit the first stream $s_1(n)$ through the first antenna and the second stream $s_2(n)$ via the second antenna. In the framework of Section 2.2 the transmitted signal is:

$$\boldsymbol{x}(n) = \begin{bmatrix} s_1(n) \\ s_2(n) \end{bmatrix}, \quad n = -N_p, \ldots, N_0 + N_p - 1 \tag{8.3.3}$$

Time-reversed burst: Following the midamble, we transmit the *time-reversed* data (including its preamble and the postamble) $s_2^*(N_0 - 1 - n)$ via the first antenna and $-s_1^*(N_0 - 1 - n)$ via the second antenna for $n = -N_p, \ldots, N_0 + N_p - 1$. Using the notation of Section 2.2:

$$\boldsymbol{x}(n) = \begin{bmatrix} s_2^*(N_0 - 1 - n) \\ -s_1^*(N_0 - 1 - n) \end{bmatrix}, \quad n = -N_p, \ldots, N_0 + N_p - 1 \tag{8.3.4}$$

8.3.2 ML Detection

For TR-OSTBC it turns out to be simpler to work with an FIR filter representation instead of the matrix algebraic formulation used so far in this chapter.

Let $\boldsymbol{y}_1(n)$, $n = 0, \ldots, N_0 + N_p - 1$, be a sequence of n_r-vectors that contain the data received during the regular burst. This data can be expressed as:

$$\boldsymbol{y}_1(n) = \begin{bmatrix} \boldsymbol{h}_1(z^{-1}) & \boldsymbol{h}_2(z^{-1}) \end{bmatrix} \begin{bmatrix} s_1(n) \\ s_2(n) \end{bmatrix} + \boldsymbol{e}_1(n) \tag{8.3.5}$$

for $n = 0, \ldots, N_0 + N_p - 1$. Note that $\boldsymbol{y}_1(n)$ depends on the data symbols as well as the preamble and postamble. Similarly, let $\boldsymbol{y}_2(n)$, $n = 0, \ldots, N_0 + N_p - 1$, be the data received during the time-reversed burst (the reason for the notation $\check{(\cdot)}$ will soon become clear):

$$\check{\boldsymbol{y}}_2(n) = \begin{bmatrix} \boldsymbol{h}_1(z^{-1}) & \boldsymbol{h}_2(z^{-1}) \end{bmatrix} \begin{bmatrix} s_2^*(N_0 - 1 - n) \\ -s_1^*(N_0 - 1 - n) \end{bmatrix} + \check{\boldsymbol{e}}_2(n) \tag{8.3.6}$$

for $n = 0, \ldots, N_0 + N_p - 1$. These data depend on the data symbols as well as the preamble and postamble (but *not* the midamble). Since $\boldsymbol{y}_1(n)$ and $\check{\boldsymbol{y}}_2(n)$ are collected during disjunct time intervals, to simplify the notation we use the same time index n for $\boldsymbol{y}_1(n)$ and for $\check{\boldsymbol{y}}_2(n)$ even though they correspond to different *absolute* time instants.

In (8.3.5) and (8.3.6), the terms $\boldsymbol{e}_1(n)$ and $\check{\boldsymbol{e}}_2(n)$ contain noise that we assume, as before, to be spatially and temporally white complex Gaussian with variance σ^2. In particular, this means that

$$E\left[\begin{bmatrix}\boldsymbol{e}_1(n)\\ \check{\boldsymbol{e}}_2(n)\end{bmatrix}\begin{bmatrix}\boldsymbol{e}_1^H(s) & \check{\boldsymbol{e}}_2^H(s)\end{bmatrix}\right] = \sigma^2\delta_{n-s}\boldsymbol{I} \tag{8.3.7}$$

Next we arrange the received data in a $2n_r$-vector as follows (somewhat analogous to what was done for the Alamouti code, the data received during the time-reversed burst is time-reversed and complex conjugated):

$$\boldsymbol{y}(n) \triangleq \begin{bmatrix}\boldsymbol{y}_1(n)\\ \boldsymbol{y}_2(n)\end{bmatrix} = \begin{bmatrix}\boldsymbol{y}_1(n)\\ \check{\boldsymbol{y}}_2^*(N_0+N_p-1-n)\end{bmatrix} = \boldsymbol{H}(z^{-1})\boldsymbol{s}(n)+\boldsymbol{e}(n) \tag{8.3.8}$$

for $n = 0,\ldots,N_0+N_p-1$ where we have implicitly defined the $2n_r \times 2$ causal matrix filter

$$\boldsymbol{H}(z^{-1}) = \begin{bmatrix}\boldsymbol{h}_1(z^{-1}) & \boldsymbol{h}_2(z^{-1})\\ -z^{-N_p}\boldsymbol{h}_2^*(z) & z^{-N_p}\boldsymbol{h}_1^*(z)\end{bmatrix} \tag{8.3.9}$$

and the $2n_r$-vector of noise

$$\boldsymbol{e}(n) = \begin{bmatrix}\boldsymbol{e}_1(n)\\ \check{\boldsymbol{e}}_2^*(N_0+N_p-1-n)\end{bmatrix} \tag{8.3.10}$$

The noise $\boldsymbol{e}(n)$ in (8.3.10) is still both spatially and temporally white. (Note the similarity between (8.3.9) and the matrix in (6.3.8) for the Alamouti code.)

To verify (8.3.8), observe from (8.3.6) that

$$\begin{aligned}\check{\boldsymbol{y}}_2^*(N_0+N_p-1-n) =& \sum_{l=0}^{L}\Big(-\boldsymbol{h}_2^{(l)*}s_1(N_0-1-[N_0+N_p-1-n-l])\\ &+\boldsymbol{h}_1^{(l)*}s_2(N_0-1-[N_0+N_p-1-n-l])\Big)\\ =&\sum_{l=0}^{L}\Big(-\boldsymbol{h}_2^{(l)*}s_1(n+l-N_p)+\boldsymbol{h}_1^{(l)*}s_2(n+l-N_p)\Big)\\ =&-z^{-N_p}\boldsymbol{h}_2^*(z)s_1(n)+z^{-N_p}\boldsymbol{h}_1^*(z)s_2(n)\end{aligned} \tag{8.3.11}$$

for $n = 0,\ldots,N_0+N_p-1$.

Let $\boldsymbol{y}(n)$ be the received data as defined in (8.3.8). Then the ML metric for symbol detection is given by:

$$\sum_{n=0}^{N_0+L-1} \|\boldsymbol{y}(n) - \boldsymbol{H}(z^{-1})\boldsymbol{s}(n)\|^2 \tag{8.3.12}$$

Apparently, the detection of $s_1(n)$ and $s_2(n)$ is coupled. While in principle a sequence detector such as the Viterbi algorithm could be applied *directly* to (8.3.12), the computational complexity of such an approach would be high. In particular, due to the coupling between $s_1(n)$ and $s_2(n)$ in (8.3.12), the number of states in the trellis diagram would increase from $|\mathcal{S}|^L$ to $|\mathcal{S}|^{2L}$. In Section 8.3.4 below we show how the ML metric in (8.3.12) can be approximately decoupled into a sum of two ML metrics associated with SISO channels. An important consequence of this decoupling is that a standard scalar Viterbi decoder (with memory length L) can be employed for equalization. However, before discussing this decoupling, we analyze the order of diversity that is achievable by TR-OSTBC using ML detection.

8.3.3 Achievable Diversity Order

We will next show that the ML detection of TR-OSTBC by minimizing the metric in (8.3.12) yields a diversity of order $2n_r(L+1)$. While this result follows easily in the framework of this chapter, the first published proof was due to [ZHOU AND GIANNAKIS, 2001].

Let us study the matrix $\boldsymbol{X}$ in (8.1.3) associated with TR-OSTBC. After appropriate permutation, this matrix can be written as:

$$\boldsymbol{X}^T \triangleq \begin{bmatrix} \boldsymbol{X}_{11} & \boldsymbol{X}_{12} \\ \boldsymbol{X}_{21} & \boldsymbol{X}_{22} \end{bmatrix} \tag{8.3.13}$$

where

$$\boldsymbol{X}_{11} = \begin{bmatrix} s_1(0) & s_1(-1) & \cdots & s_1(-L) \\ s_1(1) & s_1(0) & & s_1(1-L) \\ \vdots & & \ddots & \vdots \\ \vdots & & & s_1(0) \\ \vdots & & & \vdots \\ s_1(N_0+L-2) & s_1(N_0+L-3) & & s_1(N_0-2) \\ s_1(N_0+L-1) & s_1(N_0+L-2) & \cdots & s_1(N_0-1) \end{bmatrix} \tag{8.3.14}$$

$$\boldsymbol{X}_{12} = \begin{bmatrix} s_2^*(N_0-1) & s_2^*(N_0) & \cdots & s_2^*(N_0+L-1) \\ s_2^*(N_0-2) & s_2^*(N_0-1) & & s_2^*(N_0+L-2) \\ \vdots & & \ddots & \vdots \\ \vdots & & & s_2^*(N_0-1) \\ \vdots & & & \vdots \\ s_2^*(1-L) & s_2^*(2-L) & & s_2^*(1) \\ s_2^*(-L) & s_2^*(1-L) & \cdots & s_2^*(0) \end{bmatrix} \tag{8.3.15}$$

$$\boldsymbol{X}_{21} = \begin{bmatrix} s_2(0) & s_2(-1) & \cdots & s_2(-L) \\ s_2(1) & s_2(0) & & s_2(1-L) \\ \vdots & & \ddots & \vdots \\ \vdots & & & s_2(0) \\ \vdots & & & \vdots \\ s_2(N_0+L-2) & s_2(N_0+L-3) & & s_2(N_0-2) \\ s_2(N_0+L-1) & s_2(N_0+L-2) & \cdots & s_2(N_0-1) \end{bmatrix} \tag{8.3.16}$$

$$\boldsymbol{X}_{22} = - \begin{bmatrix} s_1^*(N_0-1) & s_1^*(N_0) & \cdots & s_1^*(N_0+L-1) \\ s_1^*(N_0-2) & s_1^*(N_0-1) & & s_1^*(N_0+L-2) \\ \vdots & & \ddots & \vdots \\ \vdots & & & s_1^*(N_0-1) \\ \vdots & & & \vdots \\ s_1^*(1-L) & s_1^*(2-L) & & s_1^*(1) \\ s_1^*(-L) & s_1^*(1-L) & \cdots & s_1^*(0) \end{bmatrix} \tag{8.3.17}$$

For a matrix $\breve{\boldsymbol{X}}$ defined as in (8.3.13) but corresponding to a hypothetical error signal $\breve{\boldsymbol{s}}(n) = \boldsymbol{s}_0(n) - \boldsymbol{s}(n)$, we have that

$$\breve{\boldsymbol{X}}^* \breve{\boldsymbol{X}}^T = \begin{bmatrix} \breve{\boldsymbol{X}}_{11}^H \breve{\boldsymbol{X}}_{11} + \breve{\boldsymbol{X}}_{21}^H \breve{\boldsymbol{X}}_{21} & \breve{\boldsymbol{X}}_{11}^H \breve{\boldsymbol{X}}_{12} + \breve{\boldsymbol{X}}_{21}^H \breve{\boldsymbol{X}}_{22} \\ \breve{\boldsymbol{X}}_{12}^H \breve{\boldsymbol{X}}_{11} + \breve{\boldsymbol{X}}_{22}^H \breve{\boldsymbol{X}}_{21} & \breve{\boldsymbol{X}}_{12}^H \breve{\boldsymbol{X}}_{12} + \breve{\boldsymbol{X}}_{22}^H \breve{\boldsymbol{X}}_{22} \end{bmatrix} \tag{8.3.18}$$

Under the assumption of known preambles and postambles, $\breve{\boldsymbol{s}}(n)$ is identical to zero for $n = -N_p, \ldots, -1$ and $n = N_0, \ldots, N_0 + N_p - 1$. By using this observation it can be shown that both diagonal blocks of $\breve{\boldsymbol{X}}\breve{\boldsymbol{X}}^H$ are positive definite and that the two off-diagonal blocks of $\breve{\boldsymbol{X}}\breve{\boldsymbol{X}}^H$ are equal to zero. To show the latter, it is useful to observe that the elements of the off-diagonal blocks can be expressed as convolution sums. Hence the matrix $\breve{\boldsymbol{X}}$ has full rank for any $\breve{\boldsymbol{s}}(n) \neq \boldsymbol{0}$. Application of Theorem 4.2 (see page 45) then shows that TR-OSTBC can achieve a diversity order equal to $2n_r(L+1)$, provided that ML decoding is used to detect the symbols.

8.3.4 Decoupling, Matched Filtering and Approximate ML Equalization

The key property of TR-OSTBC that enables a decoupled detector structure is that $\boldsymbol{H}(z^{-1})$ in (8.3.9) is "unitary":

$$\begin{aligned}\boldsymbol{H}^H(z)\boldsymbol{H}(z^{-1}) &= \begin{bmatrix} \boldsymbol{h}_1^H(z) & -z^{N_p}\boldsymbol{h}_2^T(z^{-1}) \\ \boldsymbol{h}_2^H(z) & z^{N_p}\boldsymbol{h}_1^T(z^{-1}) \end{bmatrix} \begin{bmatrix} \boldsymbol{h}_1(z^{-1}) & \boldsymbol{h}_2(z^{-1}) \\ -z^{-N_p}\boldsymbol{h}_2^*(z) & z^{-N_p}\boldsymbol{h}_1^*(z) \end{bmatrix} \\ &= (\boldsymbol{h}_1^H(z)\boldsymbol{h}_1(z^{-1}) + \boldsymbol{h}_2^H(z)\boldsymbol{h}_2(z^{-1}))\boldsymbol{I} \\ &= \gamma(z^{-1})\boldsymbol{I} = \sum_{l=-L}^{L} \gamma_l z^{-l}\boldsymbol{I} \end{aligned} \tag{8.3.19}$$

where the constants $\gamma_{-l} = \gamma_l^*$ are the coefficients of the following scalar filter:

$$\gamma(z^{-1}) \triangleq \boldsymbol{h}_1^H(z)\boldsymbol{h}_1(z^{-1}) + \boldsymbol{h}_2^H(z)\boldsymbol{h}_2(z^{-1}) \tag{8.3.20}$$

This property of $\boldsymbol{H}(z^{-1})$ can be seen as a direct extension of the orthogonality property of the Alamouti code (cf. the discussion in Section 6.3.1).

Let us now make use of (8.3.19) to study the ML metric in (8.3.12):

$$\begin{aligned}&\sum_{n=0}^{N_0+L-1} \|\boldsymbol{y}(n) - \boldsymbol{H}(z^{-1})\boldsymbol{s}(n)\|^2 \\ &= \sum_{n=0}^{N_0+L-1} \|\boldsymbol{y}(n)\|^2 - 2\mathrm{Re}\left\{ \sum_{n=0}^{N_0+L-1} \left(\boldsymbol{H}(z^{-1})\boldsymbol{s}(n)\right)^H \boldsymbol{y}(n) \right\} \\ &\quad + \sum_{n=0}^{N_0+L-1} \left(\boldsymbol{H}(z^{-1})\boldsymbol{s}(n)\right)^H \left(\boldsymbol{H}(z^{-1})\boldsymbol{s}(n)\right) \end{aligned} \tag{8.3.21}$$

The second term of (8.3.21) can be approximated in a way similar to (5.2.11):

$$\mathrm{Re}\left\{ \sum_{n=0}^{N_0+L-1} \left(\boldsymbol{H}(z^{-1})\boldsymbol{s}(n)\right)^H \boldsymbol{y}(n) \right\} \approx \mathrm{Re}\left\{ \sum_{n=0}^{N_0+L-1} \boldsymbol{s}^H(n) \cdot \boldsymbol{H}^H(z)\boldsymbol{y}(n) \right\} \tag{8.3.22}$$

where, like in Section 5.2.2, all approximations denoted by "$\approx$" are attributable to end-effects. The third term in (8.3.21) can be approximated using (8.3.19) in a manner similar to (5.2.13):

$$\begin{aligned}
&\sum_{n=0}^{N_0+L-1} \left(\boldsymbol{H}(z^{-1})\boldsymbol{s}(n)\right)^H \left(\boldsymbol{H}(z^{-1})\boldsymbol{s}(n)\right) \\
&\approx \mathrm{Re}\left\{ \sum_{n=0}^{N_0+L-1} \left(\gamma_0 \boldsymbol{s}^H(n)\boldsymbol{s}(n) + 2\sum_{l=1}^{L} \gamma_l \boldsymbol{s}^H(n)\boldsymbol{s}(n-l)\right)\right\}
\end{aligned} \tag{8.3.23}$$

Equations (8.3.22) and (8.3.23) show that minimizing the ML metric in (8.3.12) is equivalent (to within approximations attributable to end-effects) to maximizing the following metric:

$$\begin{aligned}
&\mathrm{Re}\left\{ \sum_{n=0}^{N_0+L-1} s_1^*(n)\Big(z_1(n) - \sum_{l=1}^{L} \gamma_l s_1(n-l) - \frac{1}{2}\gamma_0 s_1(n)\Big)\right\} \\
&+\mathrm{Re}\left\{ \sum_{n=0}^{N_0+L-1} s_2^*(n)\Big(z_2(n) - \sum_{l=1}^{L} \gamma_l s_2(n-l) - \frac{1}{2}\gamma_0 s_2(n)\Big)\right\}
\end{aligned} \tag{8.3.24}$$

where

$$\boldsymbol{z}(n) \triangleq \begin{bmatrix} z_1(n) \\ z_2(n) \end{bmatrix} = \boldsymbol{H}^H(z)\boldsymbol{y}(n) \tag{8.3.25}$$

From (8.3.24) we can see that the approximate ML detection of $s_1(n)$ is decoupled from the detection of $s_2(n)$ and that $s_1(n)$ and $s_2(n)$ can be detecting using a Viterbi algorithm with memory length L. Furthermore, the metric in (8.3.24) is a function of the received data $\boldsymbol{y}(n)$ only via $\boldsymbol{z}(n)$ and hence $\boldsymbol{z}(n)$ is an (approximate) sufficient statistic for detection. The decoupling of the detection of $s_1(n)$ and $s_2(n)$ follows from the filtering in (8.3.25) and the unitary property of $\boldsymbol{H}(z^{-1})$.

The filtering of the received data $\boldsymbol{y}(n)$ in (8.3.25) with $\boldsymbol{H}^H(z)$ can be interpreted as a *matched filtering*, and the output of the matched filter can be spelled out as follows:

$$\begin{aligned}
\boldsymbol{z}(n) &= \boldsymbol{H}^H(z)\boldsymbol{y}(n) \\
&= \boldsymbol{H}^H(z)\boldsymbol{H}(z^{-1})\boldsymbol{s}(n) + \boldsymbol{H}^H(z)\boldsymbol{e}(n) \\
&= \gamma(z^{-1})\boldsymbol{s}(n) + \boldsymbol{H}^H(z)\boldsymbol{e}(n)
\end{aligned} \tag{8.3.26}$$

Equation (8.3.26) can be rewritten in the equivalent form

$$\begin{aligned}
z_1(n) &= \gamma(z^{-1})s_1(n) + \boldsymbol{h}_1^H(z)e_1(n) - z^{N_p}\boldsymbol{h}_2^T(z^{-1})e_2(n) \\
z_2(n) &= \gamma(z^{-1})s_2(n) + \boldsymbol{h}_2^H(z)e_1(n) + z^{N_p}\boldsymbol{h}_1^T(z^{-1})e_2(n)
\end{aligned} \tag{8.3.27}$$

Note that the noise term in (8.3.26) is spatially white since its power spectral density is equal to

$$\sigma^2 \cdot \boldsymbol{H}^H(z)\boldsymbol{H}(z^{-1}) = \sigma^2\gamma(z^{-1})\boldsymbol{I} \tag{8.3.28}$$

Hence the noise terms of $z_1(n)$ and $z_2(n)$ in (8.3.27) are uncorrelated with one another, but they are *not* temporally white.

8.3.5 Linear Equalization

For large L and large symbol constellations, applying a Viterbi algorithm to maximize (8.3.24) may be computationally burdensome. An alternative to such a procedure is to use a linear MMSE equalizer that operates directly on the matched filtered sequence $\boldsymbol{z}(n)$ in (8.3.26). Such an MMSE equalizer can be derived following the steps in Section 5.2.1.

Without loss of generality, let us consider the detection of $s_1(n)$. Then the sequence

$$z_1(n) = \gamma(z^{-1})s_1(n) + \boldsymbol{h}_1^H(z)\boldsymbol{e}_1(n) - z^{N_p}\boldsymbol{h}_2^T(z^{-1})\boldsymbol{e}_2(n) \tag{8.3.29}$$

is approximately a sufficient statistic for detection. We seek a (scalar) linear filter $\xi(z^{-1})$ of length $2M+1$ (where $2M \geq L$) such that the output of the filtered sequence

$$\hat{s}_1(n) = \xi(z^{-1})z_1(n) \tag{8.3.30}$$

recovers $s_1(n)$ in a MMSE sense. Assume that $s_1(n)$ is statistically white with known variance σ_s^2. Paralleling the derivation in Section 5.2.1 (with some differences; for instance, the noise in the present case is colored), we find that the vector of optimal filter coefficients

$$\boldsymbol{\xi} = \begin{bmatrix} \xi_{-M} & \cdots & \xi_M \end{bmatrix}^T \tag{8.3.31}$$

is given by:

$$\boldsymbol{\xi} = \left(\boldsymbol{\Gamma}^H\boldsymbol{\Gamma} + \frac{\sigma^2}{\sigma_s^2}\boldsymbol{\Gamma}\right)^{-1}\boldsymbol{\Gamma}^H\boldsymbol{e}_0 \tag{8.3.32}$$

where $\mathbf{\Gamma}$ is a $(2M+1) \times (2M+1)$ Toeplitz matrix defined according to:

$$\mathbf{\Gamma} = \begin{bmatrix} \gamma_0 & \gamma_{-1} & \cdots & \gamma_{-L} & 0 & \cdots & 0 \\ \gamma_1 & \gamma_0 & \ddots & & \ddots & \ddots & \vdots \\ \vdots & \ddots & \ddots & & & & 0 \\ \gamma_L & & \ddots & & \ddots & & \gamma_{-L} \\ 0 & \ddots & & & \ddots & & \\ \vdots & & & & & & \vdots \\ 0 & \cdots & 0 & \gamma_L & & \cdots & \gamma_0 \end{bmatrix} \tag{8.3.33}$$

Also, $\boldsymbol{e}_0$ is a $(2M+1)$-vector whose $(M+1)$th element is equal to one and all other elements are equal to zero.

The computation of the MMSE filter coefficients in (8.3.32) requires the solution of a linear system of equations with $2M+1$ unknowns. This is far less burdensome than solving the original ML problem (8.3.12) via the Viterbi algorithm or treating it as an integer-constrained LS problem (see Section 9.1.3). However, note that no approximations are involved in the latter techniques.

8.3.6 Numerical Performance Study

In this section we present a few numerical examples (see also [Larsson et al., 2002c]) that compare the performance of TR-OSTBC with that of delay diversity (see Section 6.3.5). Whilst DD is not optimal in any known sense, we chose it for comparison because of its simplicity; DD is the simplest space-time trellis code in terms of decoding complexity. As discussed in Section 6.3.5, in a two transmit antenna system using DD a given symbol stream is transmitted through the first antenna, and the same stream delayed by D symbol intervals is transmitted via the second antenna. In this way an artificial extra delay spread of D is induced in the channel.

Note that in contrast to TR-OSTBC, the DD scheme requires a MLSD at the receiver, even if the propagation channel is not frequency-selective (i.e., $L = 0$). In general, the DD scheme requires an MLSD with memory length $L + D$ while TR-OSTBC only requires an MLSD with memory length L. As a consequence, the equalization complexity of TR-OSTBC is lower than that of the DD scheme. Also, note that TR-OSTBC achieves a diversity of order $2(L+1)$ (for $n_r = 1$), whereas DD achieves a diversity of order $L+D+1$. Hence, for DD to achieve the same diversity order as that of TR-OSTBC, we require that $L+D+1 = 2(L+1)$

	DD (delay=1)	DD (delay=$L+1$)	TR-OSTBC
Complexity order	$\|S\|^{L+1}$	$\|S\|^{2L+1}$	$\|S\|^{L}$
Diversity order	$L+2$	$2(L+1)$	$2(L+1)$

Table 8.1. Relative equalization complexity and diversity orders (for $n_r = 1$) of TR-OSTBC and DD as a function of the physical channel length $L+1$, when using an MLSD implemented by the Viterbi algorithm.

which holds if $D = L+1$ (note that the diversity order of DD cannot be increased by further increasing D). These observations are summarized in Table 8.1.

In the examples below we transmit bursts containing $N_0 = 120$ BPSK modulated data bits. The system under consideration has one receive antenna, and we consider both conventional transmission with one transmit antenna, and TR-OSTBC with two transmit antennas as well as DD with two transmit antennas and a delay of $D = 1$. For TR-OSTBC, the symbol stream is split into two substreams of length $N_0/2 = 60$ bits each, and transmitted through the two antennas as in Figure 8.3. A midamble of length $N_m = 26$ is used for channel estimation in those simulations where the channel is assumed unknown. The preambles and postambles are of length $N_p = 5$ and consist of zeros only. In the case of transmit diversity, the amplitudes of the signals fed to the two antennas are normalized such that the total transmission power remains the same as if we had one transmit antenna.

Example 8.2: TR-OSTBC vs. Delay diversity.
We compare the performance of TR-OSTBC with that of DD (with $D = 1$). In Figure 8.5 the estimated bit-error-rate (BER) is plotted versus the SNR. A standard Viterbi equalizer is employed both to maximize the metrics in (8.3.24), and to decode the transmitted message in the case of DD. We consider two channels in this simulation. The first channel ("flat") consists of a single Rayleigh fading tap and is therefore not frequency-selective (here $L = 0$). The second channel ("delay-spread") is of length $L+1 = 2$ and consists of two independent Rayleigh fading taps with the same average power. The channels are normalized such that

$$E\left[\sum_{l=0}^{L} \left\|\boldsymbol{h}_l\right\|^2\right] = 1 \tag{8.3.34}$$

and they are assumed to be known to the receiver. We can make a number of observations from Figure 8.5.

1. For the flat channel, TR-OSTBC gives a performance gain of around 5 dB

at 1% BER compared to a conventional single-antenna transmission. This difference is entirely due to the antenna diversity gain offered by the two transmit antennas (TR-OSTBC is equivalent to OSTBC in this case and hence it achieves a diversity of order 2). The performance of the DD scheme is of the same order as that of TR-OSTBC. The performance gain of DD compared to single-antenna transmission is due to the artificial multipath induced by the DD scheme; DD also achieves a diversity of order 2 in this case.

2. For the case when the channel has a delay-spread, a multipath diversity gain is obtained from the propagation channel. The BER for single-antenna transmission and that for DD with 2 transmit antennas and a flat channel are almost the same. This is not surprising: we can expect the natural multipath and the artificial multipath induced by DD to influence the performance in a similar way.

3. The TR-OSTBC scheme outperforms DD when the physical channel has a delay spread. This is expected in view of the above discussion about diversity order.

■

Example 8.3: Linear equalization vs. Viterbi algorithm.
We compare the performance of the MMSE equalizer derived above to that of the Viterbi equalizer studied in Example 8.2. The "delay-spread" channel from Example 8.2 is used. Figure 8.6 shows the results (we show only the results for single-antenna transmission and for TR-OSTBC). It is clear that the Viterbi equalizer outperforms the MMSE receiver, as expected [Proakis, 2001, Chap. 10]. The difference in this case is moderate, which can be explained by the fact that the delay spread is relatively small (and hence the transfer function of the channel is unlikely to have many points with a small value). ■

Example 8.4: TR-OSTBC for GSM Channels.
In this example we use simulation parameters that (to some degree) resemble the Global System for Mobile communications (GSM) standard [Stüber, 2001], [Mouly and Pautet, 1992]. The number of symbols per block and the midamble are as before. We use binary modulation with an ideal pulse-shape (this is a deviation from the GSM standard) and the GSM channel model "typical urban" [ETSI, 2002], which has a delay spread of around $1\mu s$ (corresponding to $\sim$0.25

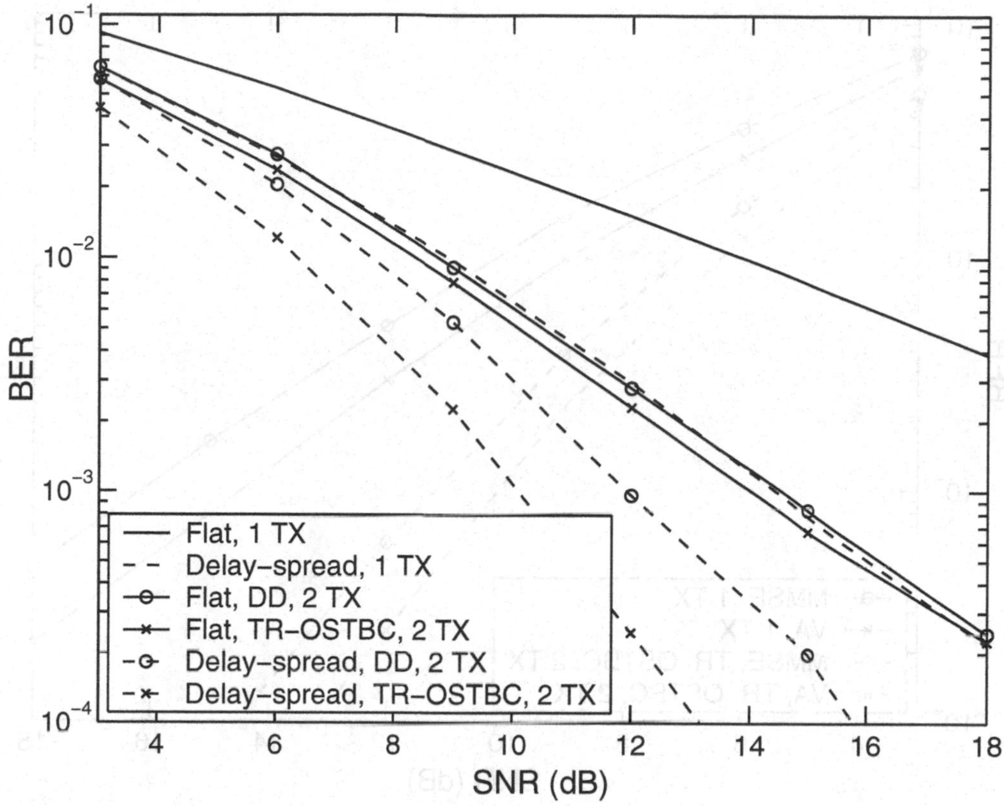

Figure 8.5. Results for Example 8.2. Comparison between TR-OSTBC and delay diversity (DD) for a flat Rayleigh fading channel and a fading multipath channel with two taps, respectively. A Viterbi equalizer was used, and the channel was assumed to be known to the receiver.

symbol intervals). This channel model is defined in continuous time and hence the delay spread of the corresponding discrete-time channel is larger than 0.25 symbol intervals. For this reason, and also to guarantee robustness to higher delay spread and synchronization errors, we use the Viterbi equalizer along with a channel with $L+1 = 3$ taps. Figure 8.7 shows the results. The performance gain we get from the extra transmit antenna is evident. If the channel is known, TR-OSTBC performs better than DD (as expected). When the channel is unknown, we use the semiblind approach to estimate it (see Section 9.5) and in this case the performances of DD and TR-OSTBC are almost the same. We attribute this behavior to the fact that for TR-OSTBC, $2(L+1) = 6$ parameters have to be estimated while for DD the channel is parameterized by only $L+2 = 4$ unknowns. In fact, the main advantage

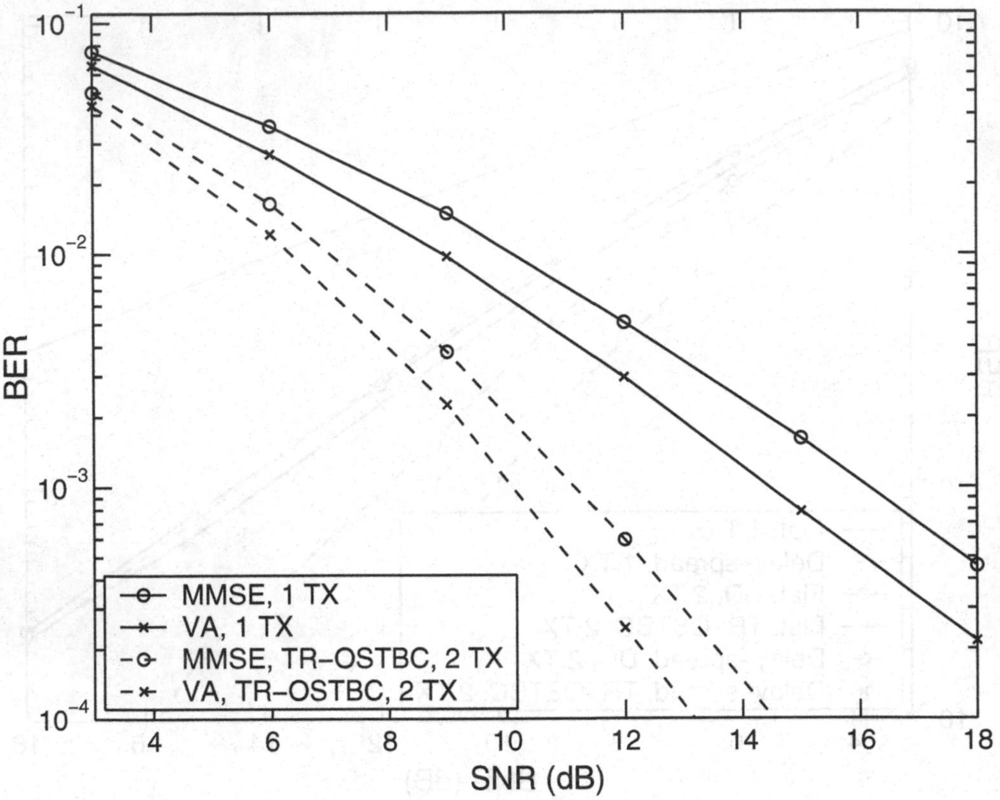

Figure 8.6. Results for Example 8.3. Comparison between MMSE and Viterbi equalizations for single-antenna transmission and for the TR-OSTBC scheme. A delay spread channel with two taps ($L+1=2$) was used. The channel was assumed to be known by the receiver.

of TR-OSTBC over DD in this case may be the lower decoding complexity. This is particularly important for systems evolving from GSM such as "enhanced data for global evolution" (EDGE) (see, e.g., [Rappaport, 2002]) which uses 8-PSK modulation; for such systems using large symbol constellations the number of states in the Viterbi decoder must be kept as small as possible. ■

8.3.7 TR-OSTBC for $n_t > 2$

Time-reversal OSTBC can be extended to more than two transmit antennas. Here we present a brief review of the results in [Stoica and Lindskog, 2002] which proposed a rate-3/4 TR-OSTBC scheme for three transmit antennas, and a rate-

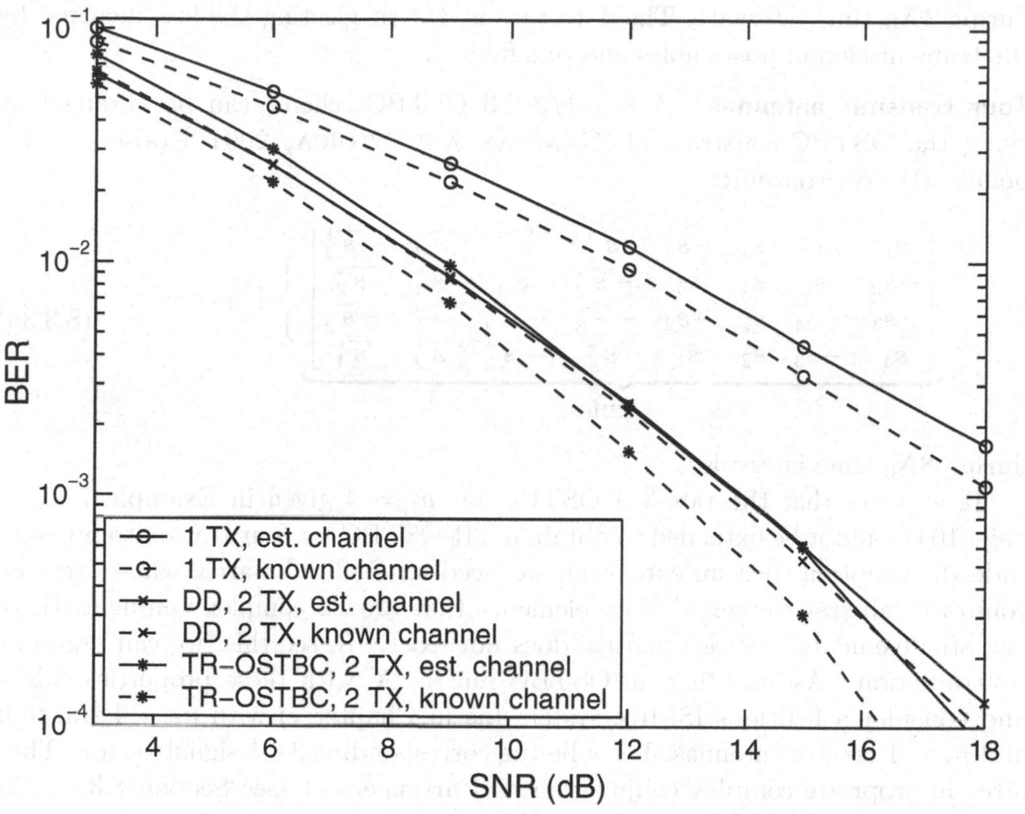

Figure 8.7. Results for Example 8.4. Comparison for a GSM-like system using the channel model "typical urban."

1/2 construct for four or more transmit antennas. We show the main results in a schematic form only, omitting the discussion about preambles and postambles, and referring the reader to the cited paper for proofs of the required properties for TR-OSTBC. In the following equations, $\boldsymbol{s}_n$ is a general symbol sequence of length N_0 and $\overleftarrow{(\cdot)}$ denotes time-reversal.

Three transmit antennas. Here a $3N_0$-symbol data sequence is split into three sub-streams of length N_0, $\{\boldsymbol{s}_n\}$ for $n = 1, \ldots, 3$, and we transmit (cf. (7.4.9))

$$\underbrace{\begin{bmatrix} \boldsymbol{s}_1 & -\overleftarrow{\boldsymbol{s}}_2^* & \overleftarrow{\boldsymbol{s}}_3^* & \mathbf{0} \\ \boldsymbol{s}_2 & \overleftarrow{\boldsymbol{s}}_1^* & \mathbf{0} & -\overleftarrow{\boldsymbol{s}}_3^* \\ \boldsymbol{s}_3 & \mathbf{0} & -\overleftarrow{\boldsymbol{s}}_1^* & \overleftarrow{\boldsymbol{s}}_2^* \end{bmatrix}}_{\rightarrow \text{ time}} \Bigg\} \downarrow \text{space} \tag{8.3.35}$$

during $4N_0$ time intervals. The data rate is 3/4 (neglecting the loss incurred by the transmission of postambles and preambles).

Four transmit antennas. A rate-1/2 TR-OSTBC scheme can be obtained by using the OSTBC construct of [Ganesan and Stoica, 2001c] (also see Appendix B). We transmit:

$$\underbrace{\begin{bmatrix} s_1 & s_2 & s_3 & -s_4 & \overleftarrow{s}_1^* & \overleftarrow{s}_2^* & \overleftarrow{s}_3^* & -\overleftarrow{s}_4^* \\ -s_2 & s_1 & s_4 & s_3 & -\overleftarrow{s}_2^* & \overleftarrow{s}_1^* & \overleftarrow{s}_4^* & \overleftarrow{s}_3^* \\ -s_3 & -s_4 & s_1 & -s_2 & -\overleftarrow{s}_3^* & -\overleftarrow{s}_4^* & \overleftarrow{s}_1^* & -\overleftarrow{s}_2^* \\ s_4 & -s_3 & s_2 & s_1 & \overleftarrow{s}_4^* & -\overleftarrow{s}_3^* & \overleftarrow{s}_2^* & \overleftarrow{s}_1^* \end{bmatrix}}_{\rightarrow \text{ time}} \Bigg\} \downarrow \text{space} \tag{8.3.36}$$

during $8N_0$ time intervals.

It appears that the rate-3/4 OSTBC for $n_t = 4$ given in Example 7.4 (see page 104) cannot be extended to obtain a TR-OSTBC for four transmit antennas. Indeed, to obtain such an extension, we need an OSTBC matrix whose rows all contains "all-star elements" (i.e., elements that are all complex conjugated), or "no-star elements." Such a matrix does not exist. To see this, we can argue by contradiction. Assume that an OSTBC matrix $\boldsymbol{X}$ with these properties exists, and consider a fictitious ISI-free system (as in Chapter 7) with $n_t = 4$ transmit and $n_r = 1$ receive antennas. Let $\boldsymbol{s}$ be the corresponding 3×1 signal vector. Then after appropriate complex conjugation and arrangement (see Section 8.3.2), the received signal can be written as:

$$\boldsymbol{y} = \boldsymbol{H}\boldsymbol{s} + \boldsymbol{e} \tag{8.3.37}$$

for some 4×3 matrix $\boldsymbol{H}$ that contains the channel gains and their complex conjugates, and for which $\boldsymbol{H}^H\boldsymbol{H} = \boldsymbol{I}$ (assuming $\|\boldsymbol{h}\|^2 = 1$). But this would imply that $\boldsymbol{H}$ is a rate-1 OSTBC matrix for three transmit antennas, and from Appendix B we know that such a matrix does not exist.

The TR-OSTBC schemes for more than two transmit antennas may in fact be of a more theoretical than practical interest (with the possible exception of three transmit antennas). First, these coding schemes do not provide full rate and hence an increase in the constellation size is necessary to achieve a high data rate (in terms of bits/sec/Hz). This may be undesirable in practice (in particular, it increases the receiver complexity). Second, if we have three transmit antennas, the TR-OSTBC scheme requires the channel to be constant during $4N_0$ time intervals, and for the case of four transmit antennas the channel must remain constant over $8N_0$ time intervals. This may be hard to achieve in practice if N_0 is large. Finally, it is known that the maximum capacity that can be achieved by OSTBC is less than the channel capacity except for in the case of two transmit antennas and one receive antenna (cf. Section 7.4.3). Hence, while OSTBC should be a good choice for improving the performance of a single-antenna communication system by employing an extra transmitting antenna, it can be argued on information theoretical grounds that OSTBC may be a less attractive choice for high-rate systems or systems with many antennas. Similar conclusions may hold for TR-OSTBC.

8.3.8 Discussion

In the TR-OSTBC scheme (for $n_t = 2$) assuming that there is no extra guard interval between the bursts, we transmit $2N_0$ symbols during $2N_0 + 4N_p + N_m$ time intervals. Since the last postamble of the transmission may coincide with the first preamble of the next block, the effective rate is $2N_0/(2N_0 + 3N_p + N_m)$. Moreover, if no midamble is present, the postamble of the regular burst and the preamble of the time-reversed burst can be merged into an N_p long midamble (see Figure 8.4) and consequently in this case the rate is the same as that of ST-OFDM:

$$R_{\text{TR-OSTBC}} = \frac{N_0}{N_0 + N_p} \leq \frac{N_0}{N_0 + L} \tag{8.3.38}$$

8.4 Summary and Discussion

This chapter has provided a discussion on linear STBC for frequency-selective channels. We started in Section 8.1 by analyzing the conditions under which a maximal diversity order, i.e., $n_r n_t(L+1)$ can be achieved. It turned out that one such condition is that the difference between any two block-Toeplitz data matrices (see (8.1.3)), corresponding to two distinct transmitted sequences, must have full rank. Given that result, we also revisited the delay diversity scheme introduced in Section 6.3.5 and re-established its diversity properties.

Next in Section 8.2 we studied space-time OFDM in some detail. ST-OFDM is a way of compromising the performance for a gain in computational efficiency; specifically, while all receiver processing can be simply done via FFT, the multipath diversity is lost and therefore ST-OFDM only achieves a diversity order equal to $n_r n_t$. ST-OFDM is reminiscent of conventional OFDM (see Section 5.2.2): by using a cyclic preamble and taking the inverse Fourier transform of the data, the frequency-selective channel is effectively transformed into parallel flat fading space-time channels. Further discussion on and developments of ST-OFDM, and combinations with other coding schemes, can be found in, for example, [AGRAWAL

ET AL., 1998], [BLUM ET AL., 2001], [BÖLCSKEI AND PAULRAJ, 2000B].

We also studied (in Section 8.3) a transmission scheme called time-reversal OSTBC (TR-OSTBC) that is related to ST-OFDM, but which does not convert the space-time channel into parallel flat fading channels. Instead, TR-OSTBC decouples (to within a certain approximation) the space-time channel into n_s parallel and independent frequency-selective SISO channels. The main appealing feature of TR-OSTBC is that standard coding and equalization procedures (such as the Viterbi equalizer) can be used on each equivalent SISO channel.

The present chapter is based on elements from Chapter 5 and Chapter 7 along with the discrete-time models in Section 2.2. Coding for frequency-selective channels will not be further discussed in this text, except in Section 9.4.1 where we treat optimal training for frequency-selective channels (both in the context of OFDM and conventional transmission).

8.5 Problems

1. Prove that the diagonal blocks of the matrix in (8.3.18) are positive definite, and that the off-diagonal blocks of this matrix are equal to zero.

2. Prove Equation (8.3.32).

3. Discuss the capacity of a ST-OFDM system (with a cyclic prefix) for frequency-selective channels, and compare it with the channel capacity. Is there any information-theoretical penalty imposed by the OFDM structure?

Hint: The capacity of OFDM and the channel capacity can be expressed by (3.2.1) using a matrix $\boldsymbol{H}$ that is diagonal, respectively Toeplitz.

4. Prove Theorems 8.1 and 8.2.

Chapter 9

COHERENT AND NON-COHERENT RECEIVERS

This chapter discusses receiver structures for systems that use transmit diversity. We begin by studying the problem of detecting linear STBC coherently, and following this we discuss noncoherent detection. We also discuss related topics including optimal training, differential detection and detection in the presence of frequency offsets. For convenience, most of our discussion focuses on flat channels, although many of the presented results extend in a straightforward manner to frequency-selective channels as well (provided that the discrete-time models in Section 2.2 are valid, and that accurate timing synchronization is possible). While Chapter 4 focused on error probabilities and code design criteria, the present chapter is primarily devoted to practical detection methods.

9.1 Coherent Detection of Linear STBC

In this chapter we assume that N_b STBC matrices (each of length N) are transmitted consecutively. Moreover, these N_b blocks are preceded by an optional known training block $\boldsymbol{X}_t$ of dimension $n_t \times N_t$. Consequently, the block of transmitted pilots and data is of dimension $n_t \times (N_t + N_b N)$ and can be written as (see Figure 9.1):

$$\boldsymbol{X} = \begin{bmatrix} \boldsymbol{X}_t & \boldsymbol{X}_1 & \cdots & \boldsymbol{X}_{N_b} \end{bmatrix} \tag{9.1.1}$$

The received data block is of dimension $n_r \times (N_t + N_b N)$ and can be expressed as follows (cf. (2.1.5)):

$$\boldsymbol{Y} = \begin{bmatrix} \boldsymbol{Y}_t & \boldsymbol{Y}_1 & \cdots & \boldsymbol{Y}_{N_b} \end{bmatrix} = \boldsymbol{H}\boldsymbol{X} + \boldsymbol{E} \tag{9.1.2}$$

where

$$\boldsymbol{E} = \begin{bmatrix} \boldsymbol{E}_t & \boldsymbol{E}_1 & \cdots & \boldsymbol{E}_{N_b} \end{bmatrix} \tag{9.1.3}$$

Training	Data 1	Data 2			

Figure 9.1. Transmission of known training data (pilots) and STBC blocks.

is an $n_r \times (N_t + N_b N)$ noise and interference matrix. The presence of the training block $\boldsymbol{X}_t$ is optional; if it is absent we simply let $N_t = 0$. Clearly, we can expect that there is a tradeoff between the amount of training used (and hence bandwidth efficiency) and the BER performance.

We have already discussed the coherent detection of STBC in Chapter 7. In the present section we extend the results therein to the block transmission model (9.1.1), and also to the case of colored noise.

9.1.1 White Noise

Assume that $\boldsymbol{X}$ is composed of matrices $\boldsymbol{X}_k$ that are formed via linear STBC using the same set of code matrices $\{\boldsymbol{A}_n, \boldsymbol{B}_n\}$. Under the assumption that the columns $\{\boldsymbol{e}_n\}$ of $\boldsymbol{E}$ are independent and complex Gaussian with covariance matrix $\sigma^2 \boldsymbol{I}$ the negative part of the log-likelihood function for detection is equal to (dropping irrelevant constants, and treating $\boldsymbol{H}$ as a deterministic variable):

$$n_r(N_t + N_b N)\log\sigma^2 + \frac{1}{\sigma^2}\|\boldsymbol{Y} - \boldsymbol{H}\boldsymbol{X}\|^2 \tag{9.1.4}$$

If $\boldsymbol{H}$ is known, the detection of $\boldsymbol{X}$ amounts to minimizing the following Euclidean distance:

$$\|\boldsymbol{Y} - \boldsymbol{H}\boldsymbol{X}\|^2 \tag{9.1.5}$$

By using the following straightforward extensions of the quantities defined in Section 7.1:

$$\begin{aligned} \boldsymbol{F}_a &= \boldsymbol{I}_{N_b} \otimes \begin{bmatrix} \operatorname{vec}(\boldsymbol{H}\boldsymbol{A}_1) & \cdots & \operatorname{vec}(\boldsymbol{H}\boldsymbol{A}_{n_s}) \end{bmatrix} \\ \boldsymbol{F}_b &= \boldsymbol{I}_{N_b} \otimes \begin{bmatrix} i\operatorname{vec}(\boldsymbol{H}\boldsymbol{B}_1) & \cdots & i\operatorname{vec}(\boldsymbol{H}\boldsymbol{B}_{n_s}) \end{bmatrix} \end{aligned} \tag{9.1.6}$$

$$\boldsymbol{F} = \begin{bmatrix} \operatorname{vec}(\boldsymbol{H}\boldsymbol{X}_t) & \boldsymbol{0} & \boldsymbol{0} \\ \boldsymbol{0} & \boldsymbol{F}_a & \boldsymbol{F}_b \end{bmatrix} \tag{9.1.7}$$

$$\boldsymbol{s}_n = \begin{bmatrix} s_1^{(n)} \\ \vdots \\ s_{n_s}^{(n)} \end{bmatrix}, \qquad n = 1, \ldots, N_b \tag{9.1.8}$$

and

$$s' = \begin{bmatrix} \bar{s}_1^T & \cdots & \bar{s}_{N_b}^T & \tilde{s}_1^T & \cdots & \tilde{s}_{N_b}^T \end{bmatrix}^T \tag{9.1.9}$$

we can write (9.1.5) as:

$$\|Y - HX\|^2 = \left\| y - F \begin{bmatrix} 1 \\ s' \end{bmatrix} \right\|^2 \tag{9.1.10}$$

where $y = \text{vec}(Y)$.

In general (9.1.10) must be minimized under the constraint that the elements of s' belong to a finite set $\mathcal{S}$. This is a fairly difficult problem, unless F has a very special structure (like for OSTBC; cf. Theorem 7.2 (see page 106)).

9.1.2 Spatially Colored Noise

A more general assumption on the noise term E is that the noise is temporally white but spatially colored, i.e., that the columns of E in (9.1.3) are independent and Gaussian with the following distribution:

$$e_n \sim N_C(\mathbf{0}, \Lambda) \tag{9.1.11}$$

where Λ is a positive definite matrix. This model may be useful in the presence of strong spatially colored interference (see Chapter 11). It can be extended to temporal correlation as well.

In the present case, the ML metric takes on the following form (treating H as a deterministic quantity):

$$(N_t + N_b N) \log |\Lambda| + \text{Tr}\left\{\Lambda^{-1}(Y - HX)(Y - HX)^H\right\} \tag{9.1.12}$$

If H and Λ are known, a coherent detection scheme similar to that in Section 9.1.1 is straightforward to obtain. The basic idea is that if the received n_r-vectors are pre-multiplied with $\Lambda^{-1/2}$, they will satisfy a pre-whitened data model to which the coherent detector for white noise can be applied. In effect, for known Λ, minimization of (9.1.12) reduces to minimization of:

$$\left\|\Lambda^{-1/2}(Y - HX)\right\|^2 = \left\|\Lambda^{-1/2}Y - \Lambda^{-1/2}HX\right\|^2 \tag{9.1.13}$$

with respect to $X \in \mathcal{X}$. By defining

$$\begin{aligned} G_a &= I_{N_b} \otimes \begin{bmatrix} \text{vec}(\Lambda^{-1/2}HA_1) & \cdots & \text{vec}(\Lambda^{-1/2}HA_{n_s}) \end{bmatrix} \\ G_b &= I_{N_b} \otimes \begin{bmatrix} i\,\text{vec}(\Lambda^{-1/2}HB_1) & \cdots & i\,\text{vec}(\Lambda^{-1/2}HB_{n_s}) \end{bmatrix} \\ G &= \begin{bmatrix} \text{vec}(\Lambda^{-1/2}HX_t) & \mathbf{0} & \mathbf{0} \\ \mathbf{0} & G_a & G_b \end{bmatrix} \end{aligned} \tag{9.1.14}$$

we find that

$$\left\|\boldsymbol{\Lambda}^{-1/2}(\boldsymbol{Y}-\boldsymbol{H}\boldsymbol{X})\right\|^2 = \left\|\check{\boldsymbol{y}} - \boldsymbol{G}\begin{bmatrix}1\\ \boldsymbol{s}'\end{bmatrix}\right\|^2 \tag{9.1.15}$$

where

$$\check{\boldsymbol{y}} = \operatorname{vec}(\boldsymbol{\Lambda}^{-1/2}\boldsymbol{Y}) \tag{9.1.16}$$

Minimizing (9.1.15) is a problem of the same form as the minimization of (9.1.10).

Example 9.1: Coherent detection of a single OSTBC matrix in colored noise. Suppose that $N_b = 1$ and that there is no training block. Then if $\{\boldsymbol{A}_n, \boldsymbol{B}_n\}$ satisfy (7.4.4), we can prove that

$$\operatorname{Re}\{\boldsymbol{G}^H\boldsymbol{G}\} = \operatorname{Tr}\{\boldsymbol{H}^H\boldsymbol{\Lambda}^{-1}\boldsymbol{H}\}\boldsymbol{I} \tag{9.1.17}$$

and that the ML metric (9.1.15) can be written as

$$\left\|\check{\boldsymbol{y}} - \boldsymbol{G}\boldsymbol{s}'\right\|^2 = \operatorname{Tr}\{\boldsymbol{H}^H\boldsymbol{\Lambda}^{-1}\boldsymbol{H}\} \cdot \left\|\boldsymbol{s} - \hat{\boldsymbol{s}}\right\|^2 + \text{const.} \tag{9.1.18}$$

where (cf. Theorem 7.3 on page 107)

$$\hat{\boldsymbol{s}} \triangleq \sqrt{\frac{n_t}{\rho^2}} \cdot \frac{\operatorname{Re}\{\boldsymbol{G}_a^H\check{\boldsymbol{y}}\} + i\operatorname{Re}\{\boldsymbol{G}_b^H\check{\boldsymbol{y}}\}}{\operatorname{Tr}\{\boldsymbol{H}^H\boldsymbol{\Lambda}^{-1}\boldsymbol{H}\}} \sim N_C\left(\boldsymbol{s}, \frac{n_t}{\rho^2\operatorname{Tr}\{\boldsymbol{H}^H\boldsymbol{\Lambda}^{-1}\boldsymbol{H}\}} \cdot \boldsymbol{I}\right) \tag{9.1.19}$$

and where, as before, ρ^2 is such that the average transmitted energy during N time intervals is equal to $\rho^2 E[\|\boldsymbol{s}\|^2]$. These equations show that $\hat{\boldsymbol{s}}$ is a sufficient statistic for detection, which can be interpreted as the output of n_s parallel and independent complex AWGN channels. The SNR on each channel is:

$$\text{SNR} = \frac{\rho^2}{n_t}\operatorname{Tr}\{\boldsymbol{H}^H\boldsymbol{\Lambda}^{-1}\boldsymbol{H}\}E\left[|s_n|^2\right] \tag{9.1.20}$$

Obviously, if $\boldsymbol{\Lambda} = \sigma^2\boldsymbol{I}$, all the results of this example reduce to those in Theorem 7.3. ■

9.1.3 The Integer-Constrained Least-Squares Problem

We have seen in the previous sections that the received data associated with a linear space-time block code can be expressed in terms of the following observation equation (after appropriate subtraction of known training data)

$$\boldsymbol{y} = \boldsymbol{F}\boldsymbol{s}' + \boldsymbol{e} \tag{9.1.21}$$

where $\boldsymbol{F}$ is a known complex matrix of dimension $m \times n$, $\boldsymbol{y}$ is a complex-valued observation vector of length m and

$$\boldsymbol{s}' = \begin{bmatrix} \bar{\boldsymbol{s}} \\ \tilde{\boldsymbol{s}} \end{bmatrix} \tag{9.1.22}$$

is a real-valued vector of length n whose elements belong to a finite alphabet $\mathcal{S}'$. Note that (9.1.21) is a generic model that includes both the model in Section 7.1 (in which case $m = Nn_r$ and $n = 2n_s$), and that in Sections 9.1.1 and 9.1.2 (in which case $m = N_bNn_r$ and $n = 2n_s$, after elimination of the training block).

If $\boldsymbol{e}$ is white and complex Gaussian noise, ML detection amounts to minimizing the following Euclidean distance:

$$\|\boldsymbol{y} - \boldsymbol{F}\boldsymbol{s}'\|^2 \tag{9.1.23}$$

with respect to $\boldsymbol{s}'$. Note that (9.1.21) can be written:

$$\boldsymbol{y}' = \boldsymbol{F}'\boldsymbol{s}' + \boldsymbol{e}' \tag{9.1.24}$$

where

$$\boldsymbol{y}' = \begin{bmatrix} \bar{\boldsymbol{y}} \\ \tilde{\boldsymbol{y}} \end{bmatrix}, \quad \boldsymbol{F}' = \begin{bmatrix} \bar{\boldsymbol{F}} \\ \tilde{\boldsymbol{F}} \end{bmatrix} \tag{9.1.25}$$

Using this notation, minimizing (9.1.23) is equivalent to minimizing:

$$\left\|\boldsymbol{y}' - \boldsymbol{F}'\boldsymbol{s}'\right\|^2 \tag{9.1.26}$$

The representation in (9.1.26) in general is easier to work with since all involved quantities are real-valued. Throughout this section, we shall assume that the constellation $\mathcal{S}$ is "separable" into two constellations $\mathcal{S}'$ for the real and imaginary parts of the symbols respectively, i.e., that these real and imaginary parts can be detected independently of each other.

The problem of minimizing (9.1.26) is sometimes called an *integer-constrained* least-squares problem and is usually quite difficult to solve, except for the case when $\text{Re}\{\boldsymbol{F}^H\boldsymbol{F}\}$ is proportional to an identity matrix (which is the case essentially only if the underlying code is an orthogonal STBC; cf. Theorem 7.2 (see page 106)). In a naive setting, finding the solution would require a brute-force search through the $|\mathcal{S}'|^{2n_s}$ possible choices of $\boldsymbol{s}'$, which may well be unfeasible. For instance, if $n_s = 4$ and the (real) constellation is 8-PAM, a brute-force ML detector would have to try $8^8 \approx 1.7 \cdot 10^7$ combinations. The main goal of this section is to review a few methods for finding exact and approximate solutions to (9.1.26).

Zero-Forcing Detection

Perhaps the simplest approach to the minimization of (9.1.26) is to relax the constraint that $s'_k \in \mathcal{S}'$ and simply minimize the criterion with respect to an unconstrained real-valued vector $\boldsymbol{s}'$. Doing so, under the weak assumption that $\boldsymbol{F}'$ has full column rank, we get:

$$\hat{\boldsymbol{s}}'_{\mathrm{ZF}} = \left(\boldsymbol{F}'^T\boldsymbol{F}'\right)^{-1}\boldsymbol{F}'^T\boldsymbol{y}' = \left(\bar{\boldsymbol{F}}^T\bar{\boldsymbol{F}} + \tilde{\boldsymbol{F}}^T\tilde{\boldsymbol{F}}\right)^{-1}\left(\bar{\boldsymbol{F}}^T\bar{\boldsymbol{y}} + \tilde{\boldsymbol{F}}^T\tilde{\boldsymbol{y}}\right) \tag{9.1.27}$$

or expressed in the original complex-valued quantities $\boldsymbol{F}$ and $\boldsymbol{y}$:

$$\hat{\boldsymbol{s}}'_{\mathrm{ZF}} = \left(\mathrm{Re}\left\{\boldsymbol{F}^H\boldsymbol{F}\right\}\right)^{-1}\mathrm{Re}\left\{\boldsymbol{F}^H\boldsymbol{y}\right\} \tag{9.1.28}$$

The zero-forcing approximation of the solution to (9.1.26) is then obtained by picking the points in the constellation $\mathcal{S}'$ that are closest to $[\hat{\boldsymbol{s}}'_{\mathrm{ZF}}]_k$. We will denote the constellation point $\hat{\hat{s}}'_k \in \mathcal{S}'$ closest to $\hat{s}'_k$ by

$$\hat{\hat{s}}'_k = \mathcal{P}(\hat{s}'_k) \tag{9.1.29}$$

where $\mathcal{P}(\cdot)$ stands for projection onto the constellation.

In general, the zero-forcing approximation does not coincide with the ML solution and hence it is not optimal, except for the case when $\mathrm{Re}\left\{\boldsymbol{F}^H\boldsymbol{F}\right\}$ is proportional to an identity matrix. Nevertheless, the zero-forcing approach is simple to implement and can perform acceptably if the matrix $\boldsymbol{F}'$ is well-conditioned.

Minimum Mean-Square Error Detection

Consider again the observation model (9.1.24):

$$\boldsymbol{y}' = \boldsymbol{F}'\boldsymbol{s}' + \boldsymbol{e}' \tag{9.1.30}$$

If we assume that $\boldsymbol{s}'$ is random with the following distribution:

$$\boldsymbol{s}' \sim N\left(\boldsymbol{0}, \frac{\rho^2}{2}\boldsymbol{I}\right) \tag{9.1.31}$$

(we assume a variance of $\rho^2/2$ because $\boldsymbol{s}'$ is real-valued) then we can compute an estimate of $\boldsymbol{s}'$ that is linear in $\boldsymbol{y}$ as follows:

$$\hat{\boldsymbol{s}}' = \boldsymbol{G}'\boldsymbol{y}' \tag{9.1.32}$$

and obtain the matrix $\boldsymbol{G}'$ that minimizes the following mean-square error:

$$E\left[\|\hat{\boldsymbol{s}}' - \boldsymbol{s}'\|^2\right] \tag{9.1.33}$$

This estimate, which is called the MMSE estimate of $\boldsymbol{s}'$, is then projected onto the constellation $\mathcal{S}'$ as in (9.1.29) to obtain the MMSE approximation of the solution to (9.1.26).

The vector $\hat{\boldsymbol{s}}'$ that minimizes the MSE in (9.1.33) can be obtained as follows. Assuming that $\boldsymbol{s}'$ and $\boldsymbol{e}'$ are independent, the MSE of $\hat{\boldsymbol{s}}'$ in (9.1.33) can be written:

$$\begin{aligned} E\left[\left\|\hat{\boldsymbol{s}}'-\boldsymbol{s}'\right\|^2\right] &= E\left[\left\|\boldsymbol{G}'\boldsymbol{y}'-\boldsymbol{s}'\right\|^2\right] \\ &= E\left[\left\|\boldsymbol{G}'\boldsymbol{F}'\boldsymbol{s}'+\boldsymbol{G}'\boldsymbol{e}'-\boldsymbol{s}'\right\|^2\right] \\ &= \frac{\rho^2}{2}\cdot\left\|\boldsymbol{G}'\boldsymbol{F}'-\boldsymbol{I}\right\|^2+\frac{\sigma^2}{2}\cdot\|\boldsymbol{G}'\|^2 \\ &= \frac{\rho^2}{2}\cdot\left\|[\boldsymbol{I}\ \ \boldsymbol{0}]-\boldsymbol{G}'\left[\boldsymbol{F}'\ \ \sqrt{\sigma^2/\rho^2}\boldsymbol{I}\right]\right\|^2 \end{aligned} \tag{9.1.34}$$

It follows that the matrix $\boldsymbol{G}'$ that minimizes the MSE in (9.1.34) is equal to:

$$\boldsymbol{G}' = \boldsymbol{F}'^T\left(\boldsymbol{F}'\boldsymbol{F}'^T+\frac{\sigma^2}{\rho^2}\boldsymbol{I}\right)^{-1} \tag{9.1.35}$$

Therefore, the resulting MMSE estimate of $\boldsymbol{s}'$ is:

$$\hat{\boldsymbol{s}}'_{\text{MMSE}} = \boldsymbol{F}'^T\left(\boldsymbol{F}'\boldsymbol{F}'^T+\frac{\sigma^2}{\rho^2}\boldsymbol{I}\right)^{-1}\boldsymbol{y}' \tag{9.1.36}$$

The MMSE technique can provide somewhat better estimates of $\boldsymbol{s}'$ than the zero-forcing algorithm, at a similar computational cost. However, the evaluation of the MMSE solution requires knowledge of the ratio σ^2/ρ^2.

The Iterative Nulling and Cancelling Algorithm

Iterative nulling and cancelling is a suboptimal algorithm that detects the elements of $\boldsymbol{s}'$ in a sequential fashion that is somewhat reminiscent of iterative interference cancellation schemes for multiuser detection [VERDÚ, 1998]. It has been an algorithm used in Bell-Labs experimental high-rate system "BLAST" [FOSCHINI, JR., 1996]. It has been shown that nulling and cancelling is closely related to a form of decision-feedback equalization [GINIS AND CIOFFI, 2001]. Many variants of this algorithm exist in the literature. We shall describe one of the simplest variants in what follows.

The main idea behind the algorithm is quite easy to understand. Consider (see (9.1.36)):

$$\hat{\boldsymbol{s}}'_{\text{MMSE}} = \boldsymbol{F}'^T\left(\boldsymbol{F}'\boldsymbol{F}'^T+\frac{\sigma^2}{\rho^2}\boldsymbol{I}\right)^{-1}\boldsymbol{y}' \tag{9.1.37}$$

which is the MMSE solution of the problem (9.1.26). The covariance matrix associated with $\hat{\boldsymbol{s}}'_{\mathrm{MMSE}}$ is:

$$\begin{aligned}\boldsymbol{Q} &= E\left[\left(\hat{\boldsymbol{s}}'_{\mathrm{MMSE}} - E\left[\hat{\boldsymbol{s}}'_{\mathrm{MMSE}}\right]\right)\left(\hat{\boldsymbol{s}}'_{\mathrm{MMSE}} - E\left[\hat{\boldsymbol{s}}'_{\mathrm{MMSE}}\right]\right)^T\right] \\ &= \frac{\sigma^2}{2} \cdot \boldsymbol{F}'^T \left(\boldsymbol{F}'\boldsymbol{F}'^T + \frac{\sigma^2}{\rho^2}\boldsymbol{I}\right)^{-2} \boldsymbol{F}'\end{aligned} \tag{9.1.38}$$

(Alternatively, we could use the MSE matrix:

$$E\left[\left(\hat{\boldsymbol{s}}'_{\mathrm{MMSE}} - \boldsymbol{s}'\right)\left(\hat{\boldsymbol{s}}'_{\mathrm{MMSE}} - \boldsymbol{s}'\right)^T\right] \tag{9.1.39}$$

the computation of which we leave as an exercise to the reader.) If $\{Q_{k,k}\}$ denote the diagonal elements of $\boldsymbol{Q}$, we would expect that the element $\hat{s}'_k$ of $\hat{\boldsymbol{s}}'_{\mathrm{MMSE}}$ for which $Q_{k,k}$ is minimal is the most accurate one. Let

$$p = \operatorname*{argmin}_k \{Q_{k,k}\} \tag{9.1.40}$$

and let us compute

$$\hat{\hat{s}}'_p = \mathcal{P}(\hat{s}'_p) \tag{9.1.41}$$

If we assume that $\hat{\hat{s}}'_p$ is correct, we can subtract its contribution from $\boldsymbol{y}'$ and obtain a problem of reduced order:

$$\min_{\boldsymbol{s}'_{\underline{p}}} \left\| \left(\boldsymbol{y}' - \hat{\hat{s}}'_p \boldsymbol{f}'_p\right) - \boldsymbol{F}'_{\underline{p}} \boldsymbol{s}'_{\underline{p}} \right\|^2 \tag{9.1.42}$$

where $\boldsymbol{f}'_p$ is the pth column of $\boldsymbol{F}'$, $\boldsymbol{s}'_{\underline{p}}$ denotes the vector $\boldsymbol{s}'$ with the pth element removed and $\boldsymbol{F}'_{\underline{p}}$ is the matrix $\boldsymbol{F}'$ with the pth column eliminated. The procedure continues iteratively by solving the reduced problem (9.1.42) until all elements of $\boldsymbol{s}'$ have been found.

The Sphere Decoding Algorithm

The task of finding the vector $\boldsymbol{s}'$ that minimizes (9.1.26) is related to that of finding the shortest vector in a finite lattice, which is a well-studied problem in number theory and computational mathematics. The work of [Fincke and Pohst, 1985] made a significant progress towards reducing the complexity of finding the solution to this minimization problem. The method proposed in the cited paper is usually referred to as "sphere decoding" and was applied in the context of digital and wireless communications in, for instance, [Viterbo and Boutros, 1999],

[Damen et al., 2000]. A complexity analysis of it, in the context of wireless communications, can be found in [Hassibi and Vikalo, 2002]. Analyzing and reducing the computational complexity of the sphere decoding algorithm and related methods is currently an active research topic (see the cited papers and also, e.g., [Agrell et al., 2002]); we shall only describe the technique in its simplest version. Like in the previous subsections, we assume that the real and imaginary parts of the symbols can be detected independently of each other.

We say that a vector $\boldsymbol{s}'$ (whose elements belong to $\mathcal{S}'$) *lies in a sphere with radius* r if it approximates the solution to (9.1.26) with a residual norm less than r:

$$\left\|\boldsymbol{y}' - \boldsymbol{F}'\boldsymbol{s}'\right\|^2 < r^2 \tag{9.1.43}$$

Let

$$\hat{\boldsymbol{s}}'_{\text{ZF}} = (\boldsymbol{F}'^T\boldsymbol{F}')^{-1}\boldsymbol{F}'^T\boldsymbol{y}' \tag{9.1.44}$$

be the zero-forcing solution (see (9.1.28)). Since

$$\hat{\boldsymbol{s}}'^T_{\text{ZF}}\boldsymbol{F}'^T\boldsymbol{F}'\boldsymbol{s}' = \boldsymbol{y}'^T\boldsymbol{F}'(\boldsymbol{F}'^T\boldsymbol{F}')^{-1}\boldsymbol{F}'^T\boldsymbol{F}'\boldsymbol{s}' = \boldsymbol{y}'^T\boldsymbol{F}'\boldsymbol{s}' \tag{9.1.45}$$

it follows that

$$\begin{aligned}\left\|\boldsymbol{F}'(\boldsymbol{s}' - \hat{\boldsymbol{s}}'_{\text{ZF}})\right\|^2 &= \left\|\boldsymbol{F}'\boldsymbol{s}'\right\|^2 + \left\|\boldsymbol{F}'\hat{\boldsymbol{s}}'_{\text{ZF}}\right\|^2 - 2\cdot\hat{\boldsymbol{s}}'^T_{\text{ZF}}\boldsymbol{F}'^T\boldsymbol{F}'\boldsymbol{s}' \\ &= \left\|\boldsymbol{F}'\boldsymbol{s}'\right\|^2 + \left\|\boldsymbol{F}'\hat{\boldsymbol{s}}'_{\text{ZF}}\right\|^2 - 2\cdot\boldsymbol{y}'^T\boldsymbol{F}'\boldsymbol{s}' \\ &= \left\|\boldsymbol{y}' - \boldsymbol{F}'\boldsymbol{s}'\right\|^2 - \|\boldsymbol{y}'\|^2 + \left\|\boldsymbol{F}'\hat{\boldsymbol{s}}'_{\text{ZF}}\right\|^2\end{aligned} \tag{9.1.46}$$

and consequently

$$\begin{aligned}\left\|\boldsymbol{y}' - \boldsymbol{F}'\boldsymbol{s}'\right\|^2 &= \left\|\boldsymbol{F}'(\boldsymbol{s}' - \hat{\boldsymbol{s}}'_{\text{ZF}})\right\|^2 + \|\boldsymbol{y}'\|^2 - \left\|\boldsymbol{F}'\hat{\boldsymbol{s}}'_{\text{ZF}}\right\|^2 \\ &= \left\|\boldsymbol{R}(\boldsymbol{s}' - \hat{\boldsymbol{s}}'_{\text{ZF}})\right\|^2 + \|\boldsymbol{y}'\|^2 - \left\|\boldsymbol{R}\hat{\boldsymbol{s}}'_{\text{ZF}}\right\|^2\end{aligned} \tag{9.1.47}$$

where $\boldsymbol{R}$ is the upper triangular matrix with positive diagonal elements that satisfies

$$\boldsymbol{R}^T\boldsymbol{R} = \boldsymbol{F}'^T\boldsymbol{F}' \tag{9.1.48}$$

In practice, $\boldsymbol{R}$ is obtained via the Cholesky factorization of the positive definite matrix $\boldsymbol{F}'^T\boldsymbol{F}'$ (see, e.g., [Golub and van Loan, 1989]). It follows that a vector $\boldsymbol{s}'$ lies in a sphere with radius r if and only if

$$\left\|\boldsymbol{R}(\boldsymbol{s}' - \hat{\boldsymbol{s}}'_{\text{ZF}})\right\|^2 < r^2 - \|\boldsymbol{y}'\|^2 + \left\|\boldsymbol{R}\hat{\boldsymbol{s}}'_{\text{ZF}}\right\|^2 \tag{9.1.49}$$

The main idea behind the sphere decoding algorithm is perhaps most easily described in the following way. Let

$$r_n'^2 = r^2 - \|\boldsymbol{y}'\|^2 + \|\boldsymbol{R}\hat{\boldsymbol{s}}'_{\text{ZF}}\|^2 \tag{9.1.50}$$

where n is the length of $\boldsymbol{s}'$. Since $\boldsymbol{R}$ is an upper-triangular matrix:

$$\begin{aligned}
&\|\boldsymbol{R}(\boldsymbol{s}' - \hat{\boldsymbol{s}}'_{\text{ZF}})\|^2 \\
=&\sum_{k=1}^{n} R_{k,k}^2 \left(s_k' - \hat{s}_k' + \sum_{l=k+1}^{n} \frac{R_{k,l}}{R_{k,k}} (s_l' - \hat{s}_l') \right)^2 \\
=&R_{n,n}^2 (s_n' - \hat{s}_n')^2 + R_{n-1,n-1}^2 \left(s_{n-1}' - \hat{s}_{n-1}' + \frac{R_{n-1,n}}{R_{n-1,n-1}} (s_n' - \hat{s}_n') \right)^2 + \cdots
\end{aligned} \tag{9.1.51}$$

(here s_k' is the kth element of $\boldsymbol{s}'$). From (9.1.51) it follows that a necessary condition for $\boldsymbol{s}'$ to lie in a sphere with radius r is that

$$R_{n,n}^2 (s_n' - \hat{s}_n')^2 \leq r_n'^2 \tag{9.1.52}$$

or equivalently that

$$\left\lceil \hat{s}_n' - \frac{r_n'}{R_{n,n}} \right\rceil \leq s_n' \leq \left\lfloor \hat{s}_n' + \frac{r_n'}{R_{n,n}} \right\rfloor \tag{9.1.53}$$

where $\lceil x \rceil$ and $\lfloor x \rfloor$ denote the integers obtained by rounding x towards $+\infty$ and $-\infty$, respectively. Note that (9.1.53) is a necessary but *not* a sufficient condition.

For a given $s_n' \in \mathcal{S}'$ that satisfies (9.1.53), let us define

$$\begin{aligned}
r_{n-1}'^2 &= r_n'^2 - R_{n,n}^2 (s_n' - \hat{s}_n')^2 \\
\hat{s}_{n-1|n}' &= \hat{s}_{n-1}' - \frac{R_{n-1,n}}{R_{n-1,n-1}} (s_n' - \hat{s}_n')
\end{aligned} \tag{9.1.54}$$

Then a further necessary condition for $\boldsymbol{s}'$ to lie in the sphere is (cf. (9.1.51)):

$$\left\lceil \hat{s}_{n-1|n}' - \frac{r_{n-1}'}{R_{n-1,n-1}} \right\rceil \leq s_{n-1}' \leq \left\lfloor \hat{s}_{n-1|n}' + \frac{r_{n-1}'}{R_{n-1,n-1}} \right\rfloor \tag{9.1.55}$$

For given $s_n' \in \mathcal{S}'$ and $s_{n-1}' \in \mathcal{S}'$ in the intervals specified by (9.1.55) we can continue to obtain a bound on s_{n-2}' that must be satisfied for $\boldsymbol{s}'$ to lie in the sphere of radius r:

$$\left\lceil \hat{s}_{n-2|n-1}' - \frac{r_{n-2}'}{R_{n-2,n-2}} \right\rceil \leq s_{n-2}' \leq \left\lfloor \hat{s}_{n-2|n-1}' + \frac{r_{n-2}'}{R_{n-2,n-2}} \right\rfloor \tag{9.1.56}$$

where

$$\begin{aligned} r'^2_{n-2} &= r'^2_{n-1} - R^2_{n-1,n-1}(s'_{n-1} - \hat{s}'_{n-1|n})^2 \\ \hat{s}'_{n-2|n-1} &= \hat{s}'_{n-2} - \frac{R_{n-2,n-1}}{R_{n-2,n-2}}(s'_{n-1} - \hat{s}'_{n-1|n}) \end{aligned} \tag{9.1.57}$$

The sphere decoder continues recursively to establish bounds on $s'_{n-3}, \ldots$. Whenever a valid value of $\hat{s}'_1$ is reached, the necessary and sufficient conditions on s' to lie in the sphere are satisfied and hence we found a vector in the sphere; otherwise, we are on a "wrong track," and we should try another combination. Each time a vector s'_* that lies in the sphere is found, the search can be restarted by restricting it to a sphere with radius $\|y' - F's'_*\|^2$. The initial r'^2_n can be obtained, for example, by "truncating" $\hat{s}'_{\text{ZF}}$ as in (9.1.29).

Note that although the procedure is recursive in nature, an iterative implementation may be advantageous from a computer implementation point of view.

9.2 Concatenation of Linear STBC with Outer Codes

In a practical system, the linear space-time block codes will be concatenated with outer error-correcting codes. In this case, in addition to solving a problem such as (9.1.26), the detector must also provide reliability information about the decisions. Note that for orthogonal STBC, the computation of optimal soft information is trivial, owing to the inherent orthogonality of OSTBC that effectively decouples the MIMO channel into a number of scalar channels (see Section 7.4). However, for general STBC, it is often necessary to resort to suboptimal techniques.

9.2.1 Optimal Information Transfer

In the general case, the optimal soft information (in a maximum-likelihood sense) is contained in the set of log-likelihood function values

$$\left\{ -n_r(N_t + N_b N)\log\sigma^2 - \frac{1}{\sigma^2}\|\boldsymbol{Y} - \boldsymbol{H}\boldsymbol{X}\|^2 \right\} \tag{9.2.1}$$

for all possible $\boldsymbol{X}$, or equivalently, in the following set:

$$\left\{ -n_r(N_t + N_b N)\log\sigma^2 - \frac{1}{\sigma^2}\|\boldsymbol{y} - \boldsymbol{F}\boldsymbol{s}'\|^2 \right\} \tag{9.2.2}$$

for all possible $\boldsymbol{s}'$. This is so since given these likelihood function values, the total likelihood function can be computed for any given hypothetical message [Larsson et al., 2002b]. Of course, computing all these metrics explicitly and passing their values to a channel decoder would violate any reasonable complexity bounds.

If the channel $\boldsymbol{H}$ and noise variance σ^2 are unknown, it was shown in [LARSSON ET AL., 2002B] that the optimal (from an ML point of view) soft information is the set of likelihood functions *concentrated with respect to* $\boldsymbol{H}, \sigma^2$:

$$\left\{ \min_{\boldsymbol{H},\sigma^2} \left\{ -n_r(N_t + N_b N) \log \sigma^2 - \frac{1}{\sigma^2} \|\boldsymbol{Y} - \boldsymbol{H}\boldsymbol{X}\|^2 \right\} \right\} \tag{9.2.3}$$

for all possible $\boldsymbol{X}$. The cited paper discussed various problems related to the computation of optimal information according to (9.2.3). For instance, since the channel $\boldsymbol{H}$ is unknown it can be assumed to be either a deterministic unknown or a stochastic variable (in the latter case, the likelihood function takes on a form different from (9.2.1); see the cited article). In fact, since $\boldsymbol{H}$ is sometimes well modeled as a random matrix with independent Gaussian elements having the same variance (see Chapter 2), one could expect that incorporation of such a priori information on the statistical distribution of $\boldsymbol{H}$ would increase the performance. In the cited reference, however, it was found that a deterministic viewpoint may have some advantages, despite the fact that the modeling of $\boldsymbol{H}$ as a deterministic quantity (as we do throughout this book) might discard some a priori information on the distribution of $\boldsymbol{H}$.

9.2.2 Bit Metric Computations

In general, it is not computationally feasible to use the optimal soft information transfer discussed in the previous Section 9.2.1 and one must resort to various approximations. In practice, it is often desirable to compute a likelihood value for each information *bit* instead of a likelihood value for each code matrix. Such bit metrics are particularly important if the code is used in concatenation with bit-interleaved coded modulation [CAIRE ET AL., 1998].

Let c be a particular information bit under study. Provided that sufficient interleaving is used, the optimal information to be forwarded to the channel decoder for each bit c is the likelihood ratio $P(c = 1)/P(c = 0)$ [LIN AND COSTELLO, 1983]. Let $\{\boldsymbol{s}' : c = 0\}$, $\{\boldsymbol{s}' : c = 1\}$ denote the set of vectors $\boldsymbol{s}'$ for which $c = 0$ respectively $c = 1$. If the information bits are equally likely and independent a priori, the likelihood ratio $P(c = 1)/P(c = 0)$ can be approximated by:

$$\frac{P(c=1)}{P(c=0)} = \frac{\sum_{\boldsymbol{s}':c=1} \exp\left(-\frac{1}{\sigma^2} \|\boldsymbol{y} - \boldsymbol{F}\boldsymbol{s}'\|^2 \right)}{\sum_{\boldsymbol{s}':c=0} \exp\left(-\frac{1}{\sigma^2} \|\boldsymbol{y} - \boldsymbol{F}\boldsymbol{s}'\|^2 \right)} \tag{9.2.4}$$

In practice, (9.2.4) can be evaluated by recursively using the fact that

$$\log\left(e^x + e^y\right) = \max(x, y) + \log\left(1 + e^{-|x-y|}\right) \tag{9.2.5}$$

The function $\log(1 + e^{-x})$ can be stored in a lookup-table, hence eliminating the need for computing exponential or logarithmic functions. If desired, the numerical calculations involved can be further simplified by using the approximation

$$\log\left(e^x + e^y\right) \approx \max(x, y) \tag{9.2.6}$$

Direct evaluation of (9.2.4) is impractical since computing the entire summation over all possible $\boldsymbol{s}'$ would take as long a time as solving the ML decoding problem (9.1.26) in a brute-force manner. This problem can be overcome, for instance, by summing only over the vectors $\boldsymbol{s}'$ that lie inside a sphere of a certain radius. The sphere decoder (see Section 9.1.3) can be used to determine those vectors $\boldsymbol{s}'$. It is also possible to solve the integer-constrained LS problem and decode the outer error-correcting code jointly in a way reminiscent of turbo coding [HOCHWALD AND TEN BRINK, 2001].

9.3 Joint ML Detection and Estimation

If the channel $\boldsymbol{H}$ is known to the receiver, the coherent detectors in Section 9.1 can be applied directly. In this section, we treat ML detection for the case that the channel is unknown to the receiver. Referring to the discussion in Section 9.2, we will assume that $\boldsymbol{H}$ and σ^2 are deterministic unknowns. Then the columns of $\boldsymbol{Y}$ are Gaussian random variables with a mean and covariance that depend on the transmitted data and hence it is, in principle, straightforward to formulate the joint likelihood function for the data $\boldsymbol{X}$, the channel $\boldsymbol{H}$ and the noise variance σ^2. The resulting joint detection (of $\boldsymbol{X}$) and estimation (of $\boldsymbol{H}$) can be seen as an application of the generalized likelihood ratio test (GRLT) principle [KAY, 1998, SEC. 6.4.2]. Another possibility is to treat $\boldsymbol{H}$ as a random variable (cf. the remark after Theorem 4.4 on page 49).

9.3.1 White Noise

Consider the logarithm of the likelihood function associated with a certain block $\boldsymbol{X}$:

$$-n_r(N_t + N_b N)\log\sigma^2 - \frac{1}{\sigma^2}\|\boldsymbol{Y} - \boldsymbol{H}\boldsymbol{X}\|^2 \tag{9.3.1}$$

Maximizing (9.3.1) with respect to $\boldsymbol{H}$ is a quadratic optimization problem. It is easy to see that the solution is given by (see, e.g., [STOICA AND MOSES, 1997, SEC. A.8.2] or [SÖDERSTRÖM AND STOICA, 1989, SEC. A.2])

$$\boldsymbol{H} = \boldsymbol{Y}\boldsymbol{X}^H(\boldsymbol{X}\boldsymbol{X}^H)^{-1} \tag{9.3.2}$$

(assuming that the above inverse exists). Insertion of (9.3.2) in (9.3.1) and dropping constant terms show that maximization of (9.3.1) is equivalent to the maximization of the following function:

$$\operatorname{Tr}\{\boldsymbol{Y}\Pi_{\boldsymbol{X}^H}\boldsymbol{Y}^H\} \tag{9.3.3}$$

with respect to the symbols in $\boldsymbol{X}$. The corresponding ML estimate of σ^2 can easily be found to be

$$\hat{\sigma}^2 = \frac{1}{n_r(N_t + N_b N)}\operatorname{Tr}\{\boldsymbol{Y}\Pi^{\perp}_{\boldsymbol{X}^H}\boldsymbol{Y}^H\} \tag{9.3.4}$$

The exact ML detection amounts to maximizing (9.3.3) *jointly* with respect to all symbols contained in $\boldsymbol{X}$. Apparently, doing so requires a search over all possible $\boldsymbol{X}$, which is clearly unfeasible if n_s and N_b are large; in particular, this would involve the evaluation of (9.3.3) for all $|\mathcal{S}|^{N_b n_s}$ possible combinations of symbols.

For the reason discussed in the previous paragraph, a suboptimal detection technique may be desirable. In the following sections, we will describe a few such suboptimal techniques. Section 9.4 describes a training-based scheme that makes use of a transmitted block of pilots. In Section 9.5 we go on to develop a technique that maximizes the likelihood function in an iterative fashion; this iterative procedure can be initialized in a blind manner, without using any training data, or by using a channel estimate obtained via training. However, before doing so we will discuss the joint ML detection and estimation problem in more detail.

Example 9.2: Joint detection/estimation for OSTBC with unitary symbols.
If $\boldsymbol{X}$ is formed using orthogonal STBC with a unitary symbol constellation $\mathcal{S}$, and if the training block $\boldsymbol{X}_t$ is semi-unitary, the problem of maximizing (9.3.3) can be cast in the form (9.1.26), and consequently the techniques in Section 9.1.3 can be used to find the ML solution. Note, however, that the main virtue of orthogonal STBC is the decoupled detection of the information symbols in the case of a known propagation channel. If the receiver can afford to use the techniques of Section 9.1.3, OSTBC may not be the preferred space-time code. Hence the following derivation may be somewhat less important from a practical point of view, although it is quite instructive.

Supposing that

$$\boldsymbol{X}_t\boldsymbol{X}_t^H = \rho^2\boldsymbol{I} \tag{9.3.5}$$

and under the assumptions made above we have that (see Theorem 7.1 on page 102):

$$\boldsymbol{X}\boldsymbol{X}^H = \left(\rho^2 + N_b n_s\right)\cdot\boldsymbol{I} \tag{9.3.6}$$

and consequently

$$\Pi_{\boldsymbol{X}^H} = \frac{1}{\rho^2 + N_b n_s} \cdot \boldsymbol{X}^H \boldsymbol{X} \tag{9.3.7}$$

Hence the maximization of (9.3.3) is equivalent to the maximization of

$$\mathrm{Tr}\{\boldsymbol{Y}\boldsymbol{X}^H\boldsymbol{X}\boldsymbol{Y}^H\} = \left\|\mathrm{vec}\,(\boldsymbol{Y}\boldsymbol{X}^H)\right\|^2 \tag{9.3.8}$$

The maximizer of (9.3.8) is the same as the maximizer of

$$[1\ \boldsymbol{s}'^T]\boldsymbol{F}^H\boldsymbol{F}\begin{bmatrix}1\\ \boldsymbol{s}'\end{bmatrix} \tag{9.3.9}$$

with respect to $\{s'_n\}$, $s'_n \in \mathcal{S}'$, where $\boldsymbol{F}$ is defined according to:

$$\begin{aligned}
\boldsymbol{F} &= \left[\mathrm{vec}(\boldsymbol{X}_t\boldsymbol{Y}^H)\ \ \boldsymbol{F}_1^{(a)}\ \ \cdots\ \ \boldsymbol{F}_{N_b}^{(a)}\ \ \boldsymbol{F}_1^{(b)}\ \ \cdots\ \ \boldsymbol{F}_{N_b}^{(b)}\right]\\
\boldsymbol{F}_k^{(a)} &= \left[\mathrm{vec}(\boldsymbol{Y}_k\boldsymbol{A}_1^H)\ \ \cdots\ \ \mathrm{vec}(\boldsymbol{Y}_k\boldsymbol{A}_{n_s}^H)\right],\quad k = 1,\ldots,N_b\\
\boldsymbol{F}_k^{(b)} &= \left[-i\,\mathrm{vec}(\boldsymbol{Y}_k\boldsymbol{B}_1^H)\ \ \cdots\ \ -i\,\mathrm{vec}(\boldsymbol{Y}_k\boldsymbol{B}_{n_s}^H)\right],\quad k = 1,\ldots,N_b
\end{aligned} \tag{9.3.10}$$

This maximization problem is exactly the one studied in Section 9.1.3. ■

9.3.2 Colored Noise

Assume that the noise is temporally white but spatially colored, i.e., that the columns of $\boldsymbol{E}$ in (9.1.3) are independent and Gaussian according to

$$\boldsymbol{e}_n \sim N_C(\boldsymbol{0}, \boldsymbol{\Lambda}) \tag{9.3.11}$$

where $\boldsymbol{\Lambda}$ is an unknown positive definite matrix. In this case, the ML metric takes on the following form (treating $\boldsymbol{H}$ as a deterministic quantity):

$$(N_t + N_b N)\log|\boldsymbol{\Lambda}| + \mathrm{Tr}\left\{\boldsymbol{\Lambda}^{-1}(\boldsymbol{Y} - \boldsymbol{H}\boldsymbol{X})(\boldsymbol{Y} - \boldsymbol{H}\boldsymbol{X})^H\right\} \tag{9.3.12}$$

It is possible to concentrate the ML metric (9.3.12) analytically with respect to $\boldsymbol{H}$ and $\boldsymbol{\Lambda}$ for a given $\boldsymbol{X}$. It is instructive to see how this can be done. Let

$$\boldsymbol{\Lambda} = \boldsymbol{T}\boldsymbol{\Delta}\boldsymbol{T}^H \tag{9.3.13}$$

be the eigenvalue-decomposition of $\boldsymbol{\Lambda}$, where

$$\boldsymbol{T}\boldsymbol{T}^H = \boldsymbol{I} \tag{9.3.14}$$

and

$$\boldsymbol{\Delta} = \begin{bmatrix} \delta_1 & 0 & \cdots & 0 \\ 0 & \delta_2 & \ddots & \vdots \\ \vdots & \ddots & \ddots & 0 \\ 0 & \cdots & 0 & \delta_{n_r} \end{bmatrix} \tag{9.3.15}$$

Then we have that

$$\begin{aligned} &(N_t + N_b N) \log |\boldsymbol{\Lambda}| + \operatorname{Tr}\left\{\boldsymbol{\Lambda}^{-1}(\boldsymbol{Y} - \boldsymbol{H}\boldsymbol{X})(\boldsymbol{Y} - \boldsymbol{H}\boldsymbol{X})^H\right\} \\ =&(N_t + N_b N) \sum_{n=1}^{n_r} \log \delta_n + \sum_{n=1}^{n_r} \frac{1}{\delta_n} \left(\boldsymbol{T}^H(\boldsymbol{Y} - \boldsymbol{H}\boldsymbol{X})(\boldsymbol{Y} - \boldsymbol{H}\boldsymbol{X})^H\boldsymbol{T}\right)_{n,n} \\ \geq&(N_t + N_b N) \sum_{n=1}^{n_r} \log \left(\boldsymbol{T}^H(\boldsymbol{Y} - \boldsymbol{H}\boldsymbol{X})(\boldsymbol{Y} - \boldsymbol{H}\boldsymbol{X})^H\boldsymbol{T}\right)_{n,n} + \text{const.} \qquad (9.3.16) \\ =&(N_t + N_b N) \log \prod_{n=1}^{n_r} \left(\boldsymbol{T}^H(\boldsymbol{Y} - \boldsymbol{H}\boldsymbol{X})(\boldsymbol{Y} - \boldsymbol{H}\boldsymbol{X})^H\boldsymbol{T}\right)_{n,n} + \text{const.} \\ \geq&(N_t + N_b N) \log |\boldsymbol{T}^H(\boldsymbol{Y} - \boldsymbol{H}\boldsymbol{X})(\boldsymbol{Y} - \boldsymbol{H}\boldsymbol{X})^H\boldsymbol{T}| + \text{const.} \end{aligned}$$

The first inequality, which is easily verified, becomes an equality if

$$\delta_n = \frac{1}{N_t + N_b N} \left(\boldsymbol{T}^H(\boldsymbol{Y} - \boldsymbol{H}\boldsymbol{X})(\boldsymbol{Y} - \boldsymbol{H}\boldsymbol{X})^H\boldsymbol{T}\right)_{n,n} \tag{9.3.17}$$

The second inequality follows from the Hadamard inequality [Horn and Johnson, 1985, Th. 7.8.1] and it becomes an equality if the matrix

$$\boldsymbol{T}^H(\boldsymbol{Y} - \boldsymbol{H}\boldsymbol{X})(\boldsymbol{Y} - \boldsymbol{H}\boldsymbol{X})^H\boldsymbol{T} \tag{9.3.18}$$

is diagonal, which, in view of (9.3.17), gives

$$\boldsymbol{T}^H(\boldsymbol{Y} - \boldsymbol{H}\boldsymbol{X})(\boldsymbol{Y} - \boldsymbol{H}\boldsymbol{X})^H\boldsymbol{T} = (N_t + N_b N)\boldsymbol{\Delta} \tag{9.3.19}$$

Hence, the ML estimate of the noise covariance matrix is given by (conditioned on $\boldsymbol{H}$):

$$\hat{\boldsymbol{\Lambda}} = \boldsymbol{T}\boldsymbol{\Delta}\boldsymbol{T}^H = \frac{1}{N_t + N_b N}(\boldsymbol{Y} - \boldsymbol{H}\boldsymbol{X})(\boldsymbol{Y} - \boldsymbol{H}\boldsymbol{X})^H \tag{9.3.20}$$

Inserting (9.3.20) in (9.3.12), we are left with minimizing:

$$\log |(\boldsymbol{Y} - \boldsymbol{H}\boldsymbol{X})(\boldsymbol{Y} - \boldsymbol{H}\boldsymbol{X})^H| \tag{9.3.21}$$

Observe that

$$
\begin{aligned}
&(\boldsymbol{Y}-\boldsymbol{H}\boldsymbol{X})(\boldsymbol{Y}-\boldsymbol{H}\boldsymbol{X})^H \\
&= \boldsymbol{Y}\boldsymbol{Y}^H - \boldsymbol{H}\boldsymbol{X}\boldsymbol{Y}^H - \boldsymbol{Y}\boldsymbol{X}^H\boldsymbol{H}^H + \boldsymbol{H}\boldsymbol{X}\boldsymbol{X}^H\boldsymbol{H}^H \\
&= (\boldsymbol{H} - \boldsymbol{Y}\boldsymbol{X}^H(\boldsymbol{X}\boldsymbol{X}^H)^{-1})\boldsymbol{X}\boldsymbol{X}^H(\boldsymbol{H} - \boldsymbol{Y}\boldsymbol{X}^H(\boldsymbol{X}\boldsymbol{X}^H)^{-1})^H \\
&\quad + \boldsymbol{Y}\boldsymbol{Y}^H - \boldsymbol{Y}\boldsymbol{X}^H(\boldsymbol{X}\boldsymbol{X}^H)^{-1}\boldsymbol{X}\boldsymbol{Y}^H \\
&\geq \boldsymbol{Y}\boldsymbol{Y}^H - \boldsymbol{Y}\boldsymbol{X}^H(\boldsymbol{X}\boldsymbol{X}^H)^{-1}\boldsymbol{X}\boldsymbol{Y}^H = \boldsymbol{Y}\Pi^{\perp}_{\boldsymbol{X}^H}\boldsymbol{Y}^H
\end{aligned}
\tag{9.3.22}
$$

where the inequality becomes an equality if and only if

$$\hat{\boldsymbol{H}} = \boldsymbol{Y}\boldsymbol{X}^H(\boldsymbol{X}\boldsymbol{X}^H)^{-1} \tag{9.3.23}$$

Hence $\hat{\boldsymbol{H}}$ in (9.3.23) minimizes all eigenvalues of $(\boldsymbol{Y}-\boldsymbol{H}\boldsymbol{X})(\boldsymbol{Y}-\boldsymbol{H}\boldsymbol{X})^H$ and therefore also the determinant in (9.3.21). Consequently $\hat{\boldsymbol{H}}$ is the ML estimate of the channel also in this case of colored noise (see the channel estimate in the case of white noise in (9.3.2)). Note that the ML estimate of $\boldsymbol{\Lambda}$ can be found by inserting (9.3.23) into (9.3.20) which gives

$$\hat{\boldsymbol{\Lambda}} = \frac{1}{N_t + N_b N}\boldsymbol{Y}\Pi^{\perp}_{\boldsymbol{X}^H}\boldsymbol{Y}^H \tag{9.3.24}$$

This formula may be of interest *per se*.

Insertion of (9.3.23) in (9.3.21) gives the following concentrated ML metric:

$$\left|\boldsymbol{Y}\Pi^{\perp}_{\boldsymbol{X}^H}\boldsymbol{Y}^H\right| \tag{9.3.25}$$

Like for the case of white noise (see (9.3.3)), minimization of (9.3.25) with respect to $\boldsymbol{X}$ requires a search through the $|\mathcal{S}|^{N_b n_s}$ possible combinations of symbols, which is often unfeasible. In the next sections, we derive suboptimal receiver structures for noncoherent detection (both for white and colored noise).

9.4 Training-Based Detection

Perhaps the most intuitive and easiest way to detect the data symbols in the case of an unknown channel is to use the received training block $\boldsymbol{Y}_t$ to estimate $\boldsymbol{H}$ and then use this channel estimate in the coherent detector (see Section 9.1). This procedure is a standard approximation to the exact ML detector and the so-obtained (suboptimal) detector will be referred to as *training-based.*

The ML estimate of the channel gain matrix $\boldsymbol{H}$ based on the received training block $\boldsymbol{Y}_t$ is, regardless of whether the receiver assumes that $\boldsymbol{\Lambda} = \sigma^2\boldsymbol{I}$, or that $\boldsymbol{\Lambda}$ is an unknown positive definite matrix:

$$\hat{\boldsymbol{H}} = \boldsymbol{Y}_t\boldsymbol{X}_t^H(\boldsymbol{X}_t\boldsymbol{X}_t^H)^{-1} \tag{9.4.1}$$

For colored noise, the noise covariance estimate is given by:

$$\hat{\boldsymbol{\Lambda}} = \frac{1}{N_t}\boldsymbol{Y}_t\Pi^{\perp}_{\boldsymbol{X}_t^H}\boldsymbol{Y}_t^H \tag{9.4.2}$$

For white noise, the noise variance estimate is:

$$\hat{\sigma}^2 = \frac{1}{N_t n_r}\operatorname{Tr}\left\{\boldsymbol{Y}_t\Pi^{\perp}_{\boldsymbol{X}_t^H}\boldsymbol{Y}_t^H\right\} \tag{9.4.3}$$

Hence, to summarize the training-based ML algorithm:

Step 1. Obtain estimates of $\boldsymbol{H}$ and of the noise parameters based on the training block $\boldsymbol{Y}_t$ using (9.4.1)–(9.4.3).

Step 2. Use the so-obtained estimates in the coherent detector (see Section 9.1) to detect the data symbols.

9.4.1 Optimal Training for White Noise

In this subsection we address the question of choosing the training block $\boldsymbol{X}_t$ so as to optimize the system performance. Finding the $\boldsymbol{X}_t$ that minimizes the BER of the system is a hard problem, but we can easily address the question of finding the $\boldsymbol{X}_t$ that minimizes the variance of the training-based channel estimate, and which should also minimize the performance difference between coherent and training-based detection.

The ML estimate of the channel $\boldsymbol{H}$ based on the received training block $\boldsymbol{Y}_t$ is given by (see (9.4.1)):

$$\hat{\boldsymbol{H}} = \boldsymbol{Y}_t\boldsymbol{X}_t^H(\boldsymbol{X}_t\boldsymbol{X}_t^H)^{-1} \tag{9.4.4}$$

Since

$$\boldsymbol{Y}_t = \boldsymbol{H}\boldsymbol{X}_t + \boldsymbol{E}_t \tag{9.4.5}$$

we have that

$$E[\hat{\boldsymbol{H}}] = \boldsymbol{H}\boldsymbol{X}_t\boldsymbol{X}_t^H(\boldsymbol{X}_t\boldsymbol{X}_t^H)^{-1} = \boldsymbol{H} \tag{9.4.6}$$

so the channel estimate in (9.4.4) is unbiased. To derive its covariance matrix, let

$$\begin{aligned} \boldsymbol{h} &= \mathrm{vec}(\boldsymbol{H}) \\ \hat{\boldsymbol{h}} &= \mathrm{vec}(\hat{\boldsymbol{H}}) \\ \boldsymbol{e}_t &= \mathrm{vec}(\boldsymbol{E}_t) \end{aligned} \tag{9.4.7}$$

Since

$$E[\boldsymbol{e}_t\boldsymbol{e}_t^H] = \sigma^2\boldsymbol{I} \tag{9.4.8}$$

it follows that

$$\begin{aligned} \boldsymbol{\Sigma} &\triangleq E[(\hat{\boldsymbol{h}} - \boldsymbol{h})(\hat{\boldsymbol{h}} - \boldsymbol{h})^H] \\ &= \left(\boldsymbol{X}_t^H(\boldsymbol{X}_t\boldsymbol{X}_t^H)^{-1} \otimes \boldsymbol{I}\right)^T E[\boldsymbol{e}_t\boldsymbol{e}_t^H]\left(\boldsymbol{X}_t^H(\boldsymbol{X}_t\boldsymbol{X}_t^H)^{-1} \otimes \boldsymbol{I}\right)^* \\ &= \sigma^2\Big(\big((\boldsymbol{X}_t\boldsymbol{X}_t^H)^{-T}(\boldsymbol{X}_t\boldsymbol{X}_t^H)^T(\boldsymbol{X}_t\boldsymbol{X}_t^H)^{-T}\big) \otimes \boldsymbol{I}\Big) \\ &= \sigma^2\Big((\boldsymbol{X}_t\boldsymbol{X}_t^H)^{-T} \otimes \boldsymbol{I}\Big) \end{aligned} \tag{9.4.9}$$

and hence that

$$\begin{aligned} \mathrm{Tr}\{\boldsymbol{\Sigma}\} &= n_r \mathrm{Tr}\{(\boldsymbol{X}_t\boldsymbol{X}_t^H)^{-1}\}\sigma^2 \\ |\boldsymbol{\Sigma}| &= |(\boldsymbol{X}_t\boldsymbol{X}_t^H)^{-1}|^{n_r}\sigma^{2n_rn_t} \end{aligned} \tag{9.4.10}$$

The covariance matrix $\boldsymbol{\Sigma}$ coincides with the Cramér-Rao bound [Söderström and Stoica, 1989, Sec. B.4] [Kay, 1993, Chap. 3], so the channel estimate in (9.4.4) is statistically efficient.

We can design the training block $\boldsymbol{X}_t$ by minimizing $\mathrm{Tr}\{\boldsymbol{\Sigma}\}$, or the corresponding determinant $|\boldsymbol{\Sigma}|$, in (9.4.10) under a power constraint

$$\mathrm{Tr}\{\boldsymbol{X}_t\boldsymbol{X}_t^H\} \leq n_t \tag{9.4.11}$$

Minimizing the trace of the error covariance has a natural physical interpretation as the trace corresponds to the sum of the variances of the error components; on the other hand, the determinant can be interpreted as the volume of a hyper-ellipsoid within which the error vector lies with a certain probability. To carry out the minimization of the error metrics, we need the following theorem.

Theorem 9.1: A result on optimal training.
Assume that $\boldsymbol{X}$ has n_t linearly independent rows and that

$$\mathrm{Tr}\{\boldsymbol{X}\boldsymbol{X}^H\} \leq n_t \tag{9.4.12}$$

Then

(i) $\mathrm{Tr}\{(\boldsymbol{X}\boldsymbol{X}^H)^{-1}\} \geq n_t$ *with equality if and only if* $\boldsymbol{X}\boldsymbol{X}^H = \boldsymbol{I}$

(ii) $|(\boldsymbol{X}\boldsymbol{X}^H)^{-1}| \geq 1$ *with equality if and only if* $\boldsymbol{X}\boldsymbol{X}^H = \boldsymbol{I}$

Proof: To show part (i), note that by the Cauchy-Schwarz inequality:

$$\begin{aligned}&\mathrm{Tr}\{\boldsymbol{X}\boldsymbol{X}^H\}\cdot\mathrm{Tr}\{(\boldsymbol{X}\boldsymbol{X}^H)^{-1}\}\\ \geq&\Big(\mathrm{Tr}\{(\boldsymbol{X}\boldsymbol{X}^H)^{1/2}\cdot(\boldsymbol{X}\boldsymbol{X}^H)^{-1/2}\}\Big)^2\\ =&(\mathrm{Tr}\{\boldsymbol{I}_{n_t}\})^2 = n_t^2\end{aligned} \tag{9.4.13}$$

It follows that

$$\mathrm{Tr}\{(\boldsymbol{X}\boldsymbol{X}^H)^{-1}\} \geq \frac{n_t^2}{\mathrm{Tr}\{\boldsymbol{X}\boldsymbol{X}^H\}} \geq \frac{n_t^2}{n_t} = n_t \tag{9.4.14}$$

with equality if and only if $\boldsymbol{X}\boldsymbol{X}^H = \boldsymbol{I}$.

To prove part (ii), let $\lambda_1, \ldots, \lambda_{n_t}$ be the eigenvalues of $\boldsymbol{X}\boldsymbol{X}^H$. By the arithmetic-geometric mean inequality,

$$\begin{aligned}|(\boldsymbol{X}\boldsymbol{X}^H)^{-1}| &= \Big(\prod_{m=1}^{n_t}\lambda_m\Big)^{-1} \geq \Big(\frac{1}{n_t}\sum_{m=1}^{n_t}\lambda_m\Big)^{-n_t}\\ &= \Big(\frac{1}{n_t}\mathrm{Tr}\{\boldsymbol{X}\boldsymbol{X}^H\}\Big)^{-n_t} \geq 1\end{aligned} \tag{9.4.15}$$

with equality if and only if $\lambda_1 = \cdots = \lambda_{n_t} = 1$. Consequently, the semi-unitary choice of $\boldsymbol{X}$ also minimizes the determinant of $(\boldsymbol{X}\boldsymbol{X}^H)^{-1}$. ■

Application of Theorem 9.1 shows that the optimal $\boldsymbol{X}_t$ must be semi-unitary:

$$\boldsymbol{X}_t\boldsymbol{X}_t^H \propto \boldsymbol{I} \tag{9.4.16}$$

A related optimality result was derived in [Hassibi and Hochwald, 2000], by considering an information theoretic criterion instead of the covariance of the channel estimate. The result (9.4.16) has the consequence that an arbitrary OSTBC matrix $\boldsymbol{X}_t$, or a concatenation of several such matrices, can be used for optimal training. The optimal training does not depend on either the true channel $\boldsymbol{H}$ or the noise variance σ^2, and this is a pleasing property.

9.4.2 Optimal Training for Colored Noise

In the case of colored noise, we should consider the accuracy of $\boldsymbol{H}$ as well as that of $\boldsymbol{\Lambda}$ when designing the optimal training block $\boldsymbol{X}_t$. Fortunately, the estimation accuracy of $\boldsymbol{\Lambda}$ does not depend on $\boldsymbol{X}_t$. This result is known in the linear regression literature [ANDERSON, 1971], but was also established in the present context in [LARSSON ET AL., 2003]. Our derivation follows that in the latter reference.

Consider the projection matrix $\Pi^{\perp}_{\boldsymbol{X}_t^H}$. Since $\Pi^{\perp}_{\boldsymbol{X}_t^H}$ has rank $N_t - n_t$, there is an $N_t \times (N_t - n_t)$ matrix $\boldsymbol{U}$ such that

$$\Pi^{\perp}_{\boldsymbol{X}_t^H} = \boldsymbol{U}\boldsymbol{U}^H \tag{9.4.17}$$

and

$$\boldsymbol{U}^H\boldsymbol{U} = \boldsymbol{I} \tag{9.4.18}$$

Let

$$\boldsymbol{Z} \triangleq \boldsymbol{U}^H\boldsymbol{E}_t^H \tag{9.4.19}$$

Then

$$\text{vec}(\boldsymbol{Z}) = (\boldsymbol{I} \otimes \boldsymbol{U}^H)\,\text{vec}(\boldsymbol{E}_t^H) \tag{9.4.20}$$

and consequently

$$\begin{aligned} E[\text{vec}(\boldsymbol{Z})] &= \boldsymbol{0} \\ E\Big[\text{vec}(\boldsymbol{Z})\big(\text{vec}(\boldsymbol{Z})\big)^H\Big] &= (\boldsymbol{I} \otimes \boldsymbol{U}^H)E[\text{vec}(\boldsymbol{E}_t^H)\big(\text{vec}(\boldsymbol{E}_t^H)\big)^H](\boldsymbol{I} \otimes \boldsymbol{U}) \\ &= (\boldsymbol{I} \otimes \boldsymbol{U}^H)(\boldsymbol{\Lambda} \otimes \boldsymbol{I})(\boldsymbol{I} \otimes \boldsymbol{U}) \\ &= (\boldsymbol{\Lambda} \otimes \boldsymbol{I}) \end{aligned} \tag{9.4.21}$$

Hence $\text{vec}(\boldsymbol{Z})$ is a zero-mean complex Gaussian vector with a covariance matrix that depends only on $\boldsymbol{\Lambda}$ (but not on $\boldsymbol{X}_t$). It follows that the statistical distribution of

$$\hat{\boldsymbol{\Lambda}} = \frac{1}{N_t}\boldsymbol{Y}_t\Pi^{\perp}_{\boldsymbol{X}_t^H}\boldsymbol{Y}_t^H = \frac{1}{N_t}\boldsymbol{E}_t\Pi^{\perp}_{\boldsymbol{X}_t^H}\boldsymbol{E}_t^H = \frac{1}{N_t}\boldsymbol{E}_t\boldsymbol{U}\boldsymbol{U}^H\boldsymbol{E}_t^H = \frac{1}{N_t}\boldsymbol{Z}^H\boldsymbol{Z} \tag{9.4.22}$$

depends only on $\boldsymbol{\Lambda}$ (but not on $\boldsymbol{X}_t$ nor on $\boldsymbol{H}$).

The LS channel estimate is given by (9.4.1):

$$\hat{\boldsymbol{H}} = \boldsymbol{Y}_t\boldsymbol{X}_t^H(\boldsymbol{X}_t\boldsymbol{X}_t^H)^{-1} \tag{9.4.23}$$

Its expectation and covariance are easily derived (cf. Section 9.4.1)

$$
\begin{aligned}
E[\hat{\boldsymbol{H}}] &= \boldsymbol{H}\boldsymbol{X}_t\boldsymbol{X}_t^H(\boldsymbol{X}_t\boldsymbol{X}_t^H)^{-1} = \boldsymbol{H} \\
\boldsymbol{\Sigma} &\triangleq E[(\hat{\boldsymbol{h}} - \boldsymbol{h})(\hat{\boldsymbol{h}} - \boldsymbol{h})^H] \\
&= (\boldsymbol{X}_t^H(\boldsymbol{X}_t\boldsymbol{X}_t^H)^{-1} \otimes \boldsymbol{I})^T E[\boldsymbol{e}_t\boldsymbol{e}_t^H](\boldsymbol{X}_t^H(\boldsymbol{X}_t\boldsymbol{X}_t^H)^{-1} \otimes \boldsymbol{I})^* \\
&= ((\boldsymbol{X}_t\boldsymbol{X}_t^H)^{-T}(\boldsymbol{X}_t\boldsymbol{X}_t^H)^T(\boldsymbol{X}_t\boldsymbol{X}_t^H)^{-T}) \otimes \boldsymbol{\Lambda} \\
&= (\boldsymbol{X}_t\boldsymbol{X}_t^H)^{-T} \otimes \boldsymbol{\Lambda}
\end{aligned}
\tag{9.4.24}
$$

It follows that

$$
\begin{aligned}
\mathrm{Tr}\{\boldsymbol{\Sigma}\} &= \mathrm{Tr}\{(\boldsymbol{X}_t\boldsymbol{X}_t^H)^{-1}\}\mathrm{Tr}\{\boldsymbol{\Lambda}\} \\
|\boldsymbol{\Sigma}| &= |(\boldsymbol{X}_t\boldsymbol{X}_t^H)^{-1}|^{n_r}|\boldsymbol{\Lambda}|^{n_t}
\end{aligned}
\tag{9.4.25}
$$

In the same way as in Section 9.4.1, we can use Theorem 9.1 (see page 176) to minimize $\mathrm{Tr}\{\boldsymbol{\Sigma}\}$ or $|\boldsymbol{\Sigma}|$ in (9.4.25) under the power constraint

$$
\mathrm{Tr}\{\boldsymbol{X}_t\boldsymbol{X}_t^H\} \leq n_t \tag{9.4.26}
$$

The result is the same as for white noise ($\boldsymbol{\Lambda} = \sigma^2\boldsymbol{I}$): the optimal training block $\boldsymbol{X}_t$ should satisfy

$$
\boldsymbol{X}_t\boldsymbol{X}_t^H \propto \boldsymbol{I} \tag{9.4.27}
$$

This is an appealing result: a semi-unitary training block is optimal regardless of the channel and the noise covariance, and regardless of whether the receiver assumes white or colored noise.

9.4.3 Training for Frequency-Selective Channels

Estimates of the parameters associated with a frequency-selective channel can be acquired by training in a manner similar to what we described for flat fading channels above. In principle, a distinction can be made between the case when the time-domain impulse response $\boldsymbol{H}(z^{-1})$ is of interest, and the case when an estimate of the transfer function $\boldsymbol{H}(\omega)$ is desired. The former case is relevant for time-domain equalization (such as for TR-OSTBC in Section 8.3), whereas the latter case is relevant for systems that process the received data in the frequency-domain (such as OFDM and ST-OFDM in Sections 5.2.2 and 8.2).

Training for Time-Domain Equalized Systems: Estimation of $\boldsymbol{H}(z^{-1})$

Let $\boldsymbol{x}_t(n)$, $n = 0, \ldots, N_t - 1$ be a set of consecutively transmitted training vectors of length n_t, and let $\boldsymbol{y}_t(n)$ be the corresponding received data. For reasons similar to those discussed in Chapter 2, the received data is well-defined only for $n = L, \ldots, N_t - 1$ where L is the maximum excess delay of the channel. Arranging the data in the same way as in Section 2.2.2 we obtain:

$$\boldsymbol{y}_t = \begin{bmatrix} \boldsymbol{y}^T(L) & \cdots & \boldsymbol{y}^T(N_t - 1) \end{bmatrix}^T = (\boldsymbol{X}_t^T \otimes \boldsymbol{I}_{n_r})\boldsymbol{h} + \boldsymbol{e}_t \tag{9.4.28}$$

where

$$\boldsymbol{e}_t = \begin{bmatrix} \boldsymbol{e}^T(L) & \cdots & \boldsymbol{e}^T(N_t - 1) \end{bmatrix}^T \tag{9.4.29}$$

is a noise vector,

$$\boldsymbol{h} = \operatorname{vec}\left(\begin{bmatrix} \boldsymbol{H}_0 & \cdots & \boldsymbol{H}_L \end{bmatrix}\right) \tag{9.4.30}$$

and the matrix $\boldsymbol{X}_t$ is built from the training data $\{\boldsymbol{x}_t(n)\}_{n=0}^{N_t-1}$ according to (cf. (2.2.11)):

$$\boldsymbol{X}_t = \begin{bmatrix} \boldsymbol{x}_t(L) & \cdots & \cdots & \cdots & \boldsymbol{x}_t(N_t - 1) \\ \boldsymbol{x}_t(L-1) & \ddots & & & \boldsymbol{x}_t(N_t - 2) \\ \vdots & & \ddots & & \vdots \\ \boldsymbol{x}_t(0) & \cdots & \cdots \quad \boldsymbol{x}_t(L) & \cdots & \boldsymbol{x}_t(N_t - L - 1) \end{bmatrix} \tag{9.4.31}$$

Then ML estimation of the channel $\boldsymbol{h}$ from the received training data $\boldsymbol{y}_t$ is equivalent to minimizing the following metric (see (8.1.4)):

$$\|\boldsymbol{y}_t - (\boldsymbol{X}_t^T \otimes \boldsymbol{I}_{n_r})\boldsymbol{h}\|^2 \tag{9.4.32}$$

The minimizer of (9.4.32) is given by:

$$\hat{\boldsymbol{h}} = ((\boldsymbol{X}_t^* \boldsymbol{X}_t^T)^{-1} \boldsymbol{X}_t^* \otimes \boldsymbol{I}_{n_r})\boldsymbol{y}_t \tag{9.4.33}$$

It is not hard to prove that, assuming that $E[\boldsymbol{e}_t \boldsymbol{e}_t^H] = \sigma^2 \boldsymbol{I}$,

$$\begin{aligned} E[\hat{\boldsymbol{h}}] &= \boldsymbol{h} \\ E[(\hat{\boldsymbol{h}} - \boldsymbol{h})(\hat{\boldsymbol{h}} - \boldsymbol{h})^H] &= \sigma^2\left((\boldsymbol{X}_t^* \boldsymbol{X}_t^T)^{-1} \otimes \boldsymbol{I}_{n_r}\right) \end{aligned} \tag{9.4.34}$$

Consequently, the channel estimate is unbiased with a covariance matrix that depends on the training data. The covariance matrix of $\hat{\boldsymbol{h}}$ is minimized when (cf. Theorem 9.1 on page 176):

$$\boldsymbol{X}_t \boldsymbol{X}_t^H \propto \boldsymbol{I} \tag{9.4.35}$$

This happens at least approximately when the sequences of training data transmitted via the different antennas are mutually uncorrelated white signals. Since $\boldsymbol{X}_t$ has a block-Toeplitz structure, it may not be possible to achieve (9.4.35) exactly, but at least the above result provides some guidelines for the design of a quasi-optimal training block.

Training for Frequency-Domain Equalized Systems: Estimation of $\boldsymbol{H}(\omega)$

For a system that uses frequency-domain equalization (with OFDM as an example), it may be desirable to obtain an estimate of the channel transfer function $\boldsymbol{H}(\omega)$ rather than the impulse response $\boldsymbol{H}(z^{-1})$. This can be done in two ways: via estimation in the frequency domain, or via estimation in the time-domain followed by transformation to the frequency domain. As we will see, the latter technique is often more advantageous.

Frequency-Domain Estimation: Let N_0 be the number of subcarriers available for training and let

$$\boldsymbol{X}_t(n) = \left[\boldsymbol{x}_t^{(1)}(n) \quad \cdots \boldsymbol{x}_t^{(N)}(n)\right], \quad n = 0, \ldots, N_0 - 1 \tag{9.4.36}$$

be an $n_t \times N$ matrix that contains the training data associated with subcarrier n. If $\boldsymbol{Y}_t(n)$ is a matrix of dimension $n_r \times N$ that contains the corresponding received data, then the ML metric for estimation of the channel transfer function associated with subcarrier n is given by (cf. Theorem 8.1 (see page 134)):

$$\left\|\boldsymbol{Y}_t(n) - \boldsymbol{H}\left(\frac{2\pi}{N_0}n\right)\boldsymbol{X}_t(n)\right\|^2 \tag{9.4.37}$$

Minimization of (9.4.37) yields

$$\hat{\boldsymbol{H}}\left(\frac{2\pi}{N_0}n\right) = \boldsymbol{Y}_t(n)\boldsymbol{X}_t^H(n)(\boldsymbol{X}_t(n)\boldsymbol{X}_t^H(n))^{-1} \tag{9.4.38}$$

Equation (9.4.38) is of the same form as (9.4.4). Therefore, we conclude that an optimal training block should satisfy:

$$\boldsymbol{X}_t(n)\boldsymbol{X}_t^H(n) \propto \boldsymbol{I} \tag{9.4.39}$$

Note that it may be the case that not all available subcarriers are being used for transmission (see, e.g., [VAN NEE ET AL., 1999]). In that case, it is not necessary to estimate the channel gain corresponding to a subcarrier that is not in use. This is easy to accomplish in the present case simply by not transmitting any pilots on these subcarriers.

Time-Domain Estimation: The procedure for estimating $\boldsymbol{H}(\omega)$ outlined in the above paragraph is not optimal in general. This is so since the physical propagation channel $\boldsymbol{H}(z^{-1})$ (in the time-domain) is determined by only $n_t n_r(L+1)$ parameters, while the method outlined above estimates $n_t n_r N_0$ parameters and typically $N_0 \gg L + 1$. According to the parsimony principle [SÖDERSTRÖM AND STOICA, 1989, SEC. 11.4], it should therefore be more advantageous to estimate the time-domain propagation coefficients and thereafter transform them to the frequency domain, instead of estimating the frequency domain propagation coefficients directly. This observation has been exploited frequently in the literature (see, e.g., [LI ET AL., 1999]).

The time-domain and the frequency-domain propagation coefficients are related in the following way:

$$\boldsymbol{H}\Big(\frac{2\pi}{N_0}k\Big) = \frac{1}{\sqrt{N_0}}\sum_{l=0}^{L}\boldsymbol{H}_l \exp\Big(-i\frac{2\pi}{N_0}kl\Big) \tag{9.4.40}$$

Let

$$\begin{aligned} \boldsymbol{h}_f &= \operatorname{vec}\Big(\Big[\boldsymbol{H}(0) \quad \cdots \quad \boldsymbol{H}\Big(\tfrac{2\pi}{N_0}(N_0-1)\Big)\Big]\Big) \\ \boldsymbol{h}_t &= \operatorname{vec}\Big(\big[\boldsymbol{H}_0 \quad \cdots \quad \boldsymbol{H}_L\big]\Big) \end{aligned} \tag{9.4.41}$$

and let $\boldsymbol{T}$ be the $N_0 \times L$ matrix whose $(k, l)th$ element is given by

$$\boldsymbol{T}_{k,l} = \frac{1}{\sqrt{N_0}}\exp\Big(-i2\pi\frac{(k-1)(l-1)}{N_0}\Big) \tag{9.4.42}$$

(note that in general $\boldsymbol{T}^H\boldsymbol{T} = \boldsymbol{I}$ but $\boldsymbol{T}\boldsymbol{T}^H \neq \boldsymbol{I}$). Then (9.4.40) can be written

$$\boldsymbol{h}_f = (\boldsymbol{T} \otimes \boldsymbol{I}_{n_r n_t})\boldsymbol{h}_t \tag{9.4.43}$$

Let

$$\boldsymbol{y}_t = \operatorname{vec}\big(\big[\boldsymbol{Y}_t(0) \quad \cdots \quad \boldsymbol{Y}_t(N_0-1)\big]\big) \tag{9.4.44}$$

The ML metric for channel estimation becomes (cf. Theorem 8.1 on page 134):

$$\begin{aligned} &\sum_{n=0}^{N_0-1}\Big\|\boldsymbol{Y}_t(n) - \boldsymbol{H}\Big(\frac{2\pi}{N_0}n\Big)\boldsymbol{X}_t(n)\Big\|^2 \\ &= \|\boldsymbol{y}_t - (\boldsymbol{X}_t \otimes \boldsymbol{I}_{n_r})\boldsymbol{h}_f\|^2 \\ &= \|\boldsymbol{y}_t - (\boldsymbol{X}_t \otimes \boldsymbol{I}_{n_r})(\boldsymbol{T} \otimes \boldsymbol{I}_{n_r n_t})\boldsymbol{h}_t\|^2 \end{aligned} \tag{9.4.45}$$

where

$$\boldsymbol{X}_t = \begin{bmatrix} \boldsymbol{X}_t^T(0) & \mathbf{0} & \cdots & \mathbf{0} \\ \mathbf{0} & \boldsymbol{X}_t^T(1) & \ddots & \vdots \\ \vdots & \ddots & \ddots & \mathbf{0} \\ \mathbf{0} & \cdots & \mathbf{0} & \boldsymbol{X}_t^T(N_0-1) \end{bmatrix} \tag{9.4.46}$$

Estimation of $\boldsymbol{h}_t$ by minimizing (9.4.45) yields

$$\begin{aligned} \hat{\boldsymbol{h}}_t &= \left[(\boldsymbol{T}^H \otimes \boldsymbol{I}_{n_r n_t})(\boldsymbol{X}_t^H \otimes \boldsymbol{I}_{n_r})(\boldsymbol{X}_t \otimes \boldsymbol{I}_{n_r})(\boldsymbol{T} \otimes \boldsymbol{I}_{n_r n_t})\right]^{-1} \\ &\quad \cdot (\boldsymbol{T}^H \otimes \boldsymbol{I}_{n_r n_t})(\boldsymbol{X}_t^H \otimes \boldsymbol{I}_{n_r})\boldsymbol{y}_t \\ &= \left[\left(\left((\boldsymbol{T}^H \otimes \boldsymbol{I}_{n_t})\boldsymbol{X}_t^H \boldsymbol{X}_t (\boldsymbol{T} \otimes \boldsymbol{I}_{n_t})\right)^{-1} (\boldsymbol{T}^H \otimes \boldsymbol{I}_{n_t})\boldsymbol{X}_t^H\right) \otimes \boldsymbol{I}_{n_r}\right] \boldsymbol{y}_t \end{aligned} \tag{9.4.47}$$

and finally the frequency-domain coefficients can be computed via

$$\begin{aligned} \hat{\boldsymbol{h}}_f &= (\boldsymbol{T} \otimes \boldsymbol{I}_{n_r n_t})\hat{\boldsymbol{h}}_t \\ &= \left[\left((\boldsymbol{T} \otimes \boldsymbol{I}_{n_t})\left((\boldsymbol{T}^H \otimes \boldsymbol{I}_{n_t})\boldsymbol{X}_t^H \boldsymbol{X}_t (\boldsymbol{T} \otimes \boldsymbol{I}_{n_t})\right)^{-1} (\boldsymbol{T}^H \otimes \boldsymbol{I}_{n_t})\boldsymbol{X}_t^H\right) \otimes \boldsymbol{I}_{n_r}\right] \boldsymbol{y}_t \end{aligned} \tag{9.4.48}$$

Example 9.3: Equivalence of time and frequency-domain estimation for $L = N_0$.
If $L = N_0$, the time-domain channel estimation and the frequency-domain channel estimation procedures become equivalent. To see this, note that in this case, $\boldsymbol{T}$ is square and unitary:

$$\boldsymbol{T}\boldsymbol{T}^H = \boldsymbol{I} \tag{9.4.49}$$

Hence, (9.4.48) becomes:

$$\hat{\boldsymbol{h}}_f = \left[\left((\boldsymbol{X}_t^H \boldsymbol{X}_t)^{-1} \boldsymbol{X}_t^H\right) \otimes \boldsymbol{I}_{n_r}\right] \boldsymbol{y}_t \tag{9.4.50}$$

or equivalently,

$$\hat{\boldsymbol{H}}\left(\frac{2\pi}{N_0}n\right) = \boldsymbol{Y}_t(n)\boldsymbol{X}_t^H(n)\left(\boldsymbol{X}_t(n)\boldsymbol{X}_t^H(n)\right)^{-1} \tag{9.4.51}$$

which is (9.4.38). ∎

Finally we note that the design of optimal training for ST-OFDM has been discussed in, for instance, [Li et al., 1999], [Tung et al., 2001], [Negi and Cioffi, 1998] and [Larsson and Li, 2001]. The first three references treat the problem in a general context, while the last paper focuses on the case when not all subcarriers are used.

9.5 Blind and Semi-Blind Detection Methods

In this section we describe two recent methods that can have some performance advantages relative to the training-based ML. These methods are called *blind* and *semi-blind*, respectively, and were proposed in [STOICA AND GANESAN, 2003], [LARSSON ET AL., 2003]; however, our presentation is slightly different from that in these articles. Related methods were also suggested in, for instance, [SWINDLEHURST AND LEUS, 2002], [MOGUEZ AND CASTEDO, 2002]. Note that the notion of blind channel estimation for communications, based on second-order data, goes back at least to [TONG ET AL., 1994] (see also the references therein).

The common core of the methods in [STOICA AND GANESAN, 2003], [LARSSON ET AL., 2003] is an iterative scheme that attempts to obtain the minimum of the ML metrics (9.1.4) and (9.1.12) by means of a cyclic minimizer; the blind and the semi-blind schemes differ simply in the way this cyclic minimizer is initialized. The complexity of the blind and semi-blind methods is relatively low and they are applicable regardless of whether the symbols belong to a unitary constellation or not. These methods are applicable to STBC in general, but the detection step in the algorithm simplifies if an OSTBC is used.

9.5.1 Cyclic Minimization of the ML Metric

The cyclic iterative scheme for minimizing the ML metric is outlined below and illustrated in Figure 9.2.

1. Obtain an initial estimate of the channel $\boldsymbol{H}$ and the noise covariance matrix $\boldsymbol{\Lambda}$, for instance blindly or by using training.

2. Use the so-obtained channel estimate in the coherent detector (see Section 9.1).

3. Re-estimate the channel using the detected symbols from Step 2, via:

$$\hat{\boldsymbol{H}} = \boldsymbol{Y}\hat{\boldsymbol{X}}(\hat{\boldsymbol{X}}\hat{\boldsymbol{X}}^H)^{-1} \tag{9.5.1}$$

(see (9.3.2)) and the noise covariance via:

$$\hat{\boldsymbol{\Lambda}} = \frac{1}{N_t + N_b N}\boldsymbol{Y}\Pi^{\perp}_{\hat{\boldsymbol{X}}^H}\boldsymbol{Y}^H \tag{9.5.2}$$

(see (9.3.24)), for colored noise, or via:

$$\hat{\sigma}^2 = \frac{1}{(N_t + N_b N)n_r}\operatorname{Tr}\left\{\boldsymbol{Y}\Pi^{\perp}_{\hat{\boldsymbol{X}}^H}\boldsymbol{Y}^H\right\} \tag{9.5.3}$$

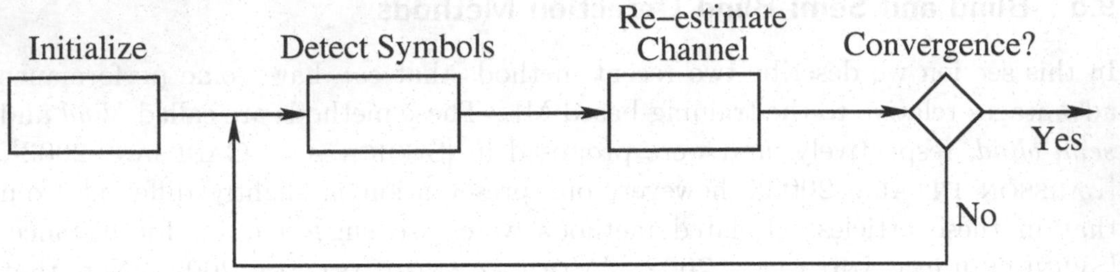

Figure 9.2. Flowchart of the cyclic detection/estimation scheme.

(see (9.3.4)), for white noise. Here $\hat{\boldsymbol{X}}$ is the concatenation of the STBC blocks (see (9.1.1)) corresponding to the most recent estimate of the symbols.

4. Iterate until convergence or until a pre-imposed number of steps have been carried out.

Note that, the initialization step aside, the above algorithm is nothing but a cyclic minimizer of the ML metric. A cyclic minimizer [ZANGWILL, 1967] minimizes an objective function by partitioning the free variables into two groups Ω_1, Ω_2 and proceeding as follows. First, minimize the objective function over Ω_1 while keeping Ω_2 fixed. Next, minimize the function over Ω_2 while keeping Ω_1 fixed at its most recently determined value. Finally, iterate these two steps until convergence. As is known, and as we explain in the following, such a cyclic minimizer will converge under general conditions.

For simplicity, let us consider the ML metric (9.1.5) associated with white noise; similar results hold in the case of colored noise. Let f_k be the value of the ML metric (9.1.5) after k iterations and let f_* be the value of it at its minimum. Of course, $f_k \geq f_*$ for all k. We have that

$$f_* = \min_{\boldsymbol{X}} \operatorname{Tr}\left\{\boldsymbol{Y}\Pi^{\perp}_{\boldsymbol{X}^H}\boldsymbol{Y}^H\right\} \tag{9.5.4}$$

and consequently $f_* > 0$ with probability one. Since each of the Steps 2 and 3 in the cyclic minimization algorithm above can only decrease the value of the objective function, we have that $f_k \leq f_{k-1}$ for all $k \geq 1$. It follows that $\{f_k\}$ forms a monotonically non-increasing sequence that must converge to

$$f_\infty \triangleq \inf\{f_k\} \geq f_* \tag{9.5.5}$$

(see, e.g., [RUDIN, 1976]). This value f_∞ corresponds to the value of the ML metric at the point of convergence, which must be at least a local minimum of the ML metric. Whether the so-obtained point of convergence is equal to the *global* minimum of the ML metric is a more difficult question — note that in general for nonlinear optimization problems the global minimum of a non-convex function is usually hard to obtain. However, since the cyclic algorithm is initialized reasonably well, and since the iteration can only decrease the value of the ML metric, we expect the cyclic algorithm to perform well in the sense that the point of convergence should at least be close to the global minimum of the ML metric.

Example 9.4: Blind detection of OSTBC.
The article [Stoica and Ganesan, 2003] suggested a *blind* approach to the detection of OSTBC, that does not require any training block. Let us define $\boldsymbol{h} = \text{vec}(\boldsymbol{H})$. Then the ML metric in (9.1.5) can be rewritten as:

$$\begin{aligned} \|\boldsymbol{Y} - \boldsymbol{H}\boldsymbol{X}\|^2 &= \text{Tr}\{\boldsymbol{Y}\boldsymbol{Y}^H\} + \|\boldsymbol{H}\|^2\|\boldsymbol{s}'\|^2 - 2\text{Re}\{\text{Tr}\{\boldsymbol{Y}^H\boldsymbol{H}\boldsymbol{X}\}\} \\ &= \|\boldsymbol{s}'\|^2\|\boldsymbol{h}\|^2 - 2\text{Re}\left\{\boldsymbol{h}^H\boldsymbol{F}\boldsymbol{s}'\right\} + \text{const.} \\ &= \|\boldsymbol{F} - \boldsymbol{h}\boldsymbol{s}'^T\|^2 + \text{const.} \end{aligned} \tag{9.5.6}$$

where $\boldsymbol{s}'$ is defined in (9.1.9) and $\boldsymbol{F}$ is defined by

$$\begin{aligned} \boldsymbol{F} &= \left[\boldsymbol{F}_1^{(a)} \;\cdots\; \boldsymbol{F}_{N_b}^{(a)} \;\; \boldsymbol{F}_1^{(b)} \;\cdots\; \boldsymbol{F}_{N_b}^{(b)}\right] \\ \boldsymbol{F}_k^{(a)} &= \left[\text{vec}(\boldsymbol{Y}_k\boldsymbol{A}_1^H) \;\cdots\; \text{vec}(\boldsymbol{Y}_k\boldsymbol{A}_{n_s}^H)\right], \quad k = 1, \ldots, N_b \\ \boldsymbol{F}_k^{(b)} &= \left[-i\,\text{vec}(\boldsymbol{Y}_k\boldsymbol{B}_1^H) \;\cdots\; -i\,\text{vec}(\boldsymbol{Y}_k\boldsymbol{B}_{n_s}^H)\right], \quad k = 1, \ldots, N_b \end{aligned} \tag{9.5.7}$$

To perform ML detection, the metric $\|\boldsymbol{F} - \boldsymbol{h}\boldsymbol{s}'^T\|^2$ must be minimized jointly with respect to $\boldsymbol{h}$ and $\boldsymbol{s}'$. If we neglect the constraint that $\boldsymbol{s}'$ belongs to a finite alphabet, this minimum is achieved by taking $\boldsymbol{h}$ and $\boldsymbol{s}'$ equal to the left and right singular vectors of the matrix $\boldsymbol{F}$, which correspond to its largest singular value. Clearly, at least one element of $\boldsymbol{s}'$ or $\boldsymbol{h}$ must be known for these two singular vectors to uniquely define $\boldsymbol{h}$ and $\boldsymbol{s}'$. If both $\boldsymbol{s}'$ and $\boldsymbol{h}$ are unconstrained vectors, an ambiguity exists since $\alpha\boldsymbol{s}'$ and $(1/\alpha)\boldsymbol{h}$ would also minimize the same criterion. Hence, the blind algorithm requires (only) one pilot symbol. The so-obtained estimate of $\boldsymbol{h}$ can be used in the cyclic minimization algorithm described above. Note that if a training block is present, the blind approach is in principle still applicable (with some algebraic modifications), even though the training-based or semi-blind methods (i.e., estimation of the channel followed by iteration as in Figure 9.2) may be preferable for several reasons.

We can make a connection between the blind algorithm outlined in this section and the exact ML detection for OSTBC discussed in Example 9.2 (see page 170).

If no training is present and the constellation $|S|$ is unitary, the concentrated ML metric is proportional to $\boldsymbol{s}'^T\boldsymbol{F}^H\boldsymbol{F}\boldsymbol{s}'$ (see (9.3.9)). If, like for the blind method above, we relax the optimization problem neglecting the constraint that $\boldsymbol{s}'$ belongs to a finite alphabet, the symbol vector $\boldsymbol{s}'$ that minimizes $\boldsymbol{s}'^T\boldsymbol{F}^H\boldsymbol{F}\boldsymbol{s}'$ is given by the right singular vector of $\boldsymbol{F}$ that corresponds to the largest singular value of this matrix. This is, however, the same solution as given by the blind algorithm. Like for the blind algorithm, for this singular vector to uniquely define $\boldsymbol{s}'$, at least one of its elements must be known.

■

Example 9.5: Blind, trained, and semi-blind detection of OSTBC.
In this example (see [Stoica and Ganesan, 2003]) we consider a system with $n_t = 3$ transmit antennas and $n_r = 2$ receive antennas using the code (7.4.8). The elements of the channel matrix $\boldsymbol{H}$ are independent and zero-mean Gaussian with variance ρ^2, and the transmitted symbols belong to a BPSK constellation. The noise is white and has variance σ^2. A variable number, N_b, of OSTBC matrices were transmitted sequentially and preceded by one training block that consisted of a known OSTBC matrix. Figure 9.3 shows the BER versus the SNR for three different values of N_b, namely $N_b = 4$, 16 and 64. From the figure, we can see that the semi-blind detector is always better than the blind detector. This is not surprising since the former makes use of the training data, while the latter does not. However, for large values of N_b, the performance of the blind detector gets somewhat closer to that of the semi-blind one. We also observe that the BER for the semi-blind detector is smaller than that corresponding to the trained detector. This may be expected in view of the fact that the training-based detector is just the first step of the semi-blind one. ■

The next two examples (see also [Larsson et al., 2003]) will illustrate the application of the training-based and semi-blind detection methods in the presence of white and colored noise, respectively.

Example 9.6: Trained and semi-blind detection for OSTBC (white noise).
We consider a system with $n_t = 2$ transmit and $n_r = 2$ receive antennas using the Alamouti code and BPSK signalling. The channel noise is spatially and temporally white Gaussian, that is, the columns of $\boldsymbol{E}$ are independent outcomes of a $N(\boldsymbol{0}, \sigma^2\boldsymbol{I})$ distribution. The elements of the channel gain matrix $\boldsymbol{H}$ in different simulation runs are independent realizations of a Gaussian random process with mean zero and variance ρ^2, and the SNR is defined as ρ^2/σ^2. Figure 9.4 shows an estimate of the BER as a function of the SNR. We note that the semi-blind ML receiver

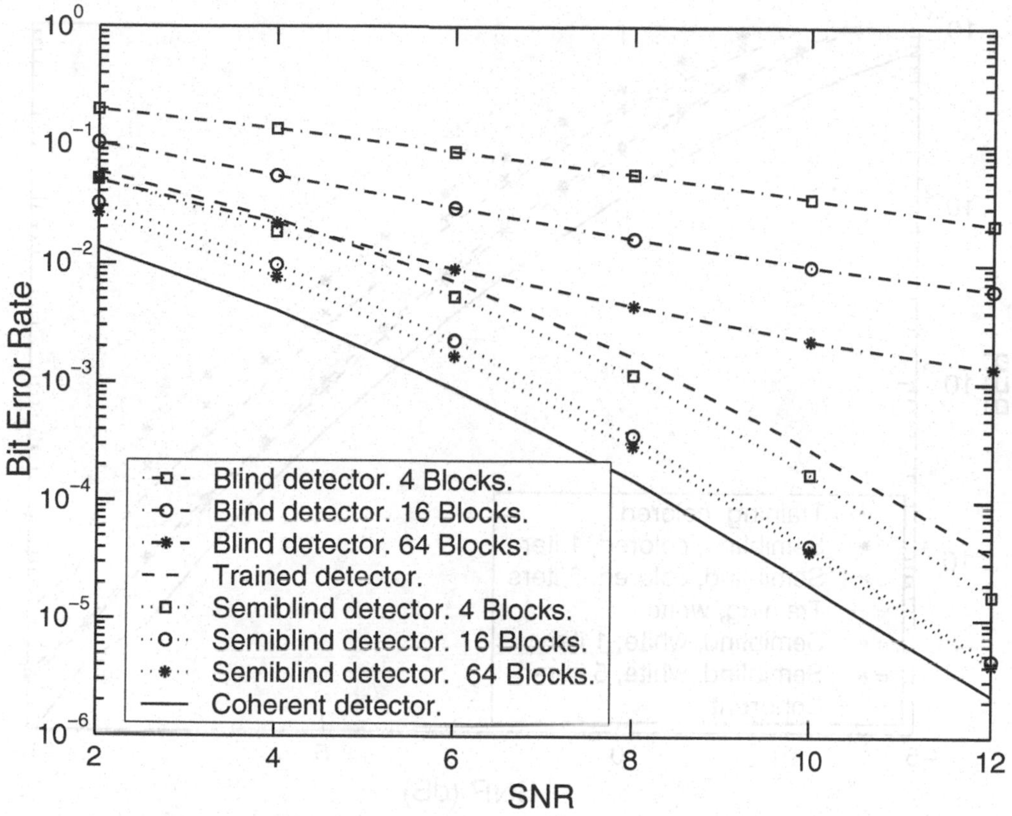

Figure 9.3. Results for Example 9.5. Coherent, training-based, semi-blind and blind detection for block lengths $N_b = 4$, 16 and 64.

outperforms the training-based algorithm by about 1 dB for the case when the receiver assumes the noise to be white, and by more than 2 dB in the case when the receiver assumes the noise to be colored. For the semi-blind ML receiver assuming white noise, the detection loss (i.e., the performance loss due to the need for estimating the channel), is less than 1 dB. This loss is somewhat higher for the receivers assuming colored noise (indeed, the receivers assuming white noise outperform those assuming colored noise by about 2 dB). The latter fact is not surprising since the receiver structures assuming colored noise are not parsimonious for this example. ■

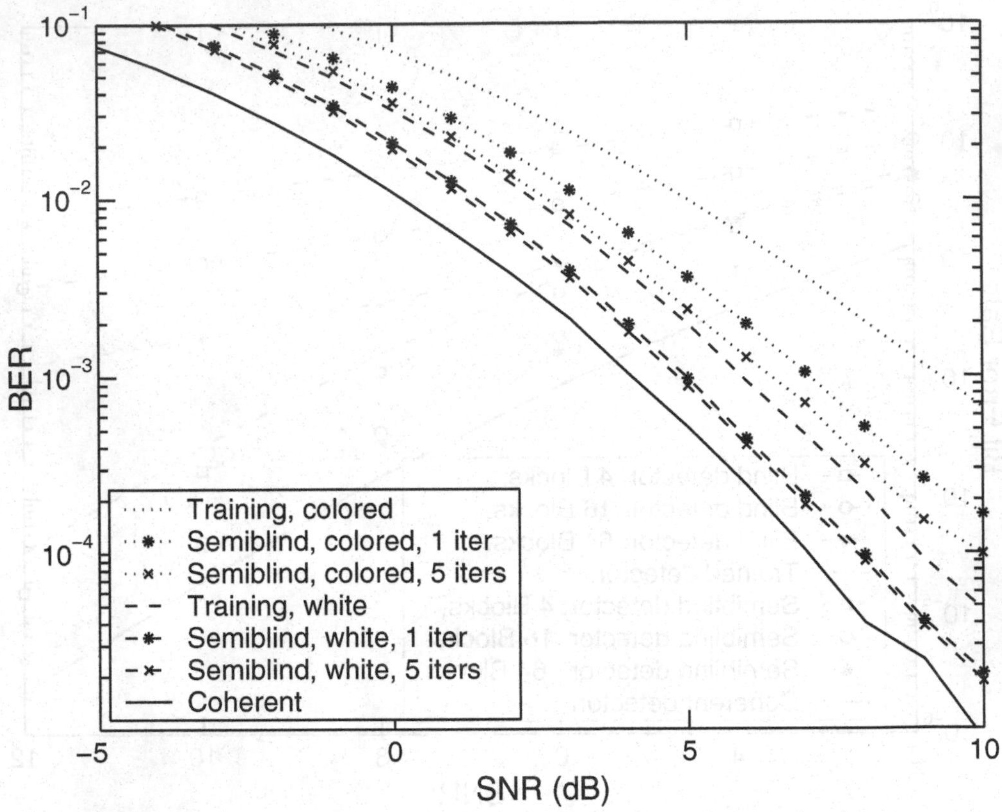

Figure 9.4. Results for Example 9.6. Only white noise is present. The figure shows the estimated bit-error-rate (BER) vs. the signal-to-noise ratio (SNR) for the coherent ML receiver (assuming perfect knowledge of $\boldsymbol{H}$ and $\boldsymbol{\Lambda}$), the training-based white and colored receivers, and the semiblind white and colored receivers using n iterations.

Example 9.7: Trained and semi-blind detection for OSTBC (spatially colored noise). In this example we consider the same system as in Example 9.6 but with one strong interference signal present. The interference signal is temporally white, has a power equal to that of the signal of interest and a channel gain vector which is also zero-mean Gaussian and independent of all other random quantities. White Gaussian noise is added to the received signal in the same way as for Example 9.6 above. Figure 9.5 shows the BER vs. the SNR for the different schemes (the SNR here is the ratio of the signal power to the power of the thermal noise, while the power of the interference signal is kept constant). We note that the semiblind algorithm outperforms the training-based method by more than 2 dB and that

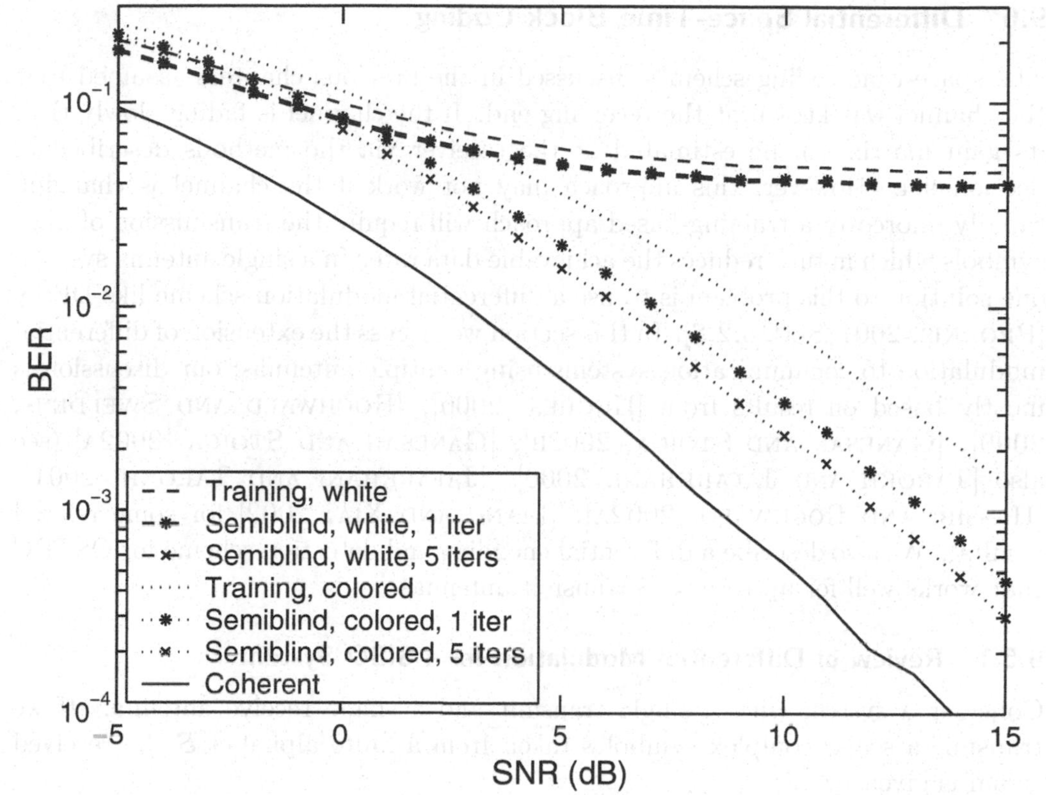

Figure 9.5. Results for Example 9.7. One strong interference signal is present.

the detection loss for the "colored" receiver is about 2 dB (which is of the same order as for Example 9.6). Note also that for high SNR, the receiver that assumes colored noise significantly outperforms the receiver that assumes white noise. In particular, the receivers relying on the white noise assumption seem to approach an error floor as SNR increases. On the other hand, when the SNR is sufficiently low, the true covariance matrix of the noise and interference is close to a scaled identity matrix, and for that reason the "white" receiver performs somewhat better than the "colored" receiver in this case. ■

9.6 Differential Space-Time Block Coding

The space-time coding schemes discussed in the previous chapters assumed that the channel was known at the receiving end. If the channel is fading slowly then its gain matrix can be estimated at the receiver via the methods described in Section 9.4. However, this approach may not work if the channel is changing rapidly; moreover a training-based approach will require the transmission of pilot symbols which in turn reduces the achievable data rate. In a single antenna system, one solution to this problem is to use a differential modulation scheme like DPSK [PROAKIS, 2001, SEC. 5.2.8]. In this section we discuss the extension of differential modulation to communication systems using multiple antennas; our discussion is mostly based on results from [HUGHES, 2000], [HOCHWALD AND SWELDENS, 2000], [GANESAN AND STOICA, 2002B], [GANESAN AND STOICA, 2002A] (see also [TAROKH AND JAFARKHANI, 2000], [JAFARKHANI AND TAROKH, 2001], [HASSIBI AND HOCHWALD, 2002A], [LIANG AND XIA, 2002] for some related results). We also describe a differential encoding and detection scheme for OSTBC that works well for up to $n_t = 8$ transmit antennas.

9.6.1 Review of Differential Modulation for a SISO System

Consider a system with a single transmit and a single receive antenna. If we transmit a scalar complex symbol s taken from a finite alphabet $\mathcal{S}$ the received signal is given by

$$y = hs + e \tag{9.6.1}$$

where h is the complex channel gain and e is noise. Assume that h is known to the receiver. Then if the noise is Gaussian, the ML detection of s given y is equivalent to minimizing:

$$|y - hs|^2 \tag{9.6.2}$$

with respect to $s \in \mathcal{S}$, which is a trivial scalar detection problem.

If h is *not* known at the receiver, we can proceed via *differential* encoding and detection as we describe next. The idea of differential encoding is to encode the information in the phase difference between two consecutively transmitted symbols, and to consider for detection a block of two consecutively received symbols at the receiver. Suppose $\{s_k\}$ is a stream of symbols to be transmitted. Instead of sending $\{s_k\}$ directly we form another set of symbols $\{w_k\}$ which is defined by:

$$\begin{aligned} w_k &= w_{k-1} s_k \\ w_0 &= 1 \end{aligned} \tag{9.6.3}$$

and transmit this sequence $\{w_k\}$ instead. At the beginning of the transmission we transmit $w_0 = 1$, and after that we transmit w_k which depends on s_k through (9.6.3). In a strict sense, differential encoding is meaningful only if the constellation $\mathcal{S}$ is unitary, i.e., $|s| = 1$ for all $s \in \mathcal{S}$, since in that case it follows immediately from (9.6.3) that $|w_k|^2 = 1$ for any k (which means that the transmit power does not vary with k). Hence we assume that $|s_k| = 1$ in the following.

To decode the symbols $\{s_k\}$ we proceed as follows. Let y_k and y_{k-1} be the received signals corresponding to w_k and w_{k-1} respectively. Then we have that

$$\begin{aligned} y_k &= hw_k + e_k \\ &= hw_{k-1}s_k + e_k \end{aligned} \tag{9.6.4}$$

and

$$y_{k-1} = hw_{k-1} + e_{k-1} \tag{9.6.5}$$

provided that h remains constant during the transmission of the two symbols. The ML detection of s_k given y_k and y_{k-1} is equivalent to minimizing

$$\begin{aligned} |y_{k-1} - hw_{k-1}|^2 + |y_k - hw_k|^2 &= |y_{k-1} - hw_{k-1}|^2 + |y_k - hw_{k-1}s_k|^2 \\ &= |y_{k-1} - g|^2 + |y_k - gs_k|^2 \\ &= |y_{k-1} - g|^2 + |s_k^* y_k - g|^2 \end{aligned} \tag{9.6.6}$$

where we defined $g = hw_{k-1}$. The minimization of (9.6.6) with respect to g gives

$$g = \frac{y_{k-1} + s_k^* y_k}{2} \tag{9.6.7}$$

Inserting (9.6.7) into (9.6.6) shows that the ML detection of s_k amounts to minimizing

$$|y_{k-1} - y_k s_k^*|^2 \tag{9.6.8}$$

Using the constraint that $|s_k| = 1$, we find that minimizing the latter function is equivalent to maximizing

$$\mathrm{Re}\,\{y_{k-1}y_k^* s_k\} = \mathrm{Re}\,\{\phi_{\text{diff}} s_k\} \tag{9.6.9}$$

where $\phi_{\text{diff}} = y_{k-1}y_k^*$, which is again a trivial scalar detection problem.

Although the differential detection scheme does not require the receiver to know the channel, the performance is worse by about 3 dB than for coherent detection. This is shown, for instance, in [Proakis, 2001, Sec. 5.2.8], but it can also be seen easily from the above equations. As $\mathcal{S}$ is a unitary constellation, it is clear from (9.6.2) that the coherent detection of s amounts to maximizing

$$\mathrm{Re}\{y^* h s\} = \mathrm{Re}\{\phi_{\text{coh}} s\} \tag{9.6.10}$$

where $\phi_{\text{coh}} = y^* h$. The SNR in ϕ_{coh} is easily found to be equal to $|h|^2/\sigma^2$, where σ^2 is the variance of the noise e. On the other hand, for (9.6.9) we have that

$$\phi_{\text{diff}} = y_{k-1} y_k^* = |h|^2 s_k^* + h w_{k-1} e_k^* + h^* w_k^* e_{k-1} + e_{k-1} e_k^* \tag{9.6.11}$$

Neglecting the last term in (9.6.11), which is of higher order, the SNR in ϕ_{diff} is found to be equal to $|h|^2/(2\sigma^2)$. Thus the differential detection method effectively trades the need for channel knowledge for a 3 dB performance loss.

9.6.2 Differential Modulation for MIMO Systems

In the case of multiple antenna systems we deal with matrices instead of scalars, and instead of unitary symbols we will work with unitary matrices [HOCHWALD AND SWELDENS, 2000], [HUGHES, 2000]. Let $\{\boldsymbol{X}_k\}$ be a set of unitary matrices to be transmitted. Equation (9.6.3) can easily be extended to this case: instead of transmitting $\{\boldsymbol{X}_k\}$ we encode the information differentially by forming a new set of matrices $\{\boldsymbol{W}_k\}$ as follows:

$$\boldsymbol{W}_k = \boldsymbol{W}_{k-1}\boldsymbol{X}_k \tag{9.6.12}$$

$$\boldsymbol{W}_0 = \boldsymbol{I} \tag{9.6.13}$$

Since $\{\boldsymbol{X}_k\}$ are unitary and $\boldsymbol{W}_0 = \boldsymbol{I}$ it follows readily that $\boldsymbol{W}_k^H \boldsymbol{W}_k = \boldsymbol{I}$ for any k. This is important as it means that the transmit power is kept constant.

If $\boldsymbol{W}_k$ is transmitted, the received $n_r \times n_t$ matrix (note that $N = n_t$) is:

$$\begin{aligned}\boldsymbol{Y}_k &= \boldsymbol{H}\boldsymbol{W}_k + \boldsymbol{E}_k \\ &= \boldsymbol{H}\boldsymbol{W}_{k-1}\boldsymbol{X}_k + \boldsymbol{E}_k\end{aligned} \tag{9.6.14}$$

where, making the same assumptions as before, the noise term $\boldsymbol{E}_k$ has independent complex Gaussian elements with variance σ^2. If $\boldsymbol{H}\boldsymbol{W}_{k-1}$ were known at the receiver, then we could simply use the coherent ML detector in Section 9.1. When the channel is unknown, we can decode $\boldsymbol{X}_k$ using two consecutively received blocks $\boldsymbol{Y}_k$ and $\boldsymbol{Y}_{k-1}$ in a fashion similar to the procedure in Section 9.6.1. Assume, as in the SISO case, that $\boldsymbol{H}$ is constant over the two blocks transmitted at time $k-1$ and k. Then

$$\boldsymbol{Y}_{k-1} = \boldsymbol{H}\boldsymbol{W}_{k-1} + \boldsymbol{E}_{k-1} \tag{9.6.15}$$

Detection of $\boldsymbol{X}_k$ from $\boldsymbol{Y}_k$ and $\boldsymbol{Y}_{k-1}$ in an ML sense amounts to minimizing the following metric:

$$\begin{aligned} &\|\boldsymbol{Y}_k - \boldsymbol{H}\boldsymbol{W}_{k-1}\boldsymbol{X}_k\|^2 + \|\boldsymbol{Y}_{k-1} - \boldsymbol{H}\boldsymbol{W}_{k-1}\|^2 \\ &= \|\boldsymbol{Y}_k\boldsymbol{X}_k^H - \boldsymbol{H}\boldsymbol{W}_{k-1}\|^2 + \|\boldsymbol{Y}_{k-1} - \boldsymbol{H}\boldsymbol{W}_{k-1}\|^2 \end{aligned} \tag{9.6.16}$$

where we have used the fact that $\boldsymbol{X}_k$ is unitary. The matrix $\boldsymbol{H}\boldsymbol{W}_{k-1}$ is a completely unknown quantity, and minimization of (9.6.16) with respect to it yields:

$$\widehat{\boldsymbol{H}\boldsymbol{W}}_{k-1} = \frac{1}{2}(\boldsymbol{Y}_k\boldsymbol{X}_k^H + \boldsymbol{Y}_{k-1}) \tag{9.6.17}$$

Substituting (9.6.17) in (9.6.16) shows that the ML detection of $\boldsymbol{X}_k$ amounts to minimizing:

$$\|\boldsymbol{Y}_k\boldsymbol{X}_k^H - \boldsymbol{Y}_{k-1}\|^2 = -2\mathrm{Re}\left\{\mathrm{Tr}\{\boldsymbol{X}_k\boldsymbol{Y}_k^H\boldsymbol{Y}_{k-1}\}\right\} + \text{const.} \tag{9.6.18}$$

with respect to $\boldsymbol{X}_k$, or equivalently, maximizing

$$\mathrm{Re}\left\{\mathrm{Tr}\{\boldsymbol{X}_k\boldsymbol{Y}_k^H\boldsymbol{Y}_{k-1}\}\right\} \tag{9.6.19}$$

with respect to $\boldsymbol{X}_k$ (cf. (9.6.9) for the SISO case). Note that the ML detector in (9.6.19) is valid for any set of unitary matrices $\{\boldsymbol{X}_k\}$.

Analysis and Design of Differential Codes

The performance of differentially detected space-time codes can be quantified by using Theorem 4.4 (see page 49). By writing (9.6.14) and (9.6.15) as

$$\begin{bmatrix}\boldsymbol{Y}_k & \boldsymbol{Y}_{k-1}\end{bmatrix} = \boldsymbol{H}\boldsymbol{W}_{k-1}\begin{bmatrix}\boldsymbol{X}_k & \boldsymbol{I}\end{bmatrix} + \begin{bmatrix}\boldsymbol{E}_k & \boldsymbol{E}_{k-1}\end{bmatrix} \tag{9.6.20}$$

we can treat differential detection as the problem of detecting $\begin{bmatrix}\boldsymbol{X}_k & \boldsymbol{I}\end{bmatrix}$ noncoherently with $\boldsymbol{H}\boldsymbol{W}_{k-1}$ as an unknown effective channel. Applying Theorem 4.4 (assuming that $\boldsymbol{H}$ has independent Gaussian elements with variance ρ^2, for simplicity) we obtain:

$$E_{\boldsymbol{H}}[P(\boldsymbol{X}_k^0 \to \boldsymbol{X}_k)] \leq \left|\boldsymbol{Z}_k^0 \Pi^{\perp}_{\boldsymbol{Z}_k^H} \boldsymbol{Z}_k^{0H}\right|^{-n_r} \cdot \left(\frac{\rho^2}{4\sigma^2}\right)^{-n_r n_t} \tag{9.6.21}$$

where $\boldsymbol{Z}_k^0 = \begin{bmatrix}\boldsymbol{X}_k^0 & \boldsymbol{I}\end{bmatrix}$ is the true transmitted matrix and $\boldsymbol{Z}_k = \begin{bmatrix}\boldsymbol{X}_k & \boldsymbol{I}\end{bmatrix}$ is any other hypothetical signal. Simplifying (9.6.21) using

$$\boldsymbol{Z}_k\boldsymbol{Z}_k^H = \begin{bmatrix}\boldsymbol{X}_k & \boldsymbol{I}\end{bmatrix}\begin{bmatrix}\boldsymbol{X}_k^H \\ \boldsymbol{I}\end{bmatrix} = 2 \cdot \boldsymbol{I} = \boldsymbol{Z}_k^0\boldsymbol{Z}_k^{0H} \tag{9.6.22}$$

yields

$$\begin{aligned} E_{\boldsymbol{H}}[P(\boldsymbol{X}_k^0 \to \boldsymbol{X}_k)] &\leq \left| \boldsymbol{Z}_k^0 \Pi^{\perp}_{\boldsymbol{Z}_k^H} \boldsymbol{Z}_k^{0H} \right|^{-n_r} \cdot \left(\frac{\rho^2}{4\sigma^2} \right)^{-n_r n_t} \\ &= \left| \begin{bmatrix} \boldsymbol{X}_k^0 & \boldsymbol{I} \end{bmatrix} \left(\boldsymbol{I} - \frac{1}{2} \cdot \begin{bmatrix} \boldsymbol{X}_k^H \\ \boldsymbol{I} \end{bmatrix} \begin{bmatrix} \boldsymbol{X}_k & \boldsymbol{I} \end{bmatrix} \right) \begin{bmatrix} \boldsymbol{X}_k^{0H} \\ \boldsymbol{I} \end{bmatrix} \right|^{-n_r} \cdot \left(\frac{\rho^2}{4\sigma^2} \right)^{-n_r n_t} \\ &= \left| 2 \cdot \boldsymbol{I} - \frac{1}{2} \cdot (\boldsymbol{X}_k^0 \boldsymbol{X}_k^H + \boldsymbol{I})(\boldsymbol{X}_k^0 \boldsymbol{X}_k^H + \boldsymbol{I})^H \right|^{-n_r} \cdot \left(\frac{\rho^2}{4\sigma^2} \right)^{-n_r n_t} \\ &= \left| \boldsymbol{I} - \frac{1}{2} \cdot (\boldsymbol{X}_k^0 \boldsymbol{X}_k^H + \boldsymbol{X}_k \boldsymbol{X}_k^{0H}) \right|^{-n_r} \cdot \left(\frac{\rho^2}{4\sigma^2} \right)^{-n_r n_t} \end{aligned} \tag{9.6.23}$$

Minimization of (9.6.23) was the approach used to design unitary matrix constellations $\{\boldsymbol{X}_k\}$ in [HUGHES, 2000]. The cited paper imposed the constraint that the set of matrices $\{\boldsymbol{X}_k\}$ should constitute a *finite* algebraic group (in other words, if $\boldsymbol{X}_k$ and $\boldsymbol{X}_l$ are two permissible code matrices then their product must also be a valid code matrix). Sometimes this is a desirable property, yet the class of code matrices that form a group is rather limited and we can expect that better codes can be obtained by abandoning the group structure. As shown in [GANESAN AND STOICA, 2002B], this is indeed the case: OSTBC can more simply and efficiently be used for differential MIMO modulation. This is the topic of the next subsection.

Using OSTBC for Differential Detection

For an arbitrary set of unitary code matrices $\{\boldsymbol{X}_k\}$ the ML detector can be computationally intensive as the minimization of (9.6.18) requires a search through all possible $\boldsymbol{X}_k$. However, as we show next, if we design $\{\boldsymbol{X}_k\}$ using OSTBC, the computational complexity can be reduced considerably. Moreover, it turns out that the codes based on OSTBC can have a better performance than the group codes of [HUGHES, 2000].

As already stated, to use differential encoding in a MIMO system, we need to implement the matrix multiplication in (9.6.12) for all k, which in turn requires that the matrices $\{\boldsymbol{X}_k\}$ are square and unitary. For OSTBC, the code matrices have a semi-unitary structure, but they are not square in general; in fact, the only instances of commonly used square OSTBC matrices, for both real and complex symbols, correspond to the cases of $n_t = 2, 4$ and 8 transmit antennas. Therefore, we consider first differential encoding for square OSTBC, and later we show how the so-obtained scheme for square OSTBC can be modified to transmit non-square blocks for up to $n_t = 8$ transmit antennas.

Square Differential OSTBC: Let us first consider that $n_t = 2$, 4 or 8, in which case the best known OSTBC matrices are square (see (7.4.6), (7.4.10) and (7.4.11)), and have rates 1, 3/4 and 1/2, respectively. Let $s_1^{(k)}, \ldots, s_{n_s}^{(k)}$ denote the symbols to be transmitted at time k. Assume that these symbols are taken from a unitary constellation $\mathcal{S}$, that is, $|s_n^{(k)}| = 1$ for all $s_n^{(k)} \in \mathcal{S}$. Let $\{\boldsymbol{A}_n\}_{n=1}^{n_s}$ and $\{\boldsymbol{B}_n\}_{n=1}^{n_s}$ be the set of $n_t \times n_t$ matrices associated with the OSTBC under consideration, and define the following normalized OSTBC matrix:

$$\boldsymbol{X}_k = \frac{1}{\sqrt{n_s}} \sum_{n=1}^{n_s} \left(\bar{s}_n^{(k)} \boldsymbol{A}_n + i \tilde{s}_n^{(k)} \boldsymbol{B}_n \right) \tag{9.6.24}$$

The normalization with the factor $1/\sqrt{n_s}$ in (9.6.24) is necessary to make $\boldsymbol{X}_k$ unitary:

$$\boldsymbol{X}_k \boldsymbol{X}_k^H = \frac{1}{n_s} \sum_{n=1}^{n_s} |s_n^{(k)}|^2 \cdot \boldsymbol{I} = \boldsymbol{I} \tag{9.6.25}$$

Hence, by using (9.6.24) in (9.6.12) we can obtain directly a scheme for differential encoding and detection of OSTBC.

The differential scheme for OSTBC can be extended to $n_t > 8$ transmit antennas, but such an extension may be of little interest since square OSTBC matrices for $n_t > 8$ have rates (much) less than 1/2 (see Appendix B).

Non-Square Differential OSTBC: In the case of $n_t = 3$, 5, 6, and 7 transmit antennas, we can modify the above scheme as follows.

The case of $n_t = 3$: Let $\boldsymbol{W}_k^{(4)}$ be the 4×4 unitary matrix at time k (after differential encoding as in (9.6.12)), designed for 4 transmit antennas. For $n_t = 3$, let us transmit the matrix $\boldsymbol{\Phi}_3 \boldsymbol{W}_k^{(4)}$ where

$$\boldsymbol{\Phi}_3 = \begin{bmatrix} 1 & 0 & 0 & 0 \\ 0 & 1 & 0 & 0 \\ 0 & 0 & 1 & 0 \end{bmatrix} \tag{9.6.26}$$

Of course, transmitting $\boldsymbol{\Phi}_3 \boldsymbol{W}_k^{(4)}$ is equivalent to transmitting the first three rows of $\boldsymbol{W}_k^{(4)}$. The received matrix $\boldsymbol{Y}_k$ can be written as:

$$\begin{aligned} \boldsymbol{Y}_k &= \boldsymbol{H} \boldsymbol{\Phi}_3 \boldsymbol{W}_k^{(4)} + \boldsymbol{E}_k \\ &= \check{\boldsymbol{H}} \boldsymbol{W}_k^{(4)} + \boldsymbol{E}_k \end{aligned} \tag{9.6.27}$$

where

$$\check{\boldsymbol{H}} = \boldsymbol{H} \boldsymbol{\Phi}_3 \tag{9.6.28}$$

and hence the multiplicative factor $\boldsymbol{\Phi}_3$ can be "absorbed" into the effective channel matrix $\check{\boldsymbol{H}}$. From (9.6.27) it follows that the receiver can use the ML detector in (9.6.18) with $\boldsymbol{X}_k$ as defined for the case of $n_t = 4$ transmit antennas.

The case of $n_t = 5, 6, 7$*:* In a fashion similar to that in the previous paragraph, let $\boldsymbol{W}_k^{(8)}$ denote the 8×8 unitary matrix corresponding to 8 transmit antennas. We transmit the first n_t rows of the matrix $\boldsymbol{W}_k^{(8)}$, or equivalently, $\boldsymbol{\Phi}_{n_t}\boldsymbol{W}_k^{(8)}$ where $\boldsymbol{\Phi}_{n_t}$ is an $n_t \times 8$ matrix given by

$$\boldsymbol{\Phi}_{n_t} = \left[\boldsymbol{I}_{n_t \times n_t}\, \boldsymbol{0}_{n_t \times (8-n_t)}\right] \tag{9.6.29}$$

Again the detector in (9.6.18) can be used directly, with $\boldsymbol{H}\boldsymbol{\Phi}_{n_t}$ in lieu of $\boldsymbol{H}$ and with $\boldsymbol{X}_k$ as defined for the case of $n_t = 8$ transmit antennas.

ML Detector for Differential OSTBC: The ML detector in (9.6.18) associated with differentially encoded OSTBC simplifies considerably compared to minimizing (9.6.18) for general unitary code matrices $\{\boldsymbol{X}_k\}$. To verify this fact, let us substitute (9.6.24) in (9.6.18). Then we obtain the following ML criterion that is to be minimized with respect to $\{s_n^{(k)} \in \mathcal{S}\}$:

$$\sum_{n=1}^{n_s}\Big(- \mathrm{Re}\left\{\mathrm{Tr}\{\boldsymbol{Y}_k^H \boldsymbol{Y}_{k-1}\boldsymbol{A}_n\}\right\}\bar{s}_n^{(k)} + \mathrm{Im}\left\{\mathrm{Tr}\{\boldsymbol{Y}_k^H \boldsymbol{Y}_{k-1}\boldsymbol{B}_n\}\right\}\tilde{s}_n^{(k)}\Big) \tag{9.6.30}$$

The above detector has the same decoupled form and computational complexity as the standard coherent detector for OSTBC in Section 7.4.

At this point, we should remark on the fact that another approach to OSTBC differential detection was proposed in [Tarokh and Jafarkhani, 2000]. While the encoding complexity of the scheme in [Tarokh and Jafarkhani, 2000] (see also [Jafarkhani and Tarokh, 2001]) is similar to that of the approach described above, the decoding associated with the approach in the cited papers (as well as the differential method in [Hughes, 2000]) is somewhat more complicated than that in (9.6.30).

SNR for Differential Detection with OSTBC: Comparing (9.6.30) with the expressions in Section 7.4.2 we see that the differential ML detector has exactly the same form as the coherent ML detector, but with $\boldsymbol{Y}_{k-1}$ instead of $\boldsymbol{H}$. In other words, the differential detector (9.6.30) can be thought of as a way of forming the decision statistic $\hat{\boldsymbol{s}}$ in Theorem 7.3 (see page 107) by using $\boldsymbol{Y}_{k-1}$ instead of the true channel matrix $\boldsymbol{H}$. Inspired by this analogy, we can calculate the SNR for the differential MIMO detector as we did for the SISO case in Section 9.6.1. Note first that

$$\begin{aligned}\boldsymbol{Y}_k^H \boldsymbol{Y}_{k-1} &= (\boldsymbol{H}\boldsymbol{W}_{k-1}\boldsymbol{X}_k + \boldsymbol{E}_k)^H (\boldsymbol{H}\boldsymbol{W}_{k-1} + \boldsymbol{E}_{k-1}) \\ &= \boldsymbol{X}_k^H \boldsymbol{W}_{k-1}^H \boldsymbol{H}^H \boldsymbol{H}\boldsymbol{W}_{k-1} + \boldsymbol{E}_k^H \boldsymbol{H}\boldsymbol{W}_{k-1} \\ &\quad + \boldsymbol{X}_k^H \boldsymbol{W}_{k-1}^H \boldsymbol{H}^H \boldsymbol{E}_{k-1} + \boldsymbol{E}_k^H \boldsymbol{E}_{k-1} \end{aligned} \tag{9.6.31}$$

Let us consider the decision statistic for $\bar{s}_n^{(k)}$ (the corresponding decision statistic for $\tilde{s}_n^{(k)}$ can be handled in a similar manner). Using (9.6.31) it is possible to show that

$$\begin{aligned}\mathrm{Re}\left\{\mathrm{Tr}\{\boldsymbol{A}_n \boldsymbol{Y}_k^H \boldsymbol{Y}_{k-1}\}\right\} &= \frac{1}{\sqrt{n_s}} \mathrm{Tr}\{\boldsymbol{H}^H \boldsymbol{H}\} \bar{s}_n^{(k)} \\ &\quad + \mathrm{Re}\left\{\mathrm{Tr}\{\boldsymbol{A}_n \boldsymbol{E}_k^H \boldsymbol{H}\boldsymbol{W}_{k-1}\}\right\} \\ &\quad + \mathrm{Re}\left\{\mathrm{Tr}\{\boldsymbol{A}_n \boldsymbol{X}_k^H \boldsymbol{W}_{k-1}^H \boldsymbol{H}^H \boldsymbol{E}_{k-1}\}\right\} \\ &\quad + \mathrm{Re}\left\{\mathrm{Tr}\{\boldsymbol{A}_n \boldsymbol{E}_k^H \boldsymbol{E}_{k-1}\}\right\} \end{aligned} \tag{9.6.32}$$

To show (9.6.32), use (7.4.4) and the fact that $\mathrm{Re}\{\mathrm{Tr}\{\boldsymbol{X}\}\} = 0$ if $\boldsymbol{X} = -\boldsymbol{X}^H$. If the noise variance σ^2 is small enough we can neglect the last term in (9.6.32). The remaining two noise terms are uncorrelated to one another and have the same variance. Using these facts we can show that approximately:

$$\sqrt{n_s} \frac{\mathrm{Re}\left\{\mathrm{Tr}\{\boldsymbol{A}_n \boldsymbol{Y}_k^H \boldsymbol{Y}_{k-1}\}\right\}}{\|\boldsymbol{H}\|^2} \sim N\left(\bar{s}_n^{(k)}, n_s \cdot \frac{\sigma^2}{\|\boldsymbol{H}\|^2}\right) \tag{9.6.33}$$

Comparing (9.6.33) with the corresponding equation for coherent detection in Theorem 7.3 (see page 107), using $\rho^2 = n_t/n_s$ to achieve the same power normalization as in the present case, we can see that the SNR in the statistic used for the differential detection of $\bar{s}_n$ is (approximately) half the SNR for coherent detection. This 3 dB SNR loss is analogous to that for conventional DPSK shown in Section 9.6.1. Note that this analysis holds true also in the case when $\boldsymbol{H}$ is replaced by $\check{\boldsymbol{H}} = \boldsymbol{H}\boldsymbol{\Phi}_{n_t}$ as required for using differential OSTBC with $n_t = 3, 5, 6$ and 7 (see (9.6.28) and the related discussion). Indeed, we have $\|\boldsymbol{H}\|^2 = \left\|\check{\boldsymbol{H}}\right\|^2$ (as $\boldsymbol{\Phi}_{n_t}\boldsymbol{\Phi}_{n_t}^H = \boldsymbol{I}$) and this is precisely what is needed for (9.6.33) to hold when the effective channel is $\check{\boldsymbol{H}}$ instead of $\boldsymbol{H}$.

Discussion: Differential detection of OSTBC yields a diversity of the same order as coherent detection. In particular, the matrix $\boldsymbol{Y}_{k-1}$ effectively serves as an effective channel estimate during the detection of $\boldsymbol{X}_k$ from $\boldsymbol{Y}_k$. Hence, application of Theorem 4.3 (see page 48) to the metric (9.6.18) shows that a diversity of order

$n_r n_t$ is achieved for differential detection as well. This observation reinforces the previously observed fact that the only difference between the performances of coherent and differential detection is a 3 dB loss in SNR.

The data rate achieved by the differential detection scheme is the same as that achieved by the coherent detection scheme. The coherent detection scheme assumed that the receiver had perfect knowledge about the channel, which is certainly not the case in practice. The channel can be estimated by the receiver via training, but this requires that the channel remains constant during the training block (see, e.g., [PEEL AND SWINDLEHURST, 2001] for a discussion of the performance tradeoffs associated with the use of training-based and differential modulation on a time-varying channel). Training also needs extra bandwidth for the transmission of pilot symbols. The differential detection scheme does not transmit any *explicit* training data and therefore the channel can change faster. If we transmit $n_t \times N$ blocks we require the channel to be practically constant for $2N$ symbol periods. Also, since there is no explicit training involved there is no loss of bandwidth. These features of the differential scheme are attractive, but they come at the cost of a 3 dB loss in performance compared to coherent detection (even though the diversity gain is the same for differential detection and for coherent detection); additionally, if the channel is slow fading the differential scheme cannot make use of that information, whereas a training-based scheme can.

Example 9.8: Differential OSTBC vs. Group Codes.
In this example (see [GANESAN AND STOICA, 2002A]), we compare differential Alamouti OSTBC with the differential technique of [HUGHES, 2000] for a system with $n_t = 2$ and $n_r = 1$. We transmit two bits per time interval; hence the differential OSTBC uses QPSK modulation whereas the code matrices for the differential group code of [HUGHES, 2000] are taken from a set of 16 possible matrices. Figure 9.6 shows the BER for the two methods, as well as the BER corresponding to coherent detection. For the group code, it is necessary to specify a mapping between each code matrix and the associated transmitted information bits; for this purpose we used a labelling technique similar to Gray coding. For this example, the differential OSTBC code outperforms the group code of [HUGHES, 2000] by about 3 dB. More numerical examples, as well as analytical comparisons of coding gains, can be found in [GANESAN AND STOICA, 2002A], [GANESAN AND STOICA, 2002B]. ■

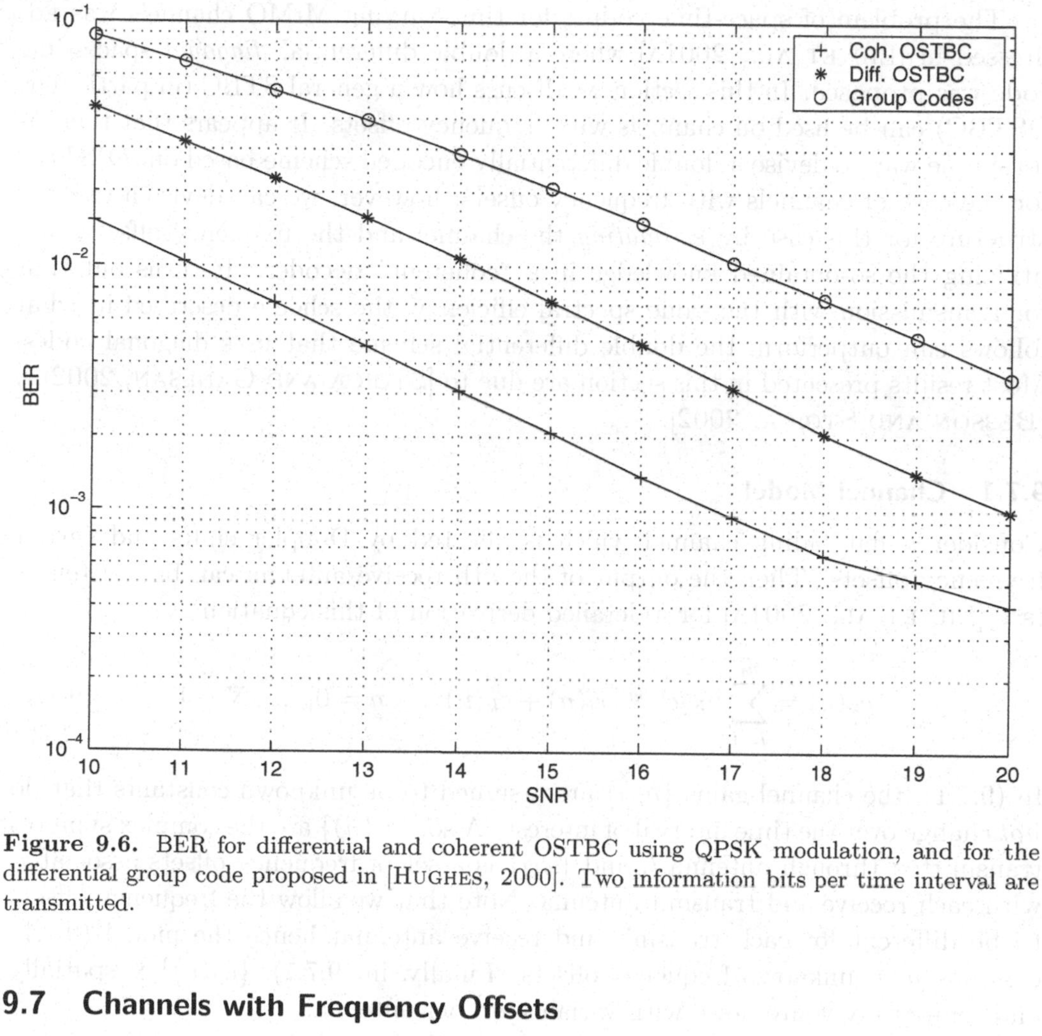

Figure 9.6. BER for differential and coherent OSTBC using QPSK modulation, and for the differential group code proposed in [HUGHES, 2000]. Two information bits per time interval are transmitted.

9.7 Channels with Frequency Offsets

All the detection methods discussed so far assume that the channel remains approximately constant for a certain number of blocks. For example, for the training-based scheme and its relatives in Section 9.4, the channel must remain unchanged over a period of time corresponding to the N_b transmitted blocks. The differential method in Section 9.6 is less restrictive as it requires that the channel is constant over only two consecutive blocks. However, in practice, there may be situations when the channel is changing so fast that even the assumptions made by the differential detector are not adequate. For instance, the channel may be affected by a frequency offset that can be either due to Doppler or to a carrier frequency mismatch between the receiver and the transmitter.

The problem of space-time coding for time-varying MIMO channels was addressed in [LIU ET AL., 2001A] where a double differential *diagonal* space-time code was proposed. In this section we discuss how a general STBC (in particular, OSTBC) can be used on channels with frequency offsets. It appears that there is no simple way to devise a double differentially encoded scheme based on (O)STBC for the case of channels with frequency offsets; however, we can devise a receiver structure for this case by *estimating* the channel and the frequency offsets, and utilizing the so-obtained knowledge in a "coherent" decoder. It turns out that for transmission with the same spectral efficiency, the scheme described in what follows can outperform the double differential scheme that uses diagonal codes. Most results presented in this section are due to [STOICA AND GANESAN, 2002B], [BESSON AND STOICA, 2002].

9.7.1 Channel Model

Consider a flat fading channel which is affected by Doppler shifts and carrier frequency offsets. Then the output of the kth receive antenna can be written as (see [LIU ET AL., 2001A] for a detailed derivation of this equation):

$$y_k(n) = \sum_{l=1}^{n_t} h_{k,l} e^{i\omega_{kl} n} x_l(n) + e_k(n) \qquad n = 0, \ldots, N-1 \tag{9.7.1}$$

In (9.7.1), the channel gains $\{h_{k,l}\}$ are assumed to be unknown constants that do *not* change over the time interval of interest. Also, $\{x_l(n)\}$ are the complex symbols transmitted through antenna l, and $\{\omega_{kl}\}$ is a set of frequency offsets associated with each receive and transmit antenna. Note that we allow the frequency offsets to be different for each transmit and receive antenna; hence the model (9.7.1) contains $n_r n_t$ unknown frequency offsets. Finally, in (9.7.1), $\{e_k(n)\}$ is spatially and temporally white noise with variance σ^2 per antenna.

9.7.2 ML Estimation of $\{\omega_{kl}\}$ and $\{h_{kl}\}$

If $\{\omega_{k,l}\}$ and $\{h_{k,l}\}$ were known, it would be easy to derive a coherent detector simply by modifying that in Section 9.1 (for general linear STBC) or that in Section 7.4.2 (for OSTBC) to compensate for the time-variation of the channel matrix; we leave doing so to the reader. Our ultimate goal is a receiver structure that estimates $\{\omega_{kl}\}$ and $\{h_{kl}\}$ via training and uses these estimates in the coherent detector in a fashion similar to Section 9.4. Hence we assume in this subsection that $\{\boldsymbol{x}_l(n)\}$ are pilots that are known to the receiver.

Introduce the following notation:

$$\begin{aligned} \boldsymbol{h}_k &= \begin{bmatrix} h_{k,1} & \cdots & h_{k,n_t} \end{bmatrix}^T \\ \boldsymbol{\omega}_k &= \begin{bmatrix} \omega_{k,1} & \cdots & \omega_{k,n_t} \end{bmatrix}^T \\ \boldsymbol{y}_k &= \begin{bmatrix} y_k(0) & \cdots & y_k(N-1) \end{bmatrix}^T \end{aligned} \tag{9.7.2}$$

and

$$\boldsymbol{X}_{\boldsymbol{\omega}_k} = \begin{bmatrix} x_1(0) & \cdots & x_{n_t}(0) \\ x_1(1)e^{i\omega_{k,1}} & \cdots & x_{n_t}(1)e^{i\omega_{k,n_t}} \\ \vdots & \vdots & \vdots \\ \vdots & \vdots & \vdots \\ x_1(N-1)e^{i(N-1)\omega_{k,1}} & \cdots & x_{n_t}(N-1)e^{i(N-1)\omega_{k,n_t}} \end{bmatrix} \tag{9.7.3}$$

Under the assumption made on the noise the ML estimation of $\{\omega_{k,l}\}$ and $\{h_{k,l}\}$, given that $\{\boldsymbol{x}_l(n)\}$ are known, amounts to minimizing the following metric:

$$\sum_{k=1}^{n_r} \|\boldsymbol{y}_k - \boldsymbol{X}_{\boldsymbol{\omega}_k}\boldsymbol{h}_k\|^2 \tag{9.7.4}$$

Each of the terms in the sum in (9.7.4) depends on one $\boldsymbol{h}_k$ and $\boldsymbol{\omega}_k$ only, and hence each term can be minimized independently. It readily follows that for a given $\boldsymbol{\omega}_k$, the minimizer of (9.7.4) with respect to $\boldsymbol{h}_k$ is:

$$\hat{\boldsymbol{h}}_k = (\boldsymbol{X}_{\boldsymbol{\omega}_k}^H \boldsymbol{X}_{\boldsymbol{\omega}_k})^{-1} \boldsymbol{X}_{\boldsymbol{\omega}_k}^H \boldsymbol{y}_k \tag{9.7.5}$$

Inserting (9.7.5) in (9.7.4) shows that the estimates of the frequency offsets are obtained as the maxima with respect to $\omega_{k,l}$ of the following function:

$$\boldsymbol{y}_k^H \boldsymbol{X}_{\boldsymbol{\omega}_k} (\boldsymbol{X}_{\boldsymbol{\omega}_k}^H \boldsymbol{X}_{\boldsymbol{\omega}_k})^{-1} \boldsymbol{X}_{\boldsymbol{\omega}_k}^H \boldsymbol{y}_k = \boldsymbol{y}_k^H \Pi_{\boldsymbol{X}_{\boldsymbol{\omega}_k}} \boldsymbol{y}_k \tag{9.7.6}$$

The matrix $\Pi_{\boldsymbol{X}_{\boldsymbol{\omega}_k}}$ depends on $\boldsymbol{\omega}_k$ in a highly nonlinear manner. Hence, in general, maximizing (9.7.6) is a hard problem that requires an n_t-dimensional search.

In the special case in which $\boldsymbol{X}_{\boldsymbol{\omega}_k}^H \boldsymbol{X}_{\boldsymbol{\omega}_k}$ is *diagonal* we can decouple the n_t-dimensional problem into n_t one-dimensional problems. The matrix $\boldsymbol{X}_{\boldsymbol{\omega}_k}^H \boldsymbol{X}_{\boldsymbol{\omega}_k}$ can be made diagonal by an *appropriate choice of the pilot symbols* $\{x_l(n)\}$. For the sake of brevity we consider the case of $n_t = 2$ transmit antennas, but the results can be easily extended to any number of transmit antennas. Let

$$\begin{aligned} \boldsymbol{x}_l &= \begin{bmatrix} x_l(0) & \cdots & x_l(N-1) \end{bmatrix}^T \\ \boldsymbol{a}_{k,l} &= \begin{bmatrix} 1 & e^{i\omega_{k,l}} & \cdots & e^{i(N-1)\omega_{k,l}} \end{bmatrix}^T \end{aligned} \tag{9.7.7}$$

and note that

$$X_{\omega_k}^H X_{\omega_k} = \begin{bmatrix} \|x_1\|^2 & a_{k,1}^H D a_{k,2} \\ a_{k,2}^H D^H a_{k,1} & \|x_2\|^2 \end{bmatrix} \tag{9.7.8}$$

where

$$D = \begin{bmatrix} x_1^*(0)x_2(0) & 0 & \cdots & 0 \\ 0 & x_1^*(1)x_2(1) & \ddots & \vdots \\ \vdots & \ddots & \ddots & 0 \\ 0 & \cdots & 0 & x_1^*(N-1)x_2(N-1) \end{bmatrix} \tag{9.7.9}$$

It is clear that $X_{\omega_k}^H X_{\omega_k}$ is a diagonal matrix for all $\omega_{k,1}, \omega_{k,2}$ if and only if

$$D = 0 \tag{9.7.10}$$

Assume for simplicity that N is even. Then the condition (9.7.10) is satisfied, for instance, if

$$\begin{array}{ll} x_1(2n) = 0 & x_1(2n+1) = s_1(n) \\ x_2(2n) = s_2(n) & x_2(2n+1) = 0 \end{array} \tag{9.7.11}$$

for $n = 0, \ldots, N/2 - 1$, where $s_1(n)$ and $s_2(n)$ are arbitrary sequences. With the choice of $x_1(n)$ and $x_2(n)$ in (9.7.11), the signal at the kth receive antenna takes on the following form

$$\begin{aligned} y_k(2n) &= h_{k,2}s_2(n)e^{i2n\omega_{k,2}} + e_k(2n) \\ y_k(2n+1) &= h_{k,1}s_1(n)e^{i(2n+1)\omega_{k,1}} + e_k(2n+1) \end{aligned} \tag{9.7.12}$$

and the corresponding ML frequency offset estimates are given by:

$$\begin{aligned} \hat{\omega}_{k,1} &= \underset{\omega}{\operatorname{argmax}} \left| \sum_{n=0}^{N/2-1} y_k(2n+1)s_1^*(n)e^{-i2\omega n} \right|^2 \\ \hat{\omega}_{k,2} &= \underset{\omega}{\operatorname{argmax}} \left| \sum_{n=0}^{N/2-1} y_k(2n)s_2^*(n)e^{-i2\omega n} \right|^2 \end{aligned} \tag{9.7.13}$$

The maxima in (9.7.13) can be found by using the fast Fourier transform. Once $\hat{\omega}_{k,1}$ and $\hat{\omega}_{k,2}$ have been calculated, the corresponding ML channel gain estimates can be computed from (9.7.5). We stress the fact that $\hat{\omega}_{k,1}$ and $\hat{\omega}_{k,2}$ above are the *true* ML estimates for the choice of pilots in (9.7.11). Note that although only $N/2$ samples are used in the estimation, the length of the "aperture" of the data string is equal to N.

9.7.3 A Simplified Channel Model

The estimation of the channel parameters can be further simplified if we assume that the frequency offset associated with a particular receive antenna is the same for all transmit antennas (i.e., $\omega_{k,l} = \omega_k$). This is true, for instance, if the frequency offset is mainly due to a carrier frequency mismatch between the transmit and receive oscillators (see [LIU ET AL., 2001A] where this assumption was made and motivated in some detail). Let ω_k denote the frequency offset associated with the kth receive antenna. Then the signal received by the kth antenna can be written as (cf. (9.7.1)):

$$y_k(n) = e^{i\omega_k n} \sum_{l=1}^{n_t} h_{k,l} x_l(n) + e_k(n) \tag{9.7.14}$$

for $n = 0, 1, \ldots, N-1$, where the different terms have the same meaning as in (9.7.1). Assume that we transmit L blocks of training symbols (each of length N), and let the $N \times n_t$ matrix $\boldsymbol{X}_l$ denote the lth pilot block (for $l = 0, \ldots, L-1$):

$$\boldsymbol{X}_l = \begin{bmatrix} x_1(Nl) & x_2(Nl) & \ldots & x_{n_t}(Nl) \\ x_1(Nl+1) & x_2(Nl+1) & \ldots & x_{n_t}(Nl+1) \\ \vdots & \vdots & & \vdots \\ x_1(Nl+N-1) & x_2(Nl+N-1) & \ldots & x_{n_t}(Nl+N-1) \end{bmatrix} \tag{9.7.15}$$

Let $\boldsymbol{h}_k$ be as defined before in (9.7.2), and let

$$\begin{aligned} \boldsymbol{y}_k(l) &= [y_k(Nl) \quad \cdots \quad y_k(Nl+N-1)]^T \\ \boldsymbol{e}_k(l) &= [e_k(Nl) \quad \cdots \quad e_k(Nl+N-1)]^T \end{aligned} \tag{9.7.16}$$

Using this notation we can write (9.7.14) as follows:

$$\boldsymbol{y}_k(l) = e^{i\omega_k Nl} \boldsymbol{\Omega}_k \boldsymbol{X}_l \boldsymbol{h}_k + \boldsymbol{e}_k(l) \quad (k = 1, \ldots, n_r) \tag{9.7.17}$$

for $l = 0, \ldots, L = 1$, where:

$$\boldsymbol{\Omega}_k = \begin{bmatrix} 1 & 0 & \cdots & 0 \\ 0 & e^{i\omega_k} & \ddots & \vdots \\ \vdots & \ddots & \ddots & 0 \\ 0 & \cdots & 0 & e^{i(N-1)\omega_k} \end{bmatrix} \tag{9.7.18}$$

We want to use the pilot blocks and the corresponding received signals, $\{\boldsymbol{y}_k(0), \ldots, \boldsymbol{y}_k(L-1)\}_{k=1}^{n_r}$ to estimate $\{\boldsymbol{h}_k\}$ and $\{\omega_k\}$.

Owing to the assumption made on the noise term in (9.7.17), the ML estimates of $\boldsymbol{h}_k$ and ω_k for given $\{\boldsymbol{X}_0, \ldots, \boldsymbol{X}_{L-1}\}$ are obtained by minimizing the following criterion:

$$\sum_{l=0}^{L-1} \left\| \boldsymbol{y}_k(l) - \boldsymbol{\Omega}_k \boldsymbol{X}_l e^{i\omega_k N l} \boldsymbol{h}_k \right\|^2 \tag{9.7.19}$$

As we will see shortly the minimization of (9.7.19) becomes simpler if the pilot blocks $\boldsymbol{X}_l$ are chosen to be square and unitary:

$$\boldsymbol{X}_l \boldsymbol{X}_l^H = \boldsymbol{I} \tag{9.7.20}$$

Under the assumption (9.7.20) we can write the criterion in (9.7.19) as:

$$\sum_{l=0}^{L-1} \left\| \boldsymbol{X}_l^H \boldsymbol{\Omega}_k^H \boldsymbol{y}_k(l) e^{-i\omega_k N l} - \boldsymbol{h}_k \right\|^2 \tag{9.7.21}$$

The minimization of (9.7.21) with respect to $\boldsymbol{h}_k$ yields:

$$\hat{\boldsymbol{h}}_k = \frac{1}{L} \sum_{l=0}^{L-1} \boldsymbol{X}_l^H \boldsymbol{\Omega}_k^H e^{-i\omega_k N l} \boldsymbol{y}_k(l) \tag{9.7.22}$$

Inserting (9.7.22) in (9.7.21) shows that the ML estimation of ω_k reduces to minimizing the following function with respect to ω_k:

$$\begin{aligned} &\left\| \left(\boldsymbol{I} - \frac{1}{L} \begin{bmatrix} \boldsymbol{I} \\ \vdots \\ e^{iN(L-1)\omega_k} \boldsymbol{I} \end{bmatrix} \begin{bmatrix} \boldsymbol{I} & \cdots & e^{-iN(L-1)\omega_k} \boldsymbol{I} \end{bmatrix} \right) \begin{bmatrix} \boldsymbol{X}_0^H \boldsymbol{\Omega}_k^H \boldsymbol{y}_k(0) \\ \vdots \\ \boldsymbol{X}_{L-1}^H \boldsymbol{\Omega}_k^H \boldsymbol{y}_k(L-1) \end{bmatrix} \right\|^2 \\ &= -\frac{1}{L} \left\| \sum_{l=0}^{L-1} e^{-iNl\omega_k} \boldsymbol{X}_l^H \boldsymbol{\Omega}_k^H \boldsymbol{y}_k(l) \right\|^2 + \text{const.} \end{aligned} \tag{9.7.23}$$

Besides choosing $\{\boldsymbol{X}_l\}$ to satisfy (9.7.20) we also choose all of these pilot blocks to be identical:

$$\boldsymbol{X}_l = \boldsymbol{X}, \quad l = 0, \ldots, L-1 \tag{9.7.24}$$

Then the function in (9.7.23) takes a simple FFT-like form, and the estimation of ω_k reduces to:

$$\max_{\omega_k} \left\| \sum_{l=0}^{L-1} e^{-iNl\omega_k} \boldsymbol{y}_k(l) \right\|^2 + \text{const.} \tag{9.7.25}$$

Note that in general we need $L \geq 2$ to guarantee the identifiability of both $\{\boldsymbol{h}_k\}$ and $\{\omega_k\}$ in (9.7.17).

In general the estimation of ω_k via (9.7.25) is still somewhat complicated as it requires a one-dimensional search. However, for the special case of $L = 2$, a *closed form* expression for ω_k can be obtained. Indeed, for $L = 2$, (9.7.25) becomes:

$$\begin{aligned} &\left\| \boldsymbol{y}_k(0) + e^{-iN\omega_k} \boldsymbol{y}_k(1) \right\|^2 \\ =&2\mathrm{Re}\left\{ e^{-iN\omega_k} \boldsymbol{y}_k^H(0)\boldsymbol{y}_k(1) \right\} + \text{const.} \end{aligned} \tag{9.7.26}$$

The value of ω_k that maximizes (9.7.26) is readily seen to be

$$\hat{\omega}_k = \frac{1}{N} \cdot \arg\left\{ \boldsymbol{y}_k^H(0)\boldsymbol{y}_k(1) \right\} \tag{9.7.27}$$

and the corresponding channel estimate is obtained from (9.7.22):

$$\hat{\boldsymbol{h}}_k = \frac{1}{2}\left(\boldsymbol{X}_0^H \hat{\boldsymbol{\Omega}}_k^H \boldsymbol{y}_k(0) + \boldsymbol{X}_1^H \hat{\boldsymbol{\Omega}}_k^H e^{-i\hat{\omega}_k N} \boldsymbol{y}_k(1) \right) \tag{9.7.28}$$

Note that the frequency estimate in (9.7.27) does not suffer from aliasing-like effects if and only if $\omega_k N \leq 2\pi$ (where $N = n_t$). This condition may well be satisfied in practice since usually ω_k is quite small.

To summarize, in a fast fading scenario we can choose $N = n_t$ and $L = 2$ (which is the smallest possible value of L) and estimate $\{\omega_k, \boldsymbol{h}_k\}$ using (9.7.27) and (9.7.28). Note that the two training blocks used for acquiring channel state information at the receiver do not have to be adjacent to one another. For instance, they can be separated by a data block as shown in Figure 9.7. The estimation formulas in (9.7.27) and (9.7.28) can be easily modified to take into account the fact that the training blocks are not consecutive. Assuming n_s symbols per block, the transmission rate of the scheme in Figure 9.7 is $n_s/(n_t+n_t)$, which for $N = n_t = n_s$ becomes 1/2. Also note that the scheme relies on the assumption that the channel (i.e, $\{\boldsymbol{h}_k\}$ and $\{\omega_k\}$) is nearly constant over three blocks (which is basically what a double differential scheme would also require; see [Liu et al., 2001a]). Finally, note that the above algorithm developed for channel estimation using pilot blocks can be used with any type of space-time code. We conclude this section with the following simulation example based on OSTBC.

Example 9.9: Trained OSTBC vs. Double Differential Diagonal Codes.
In this example (see [Stoica and Ganesan, 2002b]) we consider a system with $n_t = 2$ transmit antennas and $n_r = 1$ receive antenna. We assume that the frequency offsets $\{\omega_k\}$ are random and uniformly distributed between 0 and 0.5

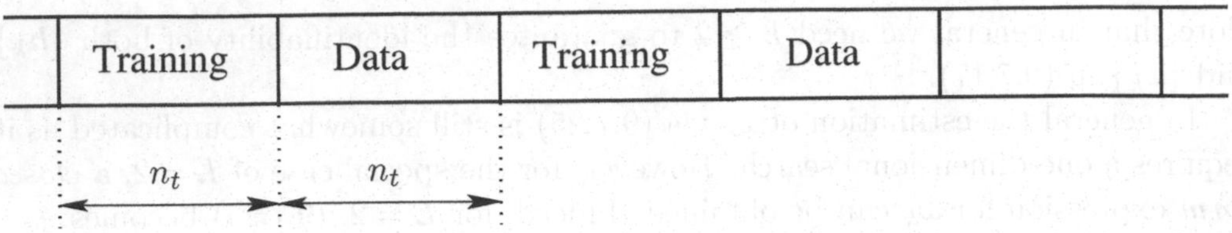

Figure 9.7. Transmission scheme for fast fading.

rad/s. The complex channel coefficients $\{h_{k,l}\}$ are assumed to be independent and Gaussian with mean zero and unity variance. The transmission scheme is the one in Figure 9.7. We transmit one 2×2 training block $\boldsymbol{X} = \boldsymbol{I}$ followed by a data block which in turn is followed by another training block and so on. The data block is an Alamouti OSTBC matrix with symbols taken from a QPSK constellation. Thus the effective data rate is $R = 1/2$. For a given data block, we use the two training blocks adjacent to it to estimate the channel and frequency offsets; doing so allows the channel to change very rapidly.

In Figure 9.8 we show the estimated BER versus the SNR. For comparison, we also plot the BER associated with the double differential scheme in [LIU ET AL., 2001A] (using the same total transmit power, and with the same effective data rate). From the figure it can be seen that the scheme using pilot blocks and OSTBC outperforms the one in [LIU ET AL., 2001A] by about 3 dB. Moreover the OSTBC is simpler to decode. ■

9.8 Summary and Discussion

The present chapter has covered several interrelated topics. We started in Section 9.1 by discussing coherent detection of linear space-time block codes. Coherent ML detection of linear STBC requires the minimization of a quadratic function subject to integer constraints. This is in general a hard problem (except for OSTBC, in which case the problem simplifies to a series of scalar detection problems), and in Section 9.1.3 we discussed some suboptimal as well as (near-)optimal approaches. The associated analysis of error probabilities was given in Chapter 4.

Section 9.2 dealt with some problems relevant to concatenation of linear STBC with an outer (for example, convolutional) error-correcting code. Expressions for bit-metrics and likelihood ratios were given, and optimal information transfer was briefly discussed.

In Sections 9.3–9.5 we treated the detection of linear STBC without channel

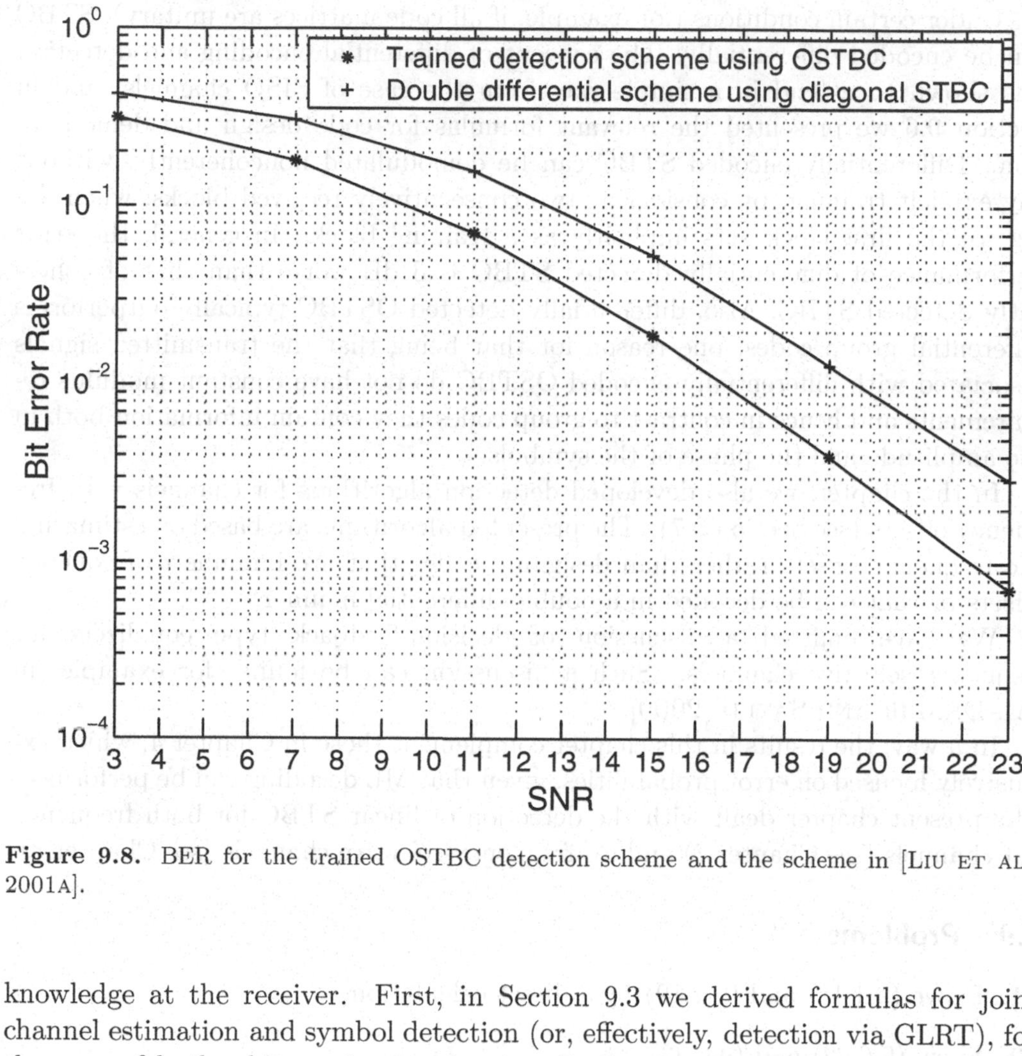

Figure 9.8. BER for the trained OSTBC detection scheme and the scheme in [Liu et al., 2001a].

knowledge at the receiver. First, in Section 9.3 we derived formulas for joint channel estimation and symbol detection (or, effectively, detection via GLRT), for the case of both white and colored noise. In Section 9.4 we discussed training-based detection in which the channel is first estimated using a pilot block known to the receiver, whereafter the coherent detector is used with the estimated channel instead of the true one. We also studied optimality conditions for the training block and found that in order to be optimal, a training block should be semi-unitary. Finally, in Section 9.5 we reviewed blind and semi-blind methods. The blind and semiblind methods presented can be seen as ways to approximate the joint estimation and detection approach of Section 9.3; although purely blind detection may have too poor a performance to be practical, semiblind methods typically outperform a training-based approach at a modest additional computational cost.

Under certain conditions (for example, if all code matrices are unitary), STBC can be encoded differentially. The concept of differential encoding and detection was shown to extend in a simple way from the case of SISO channels, and in Section 9.6 we presented the relevant formulas for code design and demodulation. Differentially encoded STBC can be demodulated noncoherently, without any explicit training, by considering two consecutively received blocks; when doing so, the first block acts implicitly as a training block. In general, the error performance of differentially detected STBC is 3 dB worse than that of coherently detected STBC. Also, differentially detected OSTBC typically outperforms differential group codes; one reason for that being that the transmitted signals associated with differentially encoded OSTBC do not have constant modulus requirements and hence in contrast to group codes they contain information both in the amplitude and the phase of the symbols.

In the chapter, we also developed detection algorithms for channels with frequency offsets (see Section 9.7). The presented algorithms are based on estimating the frequency offsets rather than designing codes that are immune to frequency offsets or that can be decoded in a double-differential manner.

We have omitted a discussion of decision-feedback type equalizers for frequency-selective channels. Such a discussion can be found, for example, in [AL-DHAHIR AND SAYED, 2000].

In a way, the results in this chapter complement those in Chapter 4, which exclusively focused on error probabilities, given that ML decoding can be performed. The present chapter dealt with the detection of linear STBC for both frequency flat channels (see Chapter 7) and for frequency-selective channels (see Chapter 8).

9.9 Problems

1. Prove (9.1.18) and (9.1.19) by a direct calculation.

2. Show (9.6.32) and (9.6.33).

3. Prove (9.1.38). Also, calculate an explicit expression for (9.1.39). Are these two expressions equal in general?

4. Discuss the validity of the interference model in Section 9.1.2. Can you give an example of interference that would be accurately described by this model? Would co-channel interference generated by another OSTBC user be described by the model?

5. Consider the channel model in Section 9.7.1. Assume that $\{\omega_{k,l}\}$ and $\{h_{k,l}\}$

are *known*. Modify the coherent detector in Section 9.1 (for general linear STBC), and that in Section 7.4.2 (for OSTBC) to this situation.

6. Derive (9.7.4).

7. Consider the model

$$\boldsymbol{Y} = \boldsymbol{H}\boldsymbol{X} + \boldsymbol{E} \tag{9.9.1}$$

Assume that $\boldsymbol{H}$ is a random matrix whose elements are Gaussian and independent with mean zero and variance ρ^2. Also, let $\boldsymbol{E}$ be white Gaussian noise with zero mean and variance σ^2, and suppose that $\boldsymbol{X}\boldsymbol{X}^H = \alpha \boldsymbol{I}$ for some constant α.

(a) Formulate the likelihood function associated with optimal detection under the previous assumptions. Prove that for known ρ^2 and σ^2, the ML rule for symbol detection takes on the same form as for the case when $\boldsymbol{H}$ is treated as an unknown *deterministic* matrix (see (9.3.3)).

(b) Try to concentrate (minimize) the likelihood function jointly with respect to ρ^2 and σ^2. Discuss the result (see also [Larsson et al., 2002b]).

8. Study a system with $n_t = 2$ transmit antennas and a frequency flat propagation channel. Suppose that instead of using the differential OSTBC scheme in Section 9.6 (which for $n_t = 2$ is based on Alamouti matrices), we first encode two streams of symbols $s_1(t), s_2(t)$ differentially:

$$\begin{aligned} x_1(t) &= x_1(t-1) \cdot s_1(t) \\ x_2(t) &= x_2(t-1) \cdot s_2(t) \end{aligned} \tag{9.9.2}$$

and then use $\{x_1(t), x_2(t)\}$ to form Alamouti matrices:

$$\begin{aligned} \boldsymbol{X}(t-1) &= \begin{bmatrix} x_1(t-1) & x_2^*(t-1) \\ x_2(t-1) & -x_1^*(t-1) \end{bmatrix} \\ \boldsymbol{X}(t) &= \begin{bmatrix} x_1(t) & x_2^*(t) \\ x_2(t) & -x_1^*(t) \end{bmatrix} = \begin{bmatrix} x_1(t-1)s_1(t) & x_2^*(t-1)s_2^*(t) \\ x_2(t-1)s_2(t) & -x_1^*(t-1)s_1^*(t) \end{bmatrix} \end{aligned} \tag{9.9.3}$$

(a) Write down an expression for the received data, given two consecutively transmitted code matrices $\boldsymbol{X}(t-1), \boldsymbol{X}(t)$.

(b) Write down the ML criterion for detection of $s_1(t), s_2(t)$, given the received matrices at time t and $t-1$.

(c) Does the ML criterion have a unique minimum with respect to $s_1(t), s_2(t)$? Can the system work?

9. Prove (9.3.20) in a "calculus-like" way, by taking derivatives of the likelihood function with respect to the elements in $\mathbf{\Lambda}$ and setting them to zero.

10. Propose an extension of the detection methods for spatially colored noise (see Sections 9.1.2 and 9.3.2), which is able to take into account the *temporal* correlation of the noise and interference as well. Discuss cases when this could be important in practice.

11. Devise an extension of the algorithms in Section 9.5 applicable to ST-OFDM by formulating a likelihood function for joint time-domain channel estimation and detection of the symbols on all subcarriers. Discuss some advantages and disadvantages of the so-obtained method.

12. Prove (9.2.5) and (9.2.6).

13. Study a system with one transmit antenna and two receive antennas, where the received data is a vector of dimension 2×1 given by:

$$\boldsymbol{y} = \boldsymbol{h}s + \boldsymbol{e} \tag{9.9.4}$$

Here s is the transmitted symbol, $\boldsymbol{h}$ is a channel vector of dimension 2×1, and $\boldsymbol{e}$ is noise.

Suppose that $\boldsymbol{h}$ is known to the receiver and that we want to determine s given $\boldsymbol{y}$. Assume that

$$\boldsymbol{e} \sim N_C(\mathbf{0}, \mathbf{\Lambda}) \tag{9.9.5}$$

where $\mathbf{\Lambda}$ is a covariance matrix that is known to the receiver. Then we know from Section 9.1.2 that ML detection amounts to minimizing:

$$\left\| \mathbf{\Lambda}^{-1/2}(\boldsymbol{y} - \boldsymbol{h}s) \right\|^2 = (\boldsymbol{y} - \boldsymbol{h}s)^H \mathbf{\Lambda}^{-1}(\boldsymbol{y} - \boldsymbol{h}s) \tag{9.9.6}$$

with respect to s.

In this problem we study the case when $\mathbf{\Lambda}$ has the following form:

$$\mathbf{\Lambda} = \boldsymbol{\lambda}\boldsymbol{\lambda}^H + \sigma^2 \boldsymbol{I} \tag{9.9.7}$$

where $\boldsymbol{\lambda}$ is some vector of dimension 2×1 and σ^2 is a very small number; hence $\mathbf{\Lambda}$ is a nearly singular matrix. Such a covariance matrix can occur, for example, in situations with very strong interference (see Chapter 11).

(a) Apply the matrix inversion lemma (see Exercise A.34 on page 246) to the matrix $\sigma^2 \boldsymbol{\Lambda}^{-1}$ and simplify the so-obtained expression.

(b) Let $\boldsymbol{\lambda}^{\perp}$ be a vector of unit norm that is orthogonal to $\boldsymbol{\lambda}$. Explain how $\sigma^2 \boldsymbol{\Lambda}^{-1}$ can be expressed in terms of $\boldsymbol{\lambda}^{\perp}$ when $\sigma^2 \to 0$.

(c) Use the result of (b) to determine a simple expression for the maximum-likelihood detector that is valid approximately in the asymptotic case when $\sigma^2 \to 0$. Simplify your result and express it using $\boldsymbol{\lambda}^{\perp}$.

14. Consider a system using linear STBC on a frequency-flat channel, and assume that the receiver knows the channel. Determine the pairwise error probability when a zero-forcing detector is used at the receiver.

Hint: Some ideas from Theorem 4.2 (see page 45) and its proof may be useful.

15. Formulate a criterion for the optimality of the training sequence in a ST-OFDM system and try to find a sequence that satisfies this criterion.

Hint: Theorem 9.1 (see page 176) may be useful.

Chapter 10

SPACE-TIME CODING FOR INFORMED TRANSMITTERS

All the discussion so far in this book (with the exception of Sections 3.3 and 6.1) has focused on the case when the channel is unknown to the transmitter. In this chapter, we will briefly study how partial channel knowledge at the transmitter can be used to improve the system performance. In particular we will study linearly precoded STBC.

10.1 Introduction

If the transmitter knows the channel, then it is optimal from an error probability point of view to use what we referred to in Section 6.1 as *beamforming*. In the case of one receive antenna ($n_r = 1$), beamforming amounts to transmitting a symbol weighted by the conjugate transpose of the channel, $\boldsymbol{h}^*$ (for $n_r = 1$, $\boldsymbol{H}$ becomes a row vector $\boldsymbol{h}^T$). Although doing so might lack the interpretation of beamforming in a strict physical sense (depending on the antenna configuration), it shares the main signal processing attributes thereof.

For a given transmit power, the performance obtained via transmit diversity using STBC is in general inferior to that of beamforming, or receive diversity, by a factor that is sometimes called the "array gain." Loosely speaking, this is so since space-time coding methods spread power uniformly in all directions in space, while beamforming uses information about the channel to steer energy in the particular direction of the receiver. The performance gap between space-time coding and transmit beamforming can be significant. For instance, it is 3 dB for a ($n_t = 2, n_r = 1$) system using the Alamouti code (see Section 6.3.1).

A feedback of partial (or full) channel state information from the receiver to the transmitter may be used to close the performance gap between STBC and beamforming. This fact has been exploited by various researchers, and a host of feedback techniques have recently appeared in the literature (see, e.g., [JÖNGREN ET AL., 2002], [MUKKAVILLI ET AL., 2001], [VISOTSKY AND MADHOW, 2001], [NARULA ET AL., 1998], [SAMPATH AND PAULRAJ, 2002], [BHASHYAM ET AL., 2002], [MOUSTAKAS AND SIMON, 2002], [SIMON AND MOUSTAKAS, 2002], [JAFAR ET AL., 2001]). In this chapter, we look at the situation of a (partly) known channel at the transmitter from a general perspective.

10.2 Information Theoretical Considerations

Let $\boldsymbol{H}$ be the channel gain matrix corresponding to a frequency independent MIMO channel as in Section 2.1, and consider a stream of transmitted zero-mean vectors $\{\boldsymbol{x}(n)\}$, each of length n_t. Let

$$\boldsymbol{P} = E\left[\boldsymbol{x}(n)\boldsymbol{x}^H(n)\right] \tag{10.2.1}$$

be the covariance matrix of these symbol vectors. Then the bandwidth-normalized information-theoretic channel capacity is given by (cf. Theorem 3.2 (see page 25)):

$$C(\boldsymbol{H}) = \log_2\left|\boldsymbol{I} + \frac{\boldsymbol{H}\boldsymbol{P}\boldsymbol{H}^H}{\sigma^2}\right| \tag{10.2.2}$$

where σ^2 is the variance of the noise that affects each receive antenna element. We know from Theorem 3.4 (see page 29) that the transmit correlation matrix $\boldsymbol{P}$ that maximizes the average capacity $E_{\boldsymbol{H}}[C(\boldsymbol{H})]$ with respect to $\boldsymbol{P}$ under the power constraint $\mathrm{Tr}\{\boldsymbol{P}\} \leq P$, if $\boldsymbol{H}$ is a Gaussian random matrix as discussed in Chapter 2, is given by:

$$\boldsymbol{P} = \frac{P}{n_t} \cdot \boldsymbol{I} \tag{10.2.3}$$

Note that if we use orthogonal STBC, we transmit matrices that are proportional to a unitary matrix (cf. (7.4.1)) and hence the associated transmit correlation matrix is given by (10.2.3), provided that the transmitted data is normalized such that the transmission power is equal to P. Therefore OSTBC satisfies the condition that the transmit correlation matrix should be proportional to a unitary matrix (which is the choice that maximizes the mutual information). However, this condition is *not sufficient* for a transmission scheme to achieve the channel capacity.

If the channel is known to the transmitter, there are in principle two different approaches. One possibility is to use beamforming (see Section 6.1) to minimize the error probability for a fixed stream of (scalar) complex symbols to be transmitted. Alternatively, we can use coding over space and time with a transmit correlation matrix $\boldsymbol{P}$ adapted such that $C(\boldsymbol{H})$ is maximized (for the channel $\boldsymbol{H}$ under consideration). We recall from Theorem 3.3 (see page 27) that in such a case the available transmit power should be distributed on the dominant eigenvectors of the MIMO channel. Not all these eigenvectors are used; depending on the SNR, between one and all of the eigen-modes of the channel are exploited. By applying Theorem 3.3 we can easily compute, for a given $\boldsymbol{H}$, the transmit correlation matrix $\boldsymbol{P}$ that maximizes the mutual information between the transmitter and receiver.

10.3 STBC with Linear Precoding

Suppose that a scheme is employed where the transmitter pre-multiplies the STBC matrix $\boldsymbol{X}$ with a weighting matrix $\boldsymbol{W}$ taken from a finite matrix constellation $\Omega = \{\boldsymbol{W}_1, \ldots, \boldsymbol{W}_K\}$; taking $K < \infty$ appears to be practical since feedback in real systems will always be of a quantized nature. Doing so we obtain the following transmission model:

$$\boldsymbol{Y} = \boldsymbol{HWX} + \boldsymbol{E} \tag{10.3.1}$$

The precoding matrix $\boldsymbol{W}$ is effectively absorbed into the channel, and the matrix $\boldsymbol{X}$ can be viewed as if it "propagates" over an effective channel $\boldsymbol{HW}$. For that reason, from a receiver point of view, the system is equivalent to an unweighted system. It is not necessary that the matrices $\{\boldsymbol{W}_k\}$ are square. For instance, an important special case of linear precoding is *antenna selection*, in which case $\boldsymbol{W}$ simply contains selected columns of the identity matrix. Antenna selection has been studied in a variety of papers, including [Sandhu et al., 2000], [Heath et al., 2001b], [Gorokhov, 2002], [Molisch et al., 2001], [Gore and Paulraj, 2002].

Example 10.1: Linear precoding for OSTBC to maximize the mutual information.
If $\boldsymbol{X}$ is an OSTBC matrix, the transmit correlation matrix associated with (10.3.1) is (cf. Theorem 7.1; see page 102), if $E[|s_n|^2] = 1$:

$$\boldsymbol{P} = \frac{P}{n_t} \cdot \boldsymbol{WW}^H \tag{10.3.2}$$

The matrix $\boldsymbol{W}$ can be chosen to maximize the mutual information between the transmitter and the receiver simply by applying Theorem 3.3 to maximize

$$\left| \boldsymbol{I} + \frac{P}{n_t \sigma^2} \boldsymbol{WW}^H \boldsymbol{H}^H \boldsymbol{H} \right| \tag{10.3.3}$$

subject to a constraint of the form $\mathrm{Tr}\{\boldsymbol{WW}^H\} \leq 1$. Doing so gives an expression for $\boldsymbol{WW}^H$, which determines the optimal precoding matrix $\boldsymbol{W}$ to within a unitary similarity transformation.

Note that if $\boldsymbol{H}$ is known, true beamforming (see Section 6.1) could be used in lieu of linear precoding as in (10.3.1). However, in particular in the presence of feedback errors, the combination of space-time coding and linear precoding appears to be a good way of guaranteeing robustness of the system (see Section 10.3.3 for an illustration of this). Also, instead of using the mutual information as an optimality criterion, we could also attempt to find $\boldsymbol{W}$ by minimizing the probability of a detection error. This will be discussed in Section 10.3.2 (for the more general case when only the *statistics* of $\boldsymbol{H}$ are known). ■

10.3.1 Quantized Feedback and Diversity

An important question is under what conditions a system with quantized feedback ($K < \infty$) achieves the same order of diversity as without feedback (i.e., when $\boldsymbol{W} = \boldsymbol{I}$). The following result (which is due to [LARSSON ET AL., 2002A]) provides a sufficient condition for this to happen.

Theorem 10.1: Quantized Feedback ($K < \infty$) and Diversity.
Suppose that $\boldsymbol{X}_k - \boldsymbol{X}_n$ has full rank for all pairs of matrices $\boldsymbol{X}_k \neq \boldsymbol{X}_n$. If all matrices $\{\boldsymbol{W}_k\}$ have rank n_t, the system achieves the same diversity order $(n_r n_t)$ as if $\boldsymbol{W} = \boldsymbol{I}$ was always chosen, regardless of the feedback rule.

Proof: Suppose for simplicity that the elements of $\boldsymbol{H}$ are independent and complex Gaussian with variance ρ^2 (we leave the extension to the general case as an exercise for the reader). Since $\boldsymbol{W}$ depends on $\boldsymbol{H}$ via the feedback rule, the statistical distribution of $\boldsymbol{W}\boldsymbol{H}$ in general is not Gaussian and hence the result does *not* follow from the analysis in Chapter 4. Instead, to prove the theorem, we proceed as follows.

First note that the behavior of the feedback rule, including any quantization effects or transmission errors associated with the feedback link, is captured by the set of conditional probabilities $\{P(\boldsymbol{W}_k|\boldsymbol{H})\}$. The probability of wrongly detecting a codeword $\boldsymbol{X}$ in lieu of a transmitted codeword $\boldsymbol{X}_0$, for a given $\boldsymbol{H}$, can be upper bounded by the Chernoff bound and Theorem 4.2 (see page 45):

$$P(\boldsymbol{X}_0 \to \boldsymbol{X}|\boldsymbol{H}, \boldsymbol{W}) \leq \exp\left(-\frac{1}{4\sigma^2}\|\boldsymbol{H}\boldsymbol{W}(\boldsymbol{X}_0 - \boldsymbol{X})\|^2\right) \tag{10.3.4}$$

To establish the diversity order, we must average (10.3.4) with respect to $\boldsymbol{H}$ and $\boldsymbol{W}$. We have that (note that $\boldsymbol{W}$ depends on $\boldsymbol{H}$):

$$P(\boldsymbol{X}_0 \to \boldsymbol{X}|\boldsymbol{H}) = \sum_{k=1}^{K} P(\boldsymbol{X}_0 \to \boldsymbol{X}|\boldsymbol{H}, \boldsymbol{W}_k)P(\boldsymbol{W}_k|\boldsymbol{H}) \tag{10.3.5}$$

Averaging (10.3.5) over $\boldsymbol{H}$ using

$$p(\boldsymbol{H}) = \frac{1}{\pi^{n_r n_t} \rho^{2 n_r n_t}} \exp\Big(-\frac{1}{\rho^2}\|\boldsymbol{H}\|^2\Big) \tag{10.3.6}$$

along with (10.3.4) and the fact that $P(\boldsymbol{W}_k|\boldsymbol{H}) \leq 1$ gives

$$\begin{aligned} P(\boldsymbol{X}_0 \to \boldsymbol{X}) &= \sum_{k=1}^{K} \int d\boldsymbol{H}\; P(\boldsymbol{X}_0 \to \boldsymbol{X}|\boldsymbol{H}, \boldsymbol{W}_k) P(\boldsymbol{W}_k|\boldsymbol{H}) p(\boldsymbol{H}) \\ &\leq (\pi\rho^2)^{-n_r n_t} \int d\boldsymbol{H} \sum_{k=1}^{K} \exp\Big(-\frac{1}{4\sigma^2}\|\boldsymbol{H}\boldsymbol{W}_k(\boldsymbol{X}_0 - \boldsymbol{X})\|^2 - \frac{1}{\rho^2}\|\boldsymbol{H}\|^2\Big) \\ &= \sum_{k=1}^{K} \Big|\frac{\rho^2}{4\sigma^2}\boldsymbol{W}_k(\boldsymbol{X}_0 - \boldsymbol{X})(\boldsymbol{X}_0 - \boldsymbol{X})^H \boldsymbol{W}_k^H + \boldsymbol{I}\Big|^{-n_r} \\ &\leq \sum_{k=1}^{K} |\boldsymbol{W}_k(\boldsymbol{X}_0 - \boldsymbol{X})(\boldsymbol{X}_0 - \boldsymbol{X})^H \boldsymbol{W}_k^H|^{-n_r} \Big(\frac{\rho^2}{4\sigma^2}\Big)^{-n_r n_t} \end{aligned} \tag{10.3.7}$$

Hence the error rate decays as $\sigma^{2n_r n_t}$ which proves the claim about diversity. ■

Although Theorem 10.1 does not fully answer the question about choosing an appropriate feedback rule, the result is encouraging as it does guarantee that for quite general feedback weights in Ω, maximal diversity is achieved no matter how "poor" the quality of the feedback is. Note also that the proof of Theorem 10.1 can be extended (under certain circumstances) to the case when $\Omega = \{\boldsymbol{W}_k\}$ is an infinite (possibly uncountable) set, although doing so may be of less interest since in practice the feedback information will likely have a quantized nature.

10.3.2 Linear Precoding for Known Fading Statistics

Recall that space-time block coding was derived with the sole purpose of being able to achieve transmit diversity in the case that the transmitter does not know the channel. Hence, if $\boldsymbol{H}$ is completely known at the transmitter, there is not much reason to use space-time coding techniques at all since in that case we can optimize the transmitted matrix $\boldsymbol{W}\boldsymbol{X}$ directly by using an optimal beamformer as in Section 6.1 (see, e.g., [SAMPATH, 2001] for further developments along these lines). Nevertheless, when $\boldsymbol{H}$ is *partially* known, we can use space-time coding and optimize only the $\boldsymbol{W}$ part of $\boldsymbol{W}\boldsymbol{X}$ according to certain criteria.

In this subsection we assume that the channel is unknown to the transmitter, but that the covariance matrix associated with the fading process (see Section 2.1.2), viz.,

$$\boldsymbol{R} = E[\boldsymbol{h}\boldsymbol{h}^H] = \boldsymbol{R}_t^T \otimes \boldsymbol{R}_r \tag{10.3.8}$$

is known (here $\boldsymbol{h} = \text{vec}(\boldsymbol{H})$). In other words, we assume that the statistics of the fading are known.

Maximizing the Mutual Information

Suppose that $\boldsymbol{R}_r = \boldsymbol{I}$ but that $\boldsymbol{R}_t$ is a general nonsingular known matrix. Then we can maximize the average mutual information between the transmitter and the receiver with respect to $\boldsymbol{W}$ in (10.3.1) (or $\boldsymbol{P}$ in (10.3.2)) simply by observing that in this case, $\boldsymbol{H} = \boldsymbol{H}_w \boldsymbol{R}_t^{1/2}$ where $\boldsymbol{H}_w$ is a matrix with i.i.d. Gaussian elements. Consequently, we need to maximize

$$\begin{aligned} E_{\boldsymbol{H}}[C(\boldsymbol{H})] &= E_{\boldsymbol{H}}\left[\log\left|\boldsymbol{I} + \frac{\boldsymbol{H}\boldsymbol{W}\boldsymbol{W}^H\boldsymbol{H}^H}{n_t\sigma^2}P\right|\right] \\ &= E_{\boldsymbol{H}}\left[\log\left|\boldsymbol{I} + \frac{\boldsymbol{H}_w\boldsymbol{R}_t^{1/2}\boldsymbol{W}\boldsymbol{W}^H\boldsymbol{R}_t^{1/2}\boldsymbol{H}_w^H}{n_t\sigma^2}P\right|\right] \end{aligned} \tag{10.3.9}$$

which is maximized if $\boldsymbol{W}\boldsymbol{W}^H = \boldsymbol{R}_t^{-1}$ (cf. Theorem 3.4 (see page 29)). The choice of $\boldsymbol{W}\boldsymbol{W}^H = \boldsymbol{R}_t^{-1}$ can be interpreted as a *pre-whitening* of the channel.

Minimizing the Error Probability

A different approach to the optimization of $\boldsymbol{W}$ is to minimize an upper bound on the average probability of a detection error. This procedure, which was proposed in [SAMPATH AND PAULRAJ, 2002], works even in the case when $\boldsymbol{R}_t$ is rank deficient, and $\boldsymbol{R}_r$ is a general positive semi-definite matrix. Moreover, interestingly enough, even though this approach does not directly relate to the information theoretical channel capacity approach, it leads to an optimization problem that can be solved using the water-filling result in Theorem 3.3 (see page 27).

If $\boldsymbol{X}_0$ is transmitted, then the average (over the channel) probability that an incorrect codeword $\boldsymbol{X}$ is decided in lieu of $\boldsymbol{X}_0$ can be upper bounded by using Theorem 4.2 (see page 45):

$$\begin{aligned} &E_{\boldsymbol{H}}[P(\boldsymbol{X}_0 \to \boldsymbol{X})] \\ \leq &\left|\boldsymbol{I} + \frac{1}{4\sigma^2}((\boldsymbol{X}_0 - \boldsymbol{X})^T\boldsymbol{W}^T \otimes \boldsymbol{I})(\boldsymbol{R}_t^T \otimes \boldsymbol{R}_r)(\boldsymbol{W}^*(\boldsymbol{X}_0 - \boldsymbol{X})^* \otimes \boldsymbol{I})\right|^{-n_r} \\ = &\left|\boldsymbol{I} + \frac{1}{4\sigma^2}((\boldsymbol{X}_0 - \boldsymbol{X})^T\boldsymbol{W}^T\boldsymbol{R}_t^T\boldsymbol{W}^*(\boldsymbol{X}_0 - \boldsymbol{X})^*) \otimes \boldsymbol{R}_r\right|^{-n_r} \end{aligned} \tag{10.3.10}$$

Application of the Schur complement formula [HORN AND JOHNSON, 1985, TH. 7.7.6] shows that the determinant of a positive semi-definite block matrix is less than the product of the determinants of its diagonal blocks (if the block size is equal to one, this relation is called the Hadamard inequality [HORN AND JOHNSON, 1985, TH. 7.8.1]). Hence, there is a constant c (that depends on $\boldsymbol{R}_r$) such that

$$E_{\boldsymbol{H}}[P(\boldsymbol{X}_0 \to \boldsymbol{X})] \leq c \cdot \left| \boldsymbol{I} + \frac{1}{4\sigma^2}(\boldsymbol{X}_0 - \boldsymbol{X})(\boldsymbol{X}_0 - \boldsymbol{X})^H \cdot \boldsymbol{W}^H \boldsymbol{R}_t \boldsymbol{W} \right|^{-n_r} \tag{10.3.11}$$

This expression can be minimized (subject to a power constraint) by using the same tools as discussed earlier in this chapter.

Example 10.2: Linear precoding for OSTBC.
For OSTBC, $(\boldsymbol{X}_0 - \boldsymbol{X})(\boldsymbol{X}_0 - \boldsymbol{X})^H$ is proportional to the identity matrix. Hence, if $E[|s_n|^2] = 1$, the $\boldsymbol{W}$ that minimizes the above bound (10.3.11) is given by:

$$\min_{\boldsymbol{W}} \left| \boldsymbol{I} + \frac{n_s}{4\sigma^2} \boldsymbol{W}^H \boldsymbol{R}_t \boldsymbol{W} \right| \tag{10.3.12}$$

which is easily found by using Theorem 3.3 (see page 27). ■

10.3.3 OSTBC with One-Bit Feedback for $n_t = 2$

An extremely simple scheme which utilizes *diagonal* weighting matrices $\boldsymbol{W}$ together with orthogonal STBC was suggested in [GANESAN ET AL., 2002]. For the case of $n_t = 2$ transmit antennas and $n_r = 1$ receive antenna, the scheme assumes that a feedback consisting of *one bit* is present, and that this information bit is used to choose one of the following two weighting matrices:

$$\boldsymbol{W}_1 = \begin{bmatrix} |a|^2 & 0 \\ 0 & 1 - |a|^2 \end{bmatrix}^{1/2}, \quad \boldsymbol{W}_2 = \begin{bmatrix} 1 - |a|^2 & 0 \\ 0 & |a|^2 \end{bmatrix}^{1/2} \tag{10.3.13}$$

for some constant a that satisfies $|a| \leq 1$.

Let h_1 and h_2 be the two elements of $\boldsymbol{H} = [h_1\ h_2] \triangleq \boldsymbol{h}^T$. Common sense suggests, and it is easily proved analytically, that provided that there is no error involved in the transmission of the feedback information the best feedback strategy, from an error probability point of view, is to use $\boldsymbol{W}_1$ if $|h_1| > |h_2|$ and $\boldsymbol{W}_2$ otherwise, where $\boldsymbol{W}_1$ and $\boldsymbol{W}_2$ are given by (10.3.13) with $|a|^2 = 1$. Apparently, since doing so amounts to using only one of the two transmit antennas at a given time, no OSTBC would be necessary. However, if errors are present in the feedback information, a value $|a|^2 \neq 1$ can be used to obtain a robust weighted transmission scheme, and moreover the use of OSTBC becomes necessary, as explained below.

We can quantify the performance of the above scheme analytically. Let P_c be the probability that the feedback bit is correct. For a given $\boldsymbol{h}$, the probability of an error can be bounded by:

$$P(\boldsymbol{X}_0 \to \boldsymbol{X}|\boldsymbol{H}, \boldsymbol{W}_k) \leq \exp\Big(- \frac{\gamma^2}{2\sigma^2}\|\boldsymbol{H}\boldsymbol{W}_k\|^2\Big) \tag{10.3.14}$$

where γ^2 is a constant that depends on the constellation. Let $h_{\max}$ and $h_{\min}$ denote the elements of $\boldsymbol{h}$ with the largest and smallest magnitude, respectively. Then the norm of the effective channel becomes, for the case that the feedback bit is correct and incorrect, respectively:

$$\begin{aligned} \|\boldsymbol{h}^T\boldsymbol{W}\|^2\Big|_{\text{correct}} &= \|[h_{\max}\ h_{\min}]\boldsymbol{W}_1\|^2 = |a|^2|h_{\max}|^2 + (1-|a|^2)|h_{\min}|^2 \\ \|\boldsymbol{h}^T\boldsymbol{W}\|^2\Big|_{\text{wrong}} &= \|[h_{\max}\ h_{\min}]\boldsymbol{W}_2\|^2 = |a|^2|h_{\min}|^2 + (1-|a|^2)|h_{\max}|^2 \end{aligned} \tag{10.3.15}$$

The associated error probability, for a given $\boldsymbol{h}$, is upper bounded by the Chernoff bound (cf. (10.3.14)):

$$\begin{aligned} P(\boldsymbol{X}_0 \to \boldsymbol{X}|\boldsymbol{h}) \leq & P_c \cdot \exp\Big(- \frac{\gamma^2}{2\sigma^2}(|a|^2|h_{\max}|^2 + (1-|a|^2)|h_{\min}|^2)\Big) \\ & + (1-P_c)\exp\Big(- \frac{\gamma^2}{2\sigma^2}(|a|^2|h_{\min}|^2 + (1-|a|^2)|h_{\max}|^2)\Big) \end{aligned} \tag{10.3.16}$$

By using order statistics [PAPOULIS, 2002, SEC. 6.2] it can be shown that the joint p.d.f. of $(|h_{\min}|^2, |h_{\max}|^2)$ is given by

$$p_{\{|h_{\min}|^2, |h_{\max}|^2\}}(x,y) = \frac{2}{\rho^4}\exp\Big(- \frac{x+y}{\rho^2}\Big) \tag{10.3.17}$$

for $x > y$ (and zero otherwise). Averaging (10.3.16) over $(|h_{\min}|^2, |h_{\max}|^2)$ gives:

$$\begin{aligned} &E_{\boldsymbol{h}}[P(\boldsymbol{X}_0 \to \boldsymbol{X})] \\ &\leq \frac{2P_c}{\rho^4}\int_0^\infty\int_y^\infty dxdy \exp\Big(-\frac{\gamma^2}{2\sigma^2}(|a|^2x + (1-|a|^2)y) - \frac{1}{\rho^2}(x+y)\Big) \\ &+ \frac{2(1-P_c)}{\rho^4}\int_0^\infty\int_y^\infty dxdy \exp\Big(-\frac{\gamma^2}{2\sigma^2}((1-|a|^2)x + |a|^2y) - \frac{1}{\rho^2}(x+y)\Big) \\ &= \frac{2}{\phi^2+2}\Big(\frac{P_c}{\phi^2|a|^2+1} + \frac{1-P_c}{\phi^2(1-|a|^2)+1}\Big) \\ &\leq \frac{2}{(\phi^2)^2}\cdot\Big(\frac{P_c}{|a|^2} + \frac{1-P_c}{1-|a|^2}\Big) \end{aligned} \tag{10.3.18}$$

where the last inequality becomes tight for large SNR=$\phi^2 = \gamma^2/\sigma^2$. Equation (10.3.18) proves, in particular, that a diversity of order 2 is achieved regardless of P_c as long as $0 \neq |a|^2 \neq 1$, which reinforces the conclusion of Theorem 10.1. In addition, the bound (10.3.18) may be useful for optimizing the system performance (see [Ganesan et al., 2002], [Larsson et al., 2002a]).

In Figure 10.1 (see [Ganesan et al., 2002]), the performance of the OSTBC scheme with feedback was compared to that of an OSTBC scheme without feedback (through Monte-Carlo simulation). The elements of $\boldsymbol{H}$ are independent complex Gaussian random variables. The receiver was assumed to have perfect channel knowledge. In the case of feedback errors it was assumed that the receiver correctly identifies the weighting matrix used by the transmitter. A system with 2 transmit antennas and 1 receive antenna was considered and the Alamouti code (see Section 6.3.1) was used, with QPSK symbols. This code has a spectral efficiency of 2 bits/sec/Hz. The SNR was varied between 4 dB and 20 dB in steps of 2 dB and the BER was measured. From Figure 10.1 it is clear that in the absence of any feedback errors, the optimal diagonal weighting improves the performance by about 1.5 dB.

10.4 Summary and Discussion

Feedback schemes can be seen as an advanced form of power control, and they are likely to find their way into future MIMO systems. The wideband CDMA standard already incorporates feedback in a simple form via an option called "closed-loop transmit diversity" [Derryberry et al., 2002]. While the currently standardized feedback schemes are fairly simple, by increasing the amount of information

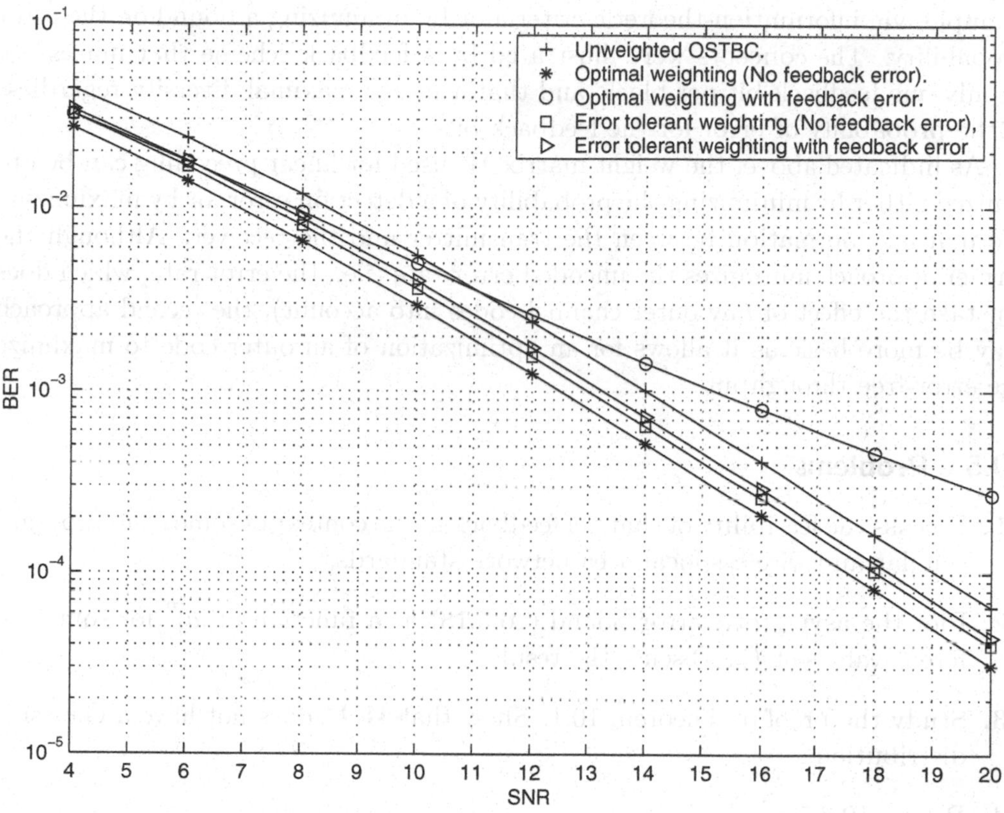

Figure 10.1. Comparison of performance for an OSTBC scheme without and with feedback. The system has two transmit antennas and one receive antenna and the spectral efficiency is 2 bits/second/Hz. The error tolerant weighting scheme was optimized for a feedback error of 5% and an SNR of 20 dB.

used at the transmitter it is possible to get closer and closer to the performance of beamforming (see Section 6.1).

This chapter has presented some fundamental principles for the design of feedback schemes. We started in Sections 10.1 and 10.2 with a general discussion from an information theoretic point of view. In Section 10.3 we introduced linearly precoded STBC. The main idea is that the transmitted STBC matrix $\boldsymbol{X}$ can be pre-multiplied (or precoded) by a weight matrix $\boldsymbol{W}$ before transmission, and that $\boldsymbol{W}$ can be optimized based on the (typically partial) channel knowledge available. We first proved a robustness result on the choice of $\boldsymbol{W}$ in the case of inaccurate or quantized feedback. Next we discussed various strategies for optimizing $\boldsymbol{W}$, for example, via information-theoretic criteria or by minimizing a bound on the error probability. The concepts were illustrated by a feedback scheme that makes use of only one feedback bit per block, and that achieves maximal diversity regardless of the probability of error for the feedback bit.

As indicated above, the weight matrix $\boldsymbol{W}$ used for linear precoding can be optimized either by minimizing the probability of a detection error, or by maximizing the mutual information between the transmitter and the receiver. Although the former approach minimizes the uncoded error rate (i.e., the error rate, which does not take the effect of any outer channel codes into account), the second approach may be more basic as it allows for an optimization of an outer code to maximize the error-free throughput.

10.5 Problems

1. Discuss the feasibility of channel feedback in the context of some contemporary cellular and wireless local area network standards.

2. Plot the asymptotic error bound (10.3.18) as a function of $|a|^2$ for some different values of P_c. Discuss the result.

3. Study the proof of Theorem 10.1. Show that $\boldsymbol{WH}$ does not have a Gaussian distribution.

4. Prove (10.3.7).

5. Extend Theorem 10.1 to channels $\boldsymbol{H}$ with a general covariance structure.

Chapter 11

SPACE-TIME CODING IN A MULTIUSER ENVIRONMENT

This short chapter is aimed at highlighting some issues related to the use of space-time coding in a multiuser environment. We will see that in general, the use of a space-time code changes the statistical properties of the transmitted signal compared with conventional transmission. We will also describe some simple techniques for multiuser interference suppression in a system that uses orthogonal STBC.

11.1 Introduction

Since the radio spectrum is a finite resource, the radio frequencies will always be shared. For this reason, all users in a system will suffer from *co-channel interference*, that is, disturbing radio signals from other users who use the same carrier frequency. The capacity of a system is related to how often, or how densely, the carrier frequencies are reused in the system. The more densely the frequencies are used, the higher the system capacity, but also the higher the level of co-channel interference. Therefore a signal processing algorithm that can suppress co-channel interference at the receiver, or maintain a functional communication link at a higher interference level, can also increase the system capacity. In a cellular system, the sharing of frequencies is usually coordinated by the network, whereas for some indoor local area networks, and so-called ad-hoc networks, it may be harder to assign radio resources in a coordinated manner; consequently the problem of mitigating co-channel interference may be even more important for such networks.

At the receiver, the co-channel interference can be mitigated in more than one way. Usually, a key element for suppressing co-channel interference is extra receive antennas. Hence, extra receive antennas can be used both for the purpose of counteracting fading (as described in the previous chapters), and to reduce the effect of co-channel interference. One approach to interference suppression is to try to decode the transmitted data corresponding to all co-channel users. This is usually referred to as *multiuser detection* (see, e.g., [VERDÚ, 1998] for a good survey of multiuser detection methods), and it is particularly important in the case when two users *intentionally* share one carrier frequency. Another approach to reducing co-channel interference is to use statistical modeling to *suppress* the interference. Since the co-channel interference consists of information-carrying signals, it has different statistical properties from those of the thermal receiver noise; moreover this interference may come from a particular direction in space. These properties of the co-channel interference can be exploited in a signal processing algorithm, and statistical interference suppression techniques have been quite successful for mitigation of interference in second generation cellular systems [KARLSSON AND HEINEGÅRD, 1996], [DAM ET AL., 1999]. Also note that although statistical interference models often do not make full use of the structure present in the co-channel signals, they have the important advantage of being able to model radio interference that originates from other external sources than co-channel users; for example, it is known that communication systems operating in the frequency band around 2.45 GHz can suffer severely from interference from a regular microwave oven.

11.2 Statistical Properties of Multiuser Interference

We discussed in Section 9.1.2 a simple algorithm for interference suppression which assumes that the sum of thermal noise and interference is temporally white but spatially colored. This assumption may approximate certain types of interference to a satisfactory degree, but in general it is not an accurate description of the interference generated by a competing user. One simple reason for this is that space-time coding in general introduces a temporal correlation (see below).

For a system that uses linear space-time block coding, we can study the statistical properties of the generated interference in the following way. Let $\boldsymbol{s}$ be an n_s-vector of complex symbols transmitted by an interferer and let

$$\boldsymbol{s}' = \begin{bmatrix} \bar{\boldsymbol{s}} \\ \tilde{\boldsymbol{s}} \end{bmatrix} \tag{11.2.1}$$

Assume that the transmitted (interfering) symbols are statistically white, i.e., that

$$E\left[\boldsymbol{s}'\boldsymbol{s}'^T\right] = \frac{\rho^2}{2} \cdot \boldsymbol{I} \tag{11.2.2}$$

(which gives a power of ρ^2 for the elements of $\boldsymbol{s}$). For linear STBC, the data matrix transmitted by the interferer is of dimension $n_t \times N$ and can be written:

$$\boldsymbol{X} = \sum_{n=1}^{n_s} (\bar{s}_n \boldsymbol{A}_n + i\tilde{s}_n \boldsymbol{B}_n) \tag{11.2.3}$$

The part of the received data that is due to co-channel interference and thermal noise can be written:

$$\boldsymbol{Y} = \boldsymbol{H}\boldsymbol{X} + \boldsymbol{E} \tag{11.2.4}$$

where all quantities were defined previously in the text. The average spatial coloration of thc interference generated by this system is essentially captured in the following covariance matrices (for a given $\boldsymbol{H}$):

$$\begin{aligned} E\left[\boldsymbol{Y}\boldsymbol{Y}^H\right] &= \boldsymbol{H}E\left[\boldsymbol{X}\boldsymbol{X}^H\right]\boldsymbol{H}^H + N \cdot \sigma^2 \boldsymbol{I} \\ E\left[\boldsymbol{Y}\boldsymbol{Y}^T\right] &= \boldsymbol{H}E\left[\boldsymbol{X}\boldsymbol{X}^T\right]\boldsymbol{H}^T \end{aligned} \tag{11.2.5}$$

Of course, the result can be extended to more than one interferer.

Example 11.1: Statistical properties of "conventional" interference.
Consider a system with $n_t = 1$ transmit antenna and $n_r \geq 1$ receive antennas. If we transmit a circularly symmetric white stream of symbols $\{s_n\}$ with variance ρ^2, the received signal has the following statistical properties:

$$\begin{aligned} E\left[\boldsymbol{Y}\boldsymbol{Y}^H\right] &= \rho^2 \cdot \boldsymbol{h}\boldsymbol{h}^H + N \cdot \sigma^2 \boldsymbol{I} \\ E\left[\boldsymbol{Y}\boldsymbol{Y}^T\right] &= \boldsymbol{0} \end{aligned} \tag{11.2.6}$$

For high SNR, the first matrix in (11.2.6) has a low-rank structure (its effective rank equals one). This is the reason why the interference suppression methods such as that in Section 9.1.2 work. ■

Example 11.2: Statistical properties of OSTBC interference.
For OSTBC, $\boldsymbol{X}\boldsymbol{X}^H = \|\boldsymbol{s}\|^2 \cdot \boldsymbol{I}$ by Theorem 7.1 (see page 102) and consequently

$$\begin{aligned} E\left[\boldsymbol{Y}\boldsymbol{Y}^H\right] &= E\left[\|\boldsymbol{s}\|^2\right] \cdot \boldsymbol{H}\boldsymbol{H}^H + N \cdot \sigma^2 \boldsymbol{I} \\ &= n_s\rho^2 \cdot \boldsymbol{H}\boldsymbol{H}^H + N \cdot \sigma^2 \boldsymbol{I} \end{aligned} \tag{11.2.7}$$

Furthermore, since $\{\boldsymbol{A}_n, \boldsymbol{B}_n\}$ are real-valued for OSTBC,

$$\begin{aligned} E\left[\boldsymbol{Y}\boldsymbol{Y}^T\right] &= \boldsymbol{H}E\left[\sum_{n=1}^{n_s}\sum_{k=1}^{n_s}\left(\bar{s}_n\bar{s}_k\boldsymbol{A}_n\boldsymbol{A}_k^T - \tilde{s}_n\tilde{s}_k\boldsymbol{B}_n\boldsymbol{B}_k^T\right.\right. \\ &\quad \left.\left. + i\bar{s}_n\tilde{s}_k\boldsymbol{A}_n\boldsymbol{B}_k^T + i\tilde{s}_n\bar{s}_k\boldsymbol{B}_n\boldsymbol{A}_k^T\right)\right]\boldsymbol{H}^T \\ &= \boldsymbol{0} \end{aligned} \tag{11.2.8}$$

under the assumption (11.2.2). Hence, unlike for conventional transmission, the spatial covariance of the interference produced by an OSTBC user does not have a low-rank structure. We can therefore expect an algorithm such as that in Section 9.1.2 to be less effective for the suppression of OSTBC multiuser interference. ■

The above formulas provide a measure of the spatial color of the signals generated by an interferer, but they do not describe the temporal coloration introduced by the coding. A more accurate description of the interference generated by linear STBC can be obtained as follows. By vectorizing the space-time received data we get (see Section 7.1):

$$\boldsymbol{y}' = \boldsymbol{F}'\boldsymbol{s}' + \boldsymbol{e}' \tag{11.2.9}$$

where

$$\boldsymbol{F} \triangleq \left[\operatorname{vec}(\boldsymbol{H}\boldsymbol{A}_1) \quad \cdots \quad \operatorname{vec}(\boldsymbol{H}\boldsymbol{A}_{n_s}) \quad i\operatorname{vec}(\boldsymbol{H}\boldsymbol{B}_1) \quad \cdots \quad i\operatorname{vec}(\boldsymbol{H}\boldsymbol{B}_{n_s})\right] \tag{11.2.10}$$

and

$$\boldsymbol{F}' = \begin{bmatrix} \bar{\boldsymbol{F}} \\ \tilde{\boldsymbol{F}} \end{bmatrix} \tag{11.2.11}$$

and where $\boldsymbol{y}'$ and $\boldsymbol{e}'$ are defined similarly to $\boldsymbol{s}'$. Using the same assumptions as before, we obtain the following covariance matrix that captures all spatial-temporal second order properties of the vector $\boldsymbol{y}'$:

$$E\left[\boldsymbol{y}'\boldsymbol{y}'^T\right] = \frac{\rho^2}{2}\begin{bmatrix} \bar{\boldsymbol{F}}\bar{\boldsymbol{F}}^T & \bar{\boldsymbol{F}}\tilde{\boldsymbol{F}}^T \\ \tilde{\boldsymbol{F}}\bar{\boldsymbol{F}}^T & \tilde{\boldsymbol{F}}\tilde{\boldsymbol{F}}^T \end{bmatrix} + \frac{\sigma^2}{2}\boldsymbol{I} \tag{11.2.12}$$

In general, this matrix is not proportional to the identity matrix. In particular, we leave to the reader the task to verify that for OSTBC the matrix in (11.2.12) is not proportional to the identity matrix. However, the analysis indicates that space-time coded signals usually have a covariance structure that can be used by an interference suppression algorithm that builds on (and extends) the ideas in Section 9.1.2. This observation motivates a more detailed study of interference cancellation for OSTBC.

11.3 OSTBC and Multiuser Interference

OSTBC has a structure that can enable co-channel interference suppression by relatively simple means, provided that the receiver knows the channel corresponding to all co-channel users.

11.3.1 The Algebraic Structure of OSTBC

Consider a system that uses the Alamouti space-time code. After appropriate complex conjugation, the received data $\boldsymbol{y}$ can be written as (see (6.3.9)):

$$\boldsymbol{y} = \boldsymbol{F}\boldsymbol{s} + \boldsymbol{e} \tag{11.3.1}$$

where $\boldsymbol{s} = [s_1 \; s_2]^T$ is a vector containing the two transmitted symbols, $\boldsymbol{e} = [e_1 \; e_2]^T$ is a noise vector and $\boldsymbol{F}$ is the following matrix:

$$\boldsymbol{F} = \begin{bmatrix} h_1 & h_2 \\ -h_2^* & h_1^* \end{bmatrix} \tag{11.3.2}$$

Such matrices have an interesting property.

Theorem 11.1: Algebraic structure of OSTBC for $n_t = 2$.
The set 2×2 *matrices with the structure (11.3.2) is closed under addition, multiplication, inversion and multiplication with a real-valued scalar.*

Proof: This result can be verified by some straightforward algebra. ■

Theorem 11.1 can be verified by the direct multiplication of relevant matrices, but we should also note that there are some more subtle connections between the set of matrices with the structure (11.3.2) and the field of complex numbers as well as the set of so-called *quaternions*. The properties of the complex number system should be well-known to the reader, but the quaternions may not be so. The quaternions were introduced by Hamilton in his study of the rotation of rigid bodies, and they have also been used in the special theory of relativity. The set of quaternions is an algebraic structure that is closed under multiplication and addition, and in which every nonzero element has an inverse. The quaternions therefore have some similarities with the real and complex numbers; however, the quaternions have four degrees of freedom compared to the complex numbers which have two and real numbers which have one, and in contrast to the real and complex numbers, the quaternions do not commute.

By using a matrix representation of complex numbers, it can be shown that the set of real-valued matrices with the structure (11.3.2) is isomorphic to the set of complex numbers. From this fact, the statement in Theorem 11.1 for real h_1,h_2 follows directly. Moreover, it can be shown that the set of *complex* matrices with the structure (11.3.2) is isomorphic to the set of quaternions. This has the implication that the set of these OSTBC matrices is closed under multiplication, addition and inversion, and from this fact the statement in Theorem 11.1 for complex matrices of the form (11.3.2) follows.

11.3.2 Suppression of Multiuser Interference in an OSTBC System

To illustrate the usefulness of Theorem 11.1, we consider for simplicity a scenario with two co-channel users A and B that are equipped with $n_t = 2$ transmit antennas each and that use the Alamouti code. We will also assume, for the simplicity of the discussion, that the receiver has $n_r = 2$ receive antennas. Let

$$\begin{aligned} \boldsymbol{\alpha} &= [\alpha_1 \; \alpha_2]^T \\ \boldsymbol{\beta} &= [\beta_1 \; \beta_2]^T \end{aligned} \tag{11.3.3}$$

be the symbol vectors corresponding to the users A and B and let

$$\begin{aligned} \boldsymbol{X}_{\boldsymbol{\alpha}} &= \begin{bmatrix} \alpha_1 & \alpha_2^* \\ \alpha_2 & -\alpha_1^* \end{bmatrix} \\ \boldsymbol{X}_{\boldsymbol{\beta}} &= \begin{bmatrix} \beta_1 & \beta_2^* \\ \beta_2 & -\beta_1^* \end{bmatrix} \end{aligned} \tag{11.3.4}$$

be the associated Alamouti matrices (cf. (6.3.1)). We use $\boldsymbol{\alpha}$ and $\boldsymbol{\beta}$ in this section instead of $\boldsymbol{s}$ as before to make the connection between the user A or B and its associated symbols. Also, let $\{\boldsymbol{A}_k, \boldsymbol{B}_k\}$ be the matrices with the following structure:

$$\begin{bmatrix} h_1 & h_2 \\ -h_2^* & h_1^* \end{bmatrix} \tag{11.3.5}$$

(see (6.3.8)) corresponding to user A and B and receive antenna number k, respectively. Furthermore, let $\boldsymbol{y}_k$ be the vector that contains the output of the kth receive antenna during the two time intervals (after complex conjugation as in Section 6.3.1). Then we can write:

$$\begin{bmatrix} \boldsymbol{y}_1 \\ \boldsymbol{y}_2 \end{bmatrix} = \begin{bmatrix} \boldsymbol{A}_1 & \boldsymbol{B}_1 \\ \boldsymbol{A}_2 & \boldsymbol{B}_2 \end{bmatrix} \begin{bmatrix} \boldsymbol{\alpha} \\ \boldsymbol{\beta} \end{bmatrix} + \begin{bmatrix} \boldsymbol{e}_1 \\ \boldsymbol{e}_2 \end{bmatrix} \tag{11.3.6}$$

where $\{\boldsymbol{e}_k\}$, $k = 1, 2$ are noise vectors.

The matrices $\{\boldsymbol{A}_k, \boldsymbol{B}_k\}$ have the structure discussed in Sections 6.3.1 and 11.3.1 and therefore they are proportional to unitary matrices:

$$\begin{aligned} \boldsymbol{A}_k^H \boldsymbol{A}_k &= \|\boldsymbol{a}_k\|^2 \boldsymbol{I} \\ \boldsymbol{B}_k^H \boldsymbol{B}_k &= \|\boldsymbol{b}_k\|^2 \boldsymbol{I} \end{aligned} \tag{11.3.7}$$

where $\boldsymbol{a}_k$ and $\boldsymbol{b}_k$ are vectors that contain the channel gains associated with user A and B, respectively. Moreover, by Theorem 11.1, the set of $\{\boldsymbol{A}_k, \boldsymbol{B}_k\}$ is closed under addition, multiplication and inversion.

Under the assumption that the elements of the noise vectors $\boldsymbol{e}_1$ and $\boldsymbol{e}_2$ in (11.3.6) are independent Gaussian random variables (which amounts to assuming that the receiver noise is spatially and temporally white), the ML metric for the joint detection of $\boldsymbol{\alpha}$ and $\boldsymbol{\beta}$ is given by:

$$\left\| \begin{bmatrix} \boldsymbol{y}_1 \\ \boldsymbol{y}_2 \end{bmatrix} - \begin{bmatrix} \boldsymbol{A}_1 & \boldsymbol{B}_1 \\ \boldsymbol{A}_2 & \boldsymbol{B}_2 \end{bmatrix} \begin{bmatrix} \boldsymbol{\alpha} \\ \boldsymbol{\beta} \end{bmatrix} \right\|^2 \tag{11.3.8}$$

Let us assume that the receiver knows $\{\boldsymbol{A}_k, \boldsymbol{B}_k\}$; this channel state information can be acquired at the receiver via training. Then the multiuser detection of $\boldsymbol{\alpha}$ and $\boldsymbol{\beta}$ (in a ML sense) amounts to minimizing (11.3.8) *jointly* with respect to $\boldsymbol{\alpha}$ and $\boldsymbol{\beta}$. Note that if the matrix

$$\begin{bmatrix} \boldsymbol{A}_1 & \boldsymbol{B}_1 \\ \boldsymbol{A}_2 & \boldsymbol{B}_2 \end{bmatrix} \tag{11.3.9}$$

had the properties of a complex orthogonal design (see Section 7.4), the detection of $\boldsymbol{\alpha}$ and $\boldsymbol{\beta}$ would decouple similarly to the detection of a single Alamouti-code user, see Section 6.3.1; however, unfortunately this is *not* the case. Consequently, the minimization of (11.3.8) may be a relatively computationally burdensome problem, even though fast search algorithms such as sphere decoding (see Section 9.1.3) can be used. Therefore, some suboptimal approaches have appeared in the literature.

For instance, [Naguib et al., 2000] (see also [Stamoulis et al., 2001]) observed that if we pre-multiply the received signals (11.3.6) with the following space-time beamforming matrix:

$$\begin{bmatrix} \boldsymbol{I} & -\frac{\boldsymbol{B}_1 \boldsymbol{B}_2^H}{\|\boldsymbol{b}_2\|^2} \end{bmatrix} \tag{11.3.10}$$

we obtain a signal vector of length two that depends only on $\boldsymbol{\alpha}$ and therefore is free of co-channel interference from the user B:

$$\check{\boldsymbol{y}} = \begin{bmatrix} \boldsymbol{I} & -\frac{\boldsymbol{B}_1\boldsymbol{B}_2^H}{\|\boldsymbol{b}_2\|^2} \end{bmatrix} \begin{bmatrix} \boldsymbol{y}_1 \\ \boldsymbol{y}_2 \end{bmatrix} = \boldsymbol{D}\boldsymbol{\alpha} + \left(\boldsymbol{e}_1 - \frac{\boldsymbol{B}_1\boldsymbol{B}_2^H}{\|\boldsymbol{b}_2\|^2}\boldsymbol{e}_2\right) \tag{11.3.11}$$

where

$$\boldsymbol{D} = \boldsymbol{A}_1 - \boldsymbol{B}_1\boldsymbol{B}_2^H\boldsymbol{A}_2/\|\boldsymbol{b}_2\|^2 \tag{11.3.12}$$

Since the set of matrices $\{\boldsymbol{A}_n, \boldsymbol{B}_n\}$ is closed under addition, multiplication and inversion, the matrix $\boldsymbol{D}$ in (11.3.12) has the same structure as (11.3.5). Furthermore, the covariance matrix of the noise vector in (11.3.11) is readily checked to be $\sigma^2(1+\|\boldsymbol{b}_1\|^2)\boldsymbol{I}$, and thus the "beamformed" noise is still spatially white. Hence the detection of $\boldsymbol{\alpha}$ from $\check{\boldsymbol{y}}$ in (11.3.11) is equivalent to the standard detection problem for a single Alamouti code user discussed in Section 6.3.1. The detector based on (11.3.11) is called a zero-forcing (ZF) detector in [Naguib et al., 2000]. The same reference also proposes a related algorithm that minimizes a certain mean-square error (MSE) criterion, a method that gives a slightly better performance than the ZF algorithm but for conciseness we will not discuss it here.

From a ML point of view, the approach of [Naguib et al., 2000], [Stamoulis et al., 2001] is not optimal. This can be intuitively understood as follows: after we pre-multiplied the received data with the matrix

$$\begin{bmatrix} \boldsymbol{I} & -\frac{\boldsymbol{B}_1\boldsymbol{B}_2^H}{\|\boldsymbol{b}_2\|^2} \end{bmatrix} \tag{11.3.13}$$

the so-obtained beamformed data are no longer a sufficient statistic for detection. In fact, the relationship between the previous technique of [Naguib et al., 2000], [Stamoulis et al., 2001] and the exact ML detection is easy to establish analytically. After some straightforward algebra, it follows that (11.3.8) is proportional to:

$$\left\| \acute{\boldsymbol{y}} - \boldsymbol{C}\boldsymbol{\alpha} - \frac{\|\boldsymbol{b}_1\|^2 + \|\boldsymbol{b}_2\|^2}{\|\boldsymbol{b}_2\|}\boldsymbol{\beta} \right\|^2 + \|\check{\boldsymbol{y}} - \boldsymbol{D}\boldsymbol{\alpha}\|^2 \tag{11.3.14}$$

where $\check{\boldsymbol{y}}$ and $\boldsymbol{D}$ are defined in (11.3.11) and (11.3.12) above, and:

$$\begin{aligned} \acute{\boldsymbol{y}} &= (\boldsymbol{B}_1^H\boldsymbol{y}_1 + \boldsymbol{B}_2^H\boldsymbol{y}_2)/\|\boldsymbol{b}_2\| \\ \boldsymbol{C} &= (\boldsymbol{B}_1^H\boldsymbol{A}_1 + \boldsymbol{B}_2^H\boldsymbol{A}_2)/\|\boldsymbol{b}_2\| \end{aligned} \tag{11.3.15}$$

In a similar manner, it can be shown that the ML metric in (11.3.8) is also proportional to:

$$\left\| \acute{z} - F\beta - \frac{\|a_1\|^2 + \|a_2\|^2}{\|a_2\|}\alpha \right\|^2 + \|\check{z} - G\beta\|^2 \tag{11.3.16}$$

where

$$\begin{aligned} \acute{z} &= (A_1^H y_1 + A_2^H y_2)/\|a_2\| \\ \check{z} &= y_1 - A_1 A_2^H y_2/\|a_2\|^2 \\ F &= (A_1^H B_1 + A_2^H B_2)/\|a_2\| \\ G &= B_1 - A_1 A_2^H B_2/\|a_2\|^2 \end{aligned} \tag{11.3.17}$$

In the light of these calculations, we can see that the ZF algorithm in [Naguib et al., 2000], [Stamoulis et al., 2001] obtains α by minimizing the second term in (11.3.14), and similarly β by minimizing the second term in (11.3.16):

$$\begin{aligned} &\min_{\alpha \in \mathcal{S}} \|\check{y} - D\alpha\|^2 \\ &\min_{\beta \in \mathcal{S}} \|\check{z} - G\beta\|^2 \end{aligned} \tag{11.3.18}$$

The detection criteria in (11.3.18) can be viewed as rough approximations of the ML criterion, obtained by neglecting a term that may be significant. This suggests that the difference in BER performance between the ZF detection scheme of [Naguib et al., 2000], [Stamoulis et al., 2001] (i.e., Equation (11.3.18)) and the optimum ML may not be marginal. In particular, it can be seen that the spatial diversity gain achieved by the methods of [Naguib et al., 2000], [Stamoulis et al., 2001] is not more than two; this should be contrasted with the fact that a single Alamouti OSTBC user transmitting over a 2×2 channel matrix achieves a spatial diversity gain of four. Also, the spatial diversity gain associated with the *joint* ML detection of the two co-channel users A and B (i.e., minimization of (11.3.8)) is equal to four. To see this, note that the samples of the (noise-free) received signal can be written as:

$$\begin{bmatrix} X_\alpha^T & 0 & X_\beta^T & 0 \\ 0 & X_\alpha^T & 0 & X_\beta^T \end{bmatrix} \begin{bmatrix} a_1 \\ a_2 \\ b_1 \\ b_2 \end{bmatrix} \triangleq \Gamma h \tag{11.3.19}$$

It holds that $\Gamma\Gamma^H = (\|\alpha\|^2 + \|\beta\|^2) I$ and consequently the rank of Γ is four for any nonzero α or β. Hence, provided that h is Gaussian distributed with a positive definite covariance matrix, Theorem 4.2 (see page 45) is applicable and it shows that the joint ML detection of the symbols corresponding to the two users gives a diversity of order four.

To summarize, the techniques of [NAGUIB ET AL., 2000], [STAMOULIS ET AL., 2001] rely on an interesting algebraic property of OSTBC (for $n_t = 2$) that we derived in Section 11.3.1. However, these techniques trade off performance for computational simplicity. We illustrate this fact in the following example.

Example 11.3: Multiuser Interference Suppression for OSTBC.
We consider a system with two co-channel (Alamouti code) users and assume that the receiver as well as each user has two antennas. The channel gains are independent circular Gaussian variables with variance ρ^2. The two users transmit QPSK symbols with unit power. The noise is spatially and temporally white circular Gaussian with variance σ^2 per antenna. We define the SNR as ρ^2/σ^2.

Figure 11.1 shows the estimated BER obtained by Monte-Carlo simulation versus the SNR for the different methods discussed above. The curve "1 user, 1 Rx" refers to the case of a single user with two transmit antennas and a receiver with one antenna. The curve "2 users, ZF" corresponds to the ZF method in [NAGUIB ET AL., 2000], [STAMOULIS ET AL., 2001] (or equivalently, Equation (9.4.9)). The curve "2 users, MMSE" corresponds to the MMSE method of [NAGUIB ET AL., 2000], [STAMOULIS ET AL., 2001] (not described in this section). The curve "2 users, ML" corresponds to the exact ML solution given by the minimizer of (11.3.8). Finally, the curve "1 user, 2 Rx" shows the BER in the case of a single user with two transmit antennas and a receiver with two transmit antennas. Note that the curves "1 user, 1 Rx" and "2 users, ZF" overlap.

Clearly, the gap between the performance of the exact ML multiuser detector and the performance of the suboptimal approaches discussed above can be quite significant. ■

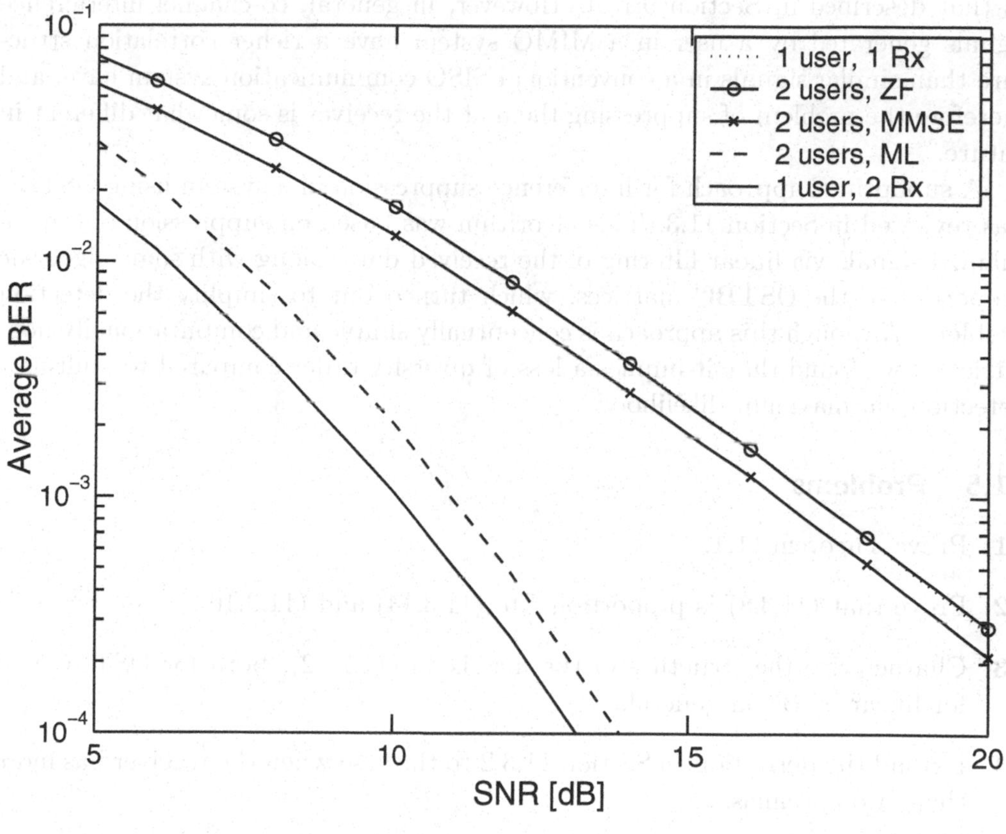

Figure 11.1. Results for Example 11.3 (multiuser detection for OSTBC).

11.4 Summary and Discussion

The goal of this chapter has been to briefly illustrate some problems that appear when multiple users share the same channel. We started in Sections 11.1 and 11.2 by discussing the statistical properties of the interference in a MIMO system. In principle, co-channel interference in a MIMO system can be suppressed via multiuser detection methods, or via statistical interference suppression methods (such as that described in Section 9.1.2). However, in general, co-channel interference signals generated by a user in a MIMO system have a richer correlation structure than similar signals in a conventional SISO communication system have, and therefore the problem of suppressing them at the receiver is somewhat different in nature.

A suboptimal approach for interference suppression in a system using OSTBC was reviewed in Section 11.3. This algorithm was based on suppression of the co-channel signals via linear filtering of the received data, along with some algebraic properties of the OSTBC matrices, which turned out to simplify the detection problem. Although this approach is conceptually simple and computationally non-intricate, we found that it implies a loss of diversity order compared to multiuser detection via maximum-likelihood.

11.5 Problems

1. Prove Theorem 11.1.

2. Prove that (11.3.8) is proportional to (11.3.14) and (11.3.16).

3. Characterize the structure of the matrix in (11.2.12), both for OSTBC and for linear STBC in general.

4. Extend the derivation in Section 11.3.2 to the case when the receiver has more than two antennas.

5. Can the results in Section 11.3 be extended to the case of more than two users?

Appendix A

SELECTED MATHEMATICAL BACKGROUND MATERIAL

This appendix is intended to help the reader by summarizing some known formulas and results from matrix algebra and probability theory. It is absolutely not a self-contained treatment of linear algebra or probability theory, nor is it a replacement for a systematic treatment or a course on these topics. However, it may be helpful as a refresher for students who have not studied these subjects for some time.

All the results presented here are well-known and their proofs can be found in many textbooks on linear algebra, probability theory and statistics. Since our discussion is somewhat scattered, and since we believe that the proof of some of the results can constitute appropriate problems for students when this book is used in a classroom setting, we present most of the results in the form of exercises.

A.1 Complex Baseband Representation of Bandpass Signals

Most signals in a wireless communication system have a narrowband character, and such narrowband (real-valued) signals can be represented with complex variables in a neat way. The general theory for doing so is treated in most digital communication textbooks (see, e.g., [PROAKIS, 2001, SEC. 4.1]), but since the complex baseband representation is a foundation for all signal models used in this book we offer a brief treatment of the topic.

Let $s(t)$ be a real-valued continuous-time signal that has a bandpass spectrum centered around $\omega_c = 2\pi f_c$ and let

$$S(f) = \int_{-\infty}^{\infty} s(t)e^{-i2\pi ft}dt \tag{A.1.1}$$

be its Fourier transform (or spectrum). Apparently,

$$S(-f) = \int_{-\infty}^{\infty} s(t)e^{i2\pi ft}dt = S^*(f) \tag{A.1.2}$$

Define the following spectrum, along with its time-domain signal:

$$\begin{aligned} S_+(f) &= \begin{cases} 2S(f) & f \geq 0 \\ 0 & \text{otherwise} \end{cases} \\ s_+(t) &= \int_{-\infty}^{\infty} S_+(f)e^{i2\pi ft}df \end{aligned} \tag{A.1.3}$$

The signal $s_+(t)$ is called the *analytic signal* and in general it is complex. By definition, its spectrum $S_+(f)$ is zero for $f \leq 0$ and narrowband around f_c. We can now define the following *complex baseband* signal $s_b(t)$ associated with $s(t)$, along with its spectrum $S_b(f)$:

$$\begin{aligned} s_b(t) &= s_+(t)e^{-i2\pi f_c t} \\ S_b(f) &= S_+(f + f_c) \end{aligned} \tag{A.1.4}$$

Although it lacks a clear physical meaning, it turns out that the complex baseband signal $s_b(t)$ is a mathematically convenient representation of $s(t)$.

We have that

$$\begin{aligned} \text{Re}\left\{s_b(t)e^{i2\pi f_c t}\right\} &= \text{Re}\left\{\int_{-\infty}^{\infty} S_+(f)e^{i2\pi ft}df\right\} \\ &= 2\text{Re}\left\{\int_{0}^{\infty} S(f)e^{i2\pi ft}df\right\} \\ &= \int_{0}^{\infty} S(f)e^{i2\pi ft}df + \int_{0}^{\infty} S^*(f)e^{-i2\pi ft}df \\ &= \int_{-\infty}^{\infty} S(f)e^{i2\pi ft}df = s(t) \end{aligned} \tag{A.1.5}$$

Hence, given the complex baseband signal $s_b(t)$ we can easily compute the corresponding narrowband signal $s(t)$. Note, however, that for a given $s(t)$, (A.1.5) does *not* define $s_b(t)$ uniquely. For instance, the complex baseband signal

$$\check{s}_b(t) = s_b(t) + ie^{-i2\pi f_c t} \tag{A.1.6}$$

yields the same $s(t)$ as does $s_b(t)$.

The quantity $|s_b(t)|$ is sometimes called the envelope of $s(t)$. Also, the signals

$$\begin{aligned} s_I(t) &= \text{Re}\{s_b(t)\} \\ s_Q(t) &= \text{Im}\{s_b(t)\} \end{aligned} \tag{A.1.7}$$

are called the in-phase and quadrature components of $s(t)$. We can see from (A.1.5) that

$$s(t) = |s_b(t)| \cdot \cos\left(2\pi f_c t + \arg[s_b(t)]\right) \tag{A.1.8}$$

and hence a phase shift of the narrowband signal $s(t)$ corresponds to a shift in the complex angle of $s_b(t)$.

Suppose that a linear system has a transfer function $H(f)$ that is narrowband around the center frequency f_c and that $h(t)$ is the associated impulse response. Assume that a narrowband (around f_c) signal is applied at the input of this linear system. Let

$$r(t) = \int_{-\infty}^{\infty} s(T)h(t-T)dT \tag{A.1.9}$$

Then it is possible to show (with obvious notation) that:

$$r_b(t) = \frac{1}{2}\int_{-\infty}^{\infty} s_b(T)h_b(t-T)dT \tag{A.1.10}$$

In other words, we can work with the complex baseband representations of both the signal and the linear system. Much of the usefulness of the complex baseband representation lies in this result.

A.2 Review of Some Concepts in Matrix Algebra

1. Trace of a matrix: The trace of a (square) matrix, $\text{Tr}\{\boldsymbol{X}\}$, is the sum of the diagonal elements of $\boldsymbol{X}$.

2. Frobenius norm of a matrix: The Frobenius norm of an $m \times n$ matrix $\boldsymbol{X}$ is defined as:

$$\|\boldsymbol{X}\|^2 = \sum_{s=1}^{m}\sum_{t=1}^{n} |X_{s,t}|^2 = \text{Tr}\{\boldsymbol{X}\boldsymbol{X}^H\} = \text{Tr}\{\boldsymbol{X}^H\boldsymbol{X}\} \tag{A.2.1}$$

   For vectors, the Frobenius norm becomes identical to the Euclidean vector norm.

3. Determinant of a matrix: The determinant $|\boldsymbol{X}|$ of a (square) matrix $\boldsymbol{X}$ can be defined as a linear combination of certain monoms involving different elements of $\boldsymbol{X}$. However, it can be shown that the determinant is also equal to the product of the eigenvalues of $\boldsymbol{X}$. For our purposes, this can be taken as the definition of the determinant.

4. Rank of a matrix: The column rank of a matrix is the number of linearly independent columns. The row rank of a matrix is the number of linearly independent rows. It turns out that the column rank and the row rank are always equal. Hence we often simply say the rank of a matrix.

5. Eigenvalue and eigenvector of a matrix: Let $\boldsymbol{X}$ be an arbitrary matrix. If the vector $\boldsymbol{u}$ and scalar λ are such that $\boldsymbol{X}\boldsymbol{u} = \lambda\boldsymbol{u}$, then $\boldsymbol{u}$ is called an eigenvector of $\boldsymbol{X}$, and the scalar λ is the corresponding eigenvalue.

6. Positive definite matrix: A matrix is positive definite if all its eigenvalues are positive. It is non-negative definite, or positive semi-definite, if all eigenvalues are non-negative.

7. Eigenvalue-decomposition: Suppose that $\boldsymbol{X}$ is a Hermitian matrix of dimension $m \times m$, i.e., $\boldsymbol{H}^H = \boldsymbol{H}$. Then there is an $m \times m$ diagonal matrix $\boldsymbol{\Delta}$ and a unitary matrix $\boldsymbol{T}$ such that:

$$\boldsymbol{X} = \boldsymbol{T}\boldsymbol{\Delta}\boldsymbol{T}^H \tag{A.2.2}$$

The pair $\{\boldsymbol{T}, \boldsymbol{\Delta}\}$ is called the eigenvalue-decomposition of $\boldsymbol{X}$. The diagonal elements of $\boldsymbol{\Delta}$ contain the (real-valued) eigenvalues of $\boldsymbol{X}$.

8. Singular value decomposition: Suppose that $\boldsymbol{X}$ is a general $m \times n$ matrix of rank r. Then there is an $r \times r$ diagonal matrix $\boldsymbol{\Delta}$, with positive diagonal elements, as well as two matrices $\boldsymbol{V}$ and $\boldsymbol{U}$ such that:

$$\begin{aligned} \boldsymbol{X} &= \boldsymbol{U}\boldsymbol{\Delta}\boldsymbol{V}^H \\ \boldsymbol{V}^H\boldsymbol{V} &= \boldsymbol{I} \\ \boldsymbol{U}^H\boldsymbol{U} &= \boldsymbol{I} \end{aligned} \tag{A.2.3}$$

The triplet $\{\boldsymbol{U}, \boldsymbol{V}, \boldsymbol{\Delta}\}$ is called the singular value decomposition of $\boldsymbol{X}$.

9. Vectorization of a matrix: Let $\boldsymbol{A}$ be an arbitrary $m \times n$ matrix. Then the vector $\mathrm{vec}(\boldsymbol{A})$ is obtained by stacking the columns of $\boldsymbol{A}$ on top of each other:

$$\mathrm{vec}(\boldsymbol{A}) = [A_{1,1}\ A_{2,1}\ \cdots\ A_{m,1}\ \cdots A_{m,n}]^T \tag{A.2.4}$$

10. **Kronecker product:** Let $\boldsymbol{A}$ be an arbitrary $m \times n$ matrix and let $\boldsymbol{B}$ be an arbitrary $\bar{m} \times \bar{n}$ matrix. Then the Kronecker product $\boldsymbol{A} \otimes \boldsymbol{B}$ is of dimension $m\bar{m} \times n\bar{n}$ and is defined as:

$$\boldsymbol{A} \otimes \boldsymbol{B} = \begin{bmatrix} A_{1,1}\boldsymbol{B} & \cdots & A_{1,n}\boldsymbol{B} \\ \vdots & & \vdots \\ A_{m,1}\boldsymbol{B} & \cdots & A_{m,n}\boldsymbol{B} \end{bmatrix} \tag{A.2.5}$$

11. **Projection Matrices:** Let $\boldsymbol{X}$ be a matrix of dimension $m \times n$, where $m \geq n$. If $\boldsymbol{X}$ has full column rank, the *orthogonal projector* onto the range space of $\boldsymbol{X}$ is defined as follows:

$$\Pi_{\boldsymbol{X}} = \boldsymbol{X}(\boldsymbol{X}^H\boldsymbol{X})^{-1}\boldsymbol{X}^H \tag{A.2.6}$$

The projection matrix $\Pi_{\boldsymbol{X}}$ has the following properties:

$$\begin{aligned} \Pi_{\boldsymbol{X}}\Pi_{\boldsymbol{X}} &= \Pi_{\boldsymbol{X}} \\ \Pi_{\boldsymbol{X}}\boldsymbol{X} &= \boldsymbol{X} \end{aligned} \tag{A.2.7}$$

Also, the orthogonal projector onto the complement of the range space of $\boldsymbol{X}$ is defined as:

$$\Pi_{\boldsymbol{X}}^{\perp} = \boldsymbol{I} - \Pi_{\boldsymbol{X}} \tag{A.2.8}$$

12. **Circulant Matrices:** A *right* circulant matrix is a matrix obtained by stacking cyclically right-shifted versions of a row vector on top of each other:

$$\boldsymbol{X} = \begin{bmatrix} x_0 & x_{N-1} & \cdots & x_2 & x_1 \\ x_1 & x_0 & & x_3 & x_2 \\ \vdots & & \ddots & & \vdots \\ x_{N-2} & x_{N-3} & & x_0 & x_{N-1} \\ x_{N-1} & x_{N-2} & \cdots & x_1 & x_0 \end{bmatrix} \tag{A.2.9}$$

A *left* circulant matrix is obtained in a similar way, by instead cyclically shifting the row vector $\boldsymbol{x}$ leftwards. Circulant matrices have appealing properties. For instance, it can be shown that $\boldsymbol{X}$ in (A.2.9) enjoys the following eigenvalue-decomposition:

$$\boldsymbol{X} = \boldsymbol{T}\boldsymbol{\Delta}\boldsymbol{T}^H \tag{A.2.10}$$

where

$$\boldsymbol{\Delta} = \begin{bmatrix} \delta_0 & 0 & \cdots & 0 \\ 0 & \delta_1 & \ddots & \vdots \\ \vdots & \ddots & \ddots & 0 \\ 0 & \cdots & 0 & \delta_{N-1} \end{bmatrix} \tag{A.2.11}$$

$$\delta_n = \sum_{l=0}^{N-1} x_l e^{-i2\pi \frac{ln}{N}}$$

and $\boldsymbol{T}$ is the unitary matrix whose (k,l)th element is given by:

$$\boldsymbol{T}_{k,l} = \frac{1}{\sqrt{N}} \exp\Big(- i2\pi \frac{(k-1)(l-1)}{N} \Big) \tag{A.2.12}$$

Similar results hold for left circulant matrices, and for block-circulant matrices as well.

13. Isomorphism between complex-valued n-vectors and real-valued $2n$-vectors: Consider the matrix-vector product

$$\boldsymbol{y} = \boldsymbol{A}\boldsymbol{x} \tag{A.2.13}$$

where $\boldsymbol{A}$ is a complex-valued matrix of dimension $m \times n$, $\boldsymbol{x}$ is a complex-valued vector of length n and $\boldsymbol{y}$ is a complex-valued vector of dimension m. This product can be equivalently expressed as follows:

$$\begin{bmatrix} \bar{\boldsymbol{y}} \\ \tilde{\boldsymbol{y}} \end{bmatrix} = \begin{bmatrix} \bar{\boldsymbol{A}} & -\tilde{\boldsymbol{A}} \\ \tilde{\boldsymbol{A}} & \bar{\boldsymbol{A}} \end{bmatrix} \begin{bmatrix} \bar{\boldsymbol{x}} \\ \tilde{\boldsymbol{x}} \end{bmatrix} \tag{A.2.14}$$

where all involved matrices and vectors are real-valued. Let us introduce the following notation

$$\boldsymbol{y}' = \begin{bmatrix} \bar{\boldsymbol{y}} \\ \tilde{\boldsymbol{y}} \end{bmatrix}, \quad \boldsymbol{x}' = \begin{bmatrix} \bar{\boldsymbol{x}} \\ \tilde{\boldsymbol{x}} \end{bmatrix}$$
$$\boldsymbol{A}' = \begin{bmatrix} \bar{\boldsymbol{A}} & -\tilde{\boldsymbol{A}} \\ \tilde{\boldsymbol{A}} & \bar{\boldsymbol{A}} \end{bmatrix} \tag{A.2.15}$$

(the notation $(\cdot)'$ is used in many places in this book). Then we can write (A.2.14) as:

$$\boldsymbol{y}' = \boldsymbol{A}'\boldsymbol{x}' \tag{A.2.16}$$

The representation in (A.2.14) is an often useful manifestation of the isomorphism between complex numbers and real-valued matrices of dimension 2×2.

A.3 Selected Concepts from Probability Theory

1. **Multivariate real-valued Gaussian random variables:** A real-valued vector $\boldsymbol{x}$ of length m is Gaussian with mean $\boldsymbol{\mu}$ and covariance matrix $\boldsymbol{\Upsilon}$ if its p.d.f. can be written

$$p(\boldsymbol{x}) = (2\pi)^{-m/2}|\boldsymbol{\Upsilon}|^{-1/2}\exp\Big(-\frac{1}{2}(\boldsymbol{x}-\boldsymbol{\mu})^T\boldsymbol{\Upsilon}^{-1}(\boldsymbol{x}-\boldsymbol{\mu})\Big) \tag{A.3.1}$$

This means that

$$P(\boldsymbol{x} \in \Omega) = \int_{\boldsymbol{x}\in\Omega} p(\boldsymbol{x})d\boldsymbol{x} \tag{A.3.2}$$

Notationally we write:

$$\boldsymbol{x} \sim N(\boldsymbol{\mu}, \boldsymbol{\Upsilon}) \tag{A.3.3}$$

2. **Complex-valued Gaussian random vectors:** A complex-valued vector $\boldsymbol{x} = \bar{\boldsymbol{x}} + i\tilde{\boldsymbol{x}}$ is Gaussian if the real and imaginary parts of all its elements are jointly Gaussian:

$$\begin{bmatrix}\bar{\boldsymbol{x}}\\ \tilde{\boldsymbol{x}}\end{bmatrix} \sim N(\boldsymbol{\mu}', \boldsymbol{\Upsilon}') \tag{A.3.4}$$

for some vector $\boldsymbol{\mu}'$ and a positive definite matrix $\boldsymbol{\Upsilon}'$ of dimension $2m \times 2m$.

3. **Circularly symmetric Gaussian random vectors:** An important subclass of complex Gaussian random vectors is that of *circularly symmetric* complex Gaussian variables. A vector $\boldsymbol{x} = \bar{\boldsymbol{x}} + i\tilde{\boldsymbol{x}}$ is circularly symmetric Gaussian if its real and imaginary parts are jointly Gaussian with mean $\boldsymbol{\mu}'$ and covariance matrix $\boldsymbol{\Upsilon}'$:

$$\boldsymbol{x}' = \begin{bmatrix}\bar{\boldsymbol{x}}\\ \tilde{\boldsymbol{x}}\end{bmatrix} \sim N(\boldsymbol{\mu}', \boldsymbol{\Upsilon}') \tag{A.3.5}$$

and, in addition, $\boldsymbol{\Upsilon}'$ has the following structure:

$$\boldsymbol{\Upsilon}' = \begin{bmatrix}\boldsymbol{\Upsilon}_1 & \boldsymbol{\Upsilon}_2\\ -\boldsymbol{\Upsilon}_2 & \boldsymbol{\Upsilon}_1\end{bmatrix} \tag{A.3.6}$$

(Note that $\boldsymbol{\Upsilon}_1^T = \boldsymbol{\Upsilon}_1$ and $\boldsymbol{\Upsilon}_2^T = -\boldsymbol{\Upsilon}_2$ by construction.)

Such a random vector $\boldsymbol{x}$ has the following properties:

$$\begin{aligned} E[\boldsymbol{x}] &\triangleq \boldsymbol{\mu} = \bar{\boldsymbol{\mu}} + i\tilde{\boldsymbol{\mu}}\\ E\left[(\boldsymbol{x}-\boldsymbol{\mu})(\boldsymbol{x}-\boldsymbol{\mu})^H\right] &\triangleq \boldsymbol{\Upsilon} = 2(\boldsymbol{\Upsilon}_1 - i\boldsymbol{\Upsilon}_2)\\ E\left[(\boldsymbol{x}-\boldsymbol{\mu})(\boldsymbol{x}-\boldsymbol{\mu})^T\right] &= \boldsymbol{0} \end{aligned} \tag{A.3.7}$$

and its p.d.f. can be written

$$p(\boldsymbol{x}) = \pi^{-m}|\boldsymbol{\Upsilon}|^{-1}\exp\Big(-(\boldsymbol{x}-\boldsymbol{\mu})^H\boldsymbol{\Upsilon}^{-1}(\boldsymbol{x}-\boldsymbol{\mu})\Big) \tag{A.3.8}$$

For short, we write

$$\boldsymbol{x} \sim N_C(\boldsymbol{\mu}, \boldsymbol{\Upsilon}) \tag{A.3.9}$$

to denote that $\boldsymbol{x}$ is a circularly symmetric Gaussian random vector.

4. **The Q-function**: Let x be a real-valued Gaussian random variable with zero mean and unit variance. Then the integrated tail of the p.d.f. of x is often denoted by the following function:

$$Q(t) \triangleq P(x \geq t) = \frac{1}{\sqrt{2\pi}}\int_t^\infty \exp\Big(-\frac{x^2}{2}\Big)dx \tag{A.3.10}$$

5. **Chernoff bound**: The value of $Q(t)$ in (A.3.10) can be upper-bounded by the Chernoff bound [Proakis, 2001, Sec. 2.1.5]:

$$Q(t) \leq \exp\Big(-\frac{t^2}{2}\Big) \tag{A.3.11}$$

6. **Rayleigh and Exponentially Distributed Variables**: If x and y are independent zero-mean real-valued Gaussian random variables with variance $\sigma^2/2$, then the following variable

$$z = \sqrt{x^2 + y^2} \tag{A.3.12}$$

has a *Rayleigh* distribution, and the variable z^2 has an *exponential* distribution. The probability density functions (p.d.f.s) of z and z^2 are:

$$\begin{aligned} p_z(t) &= \frac{2t}{\sigma^2}\cdot\exp\left(-\frac{t^2}{\sigma^2}\right) \\ p_{z^2}(t) &= \frac{1}{\sigma^2}\cdot\exp\left(-\frac{t}{\sigma^2}\right) \end{aligned} \tag{A.3.13}$$

A.4 Selected Problems

The following exercises contain useful formulas and results from matrix algebra and probability theory. The level of difficulty of these problems is quite nonuniform: some of them are very easy and some of them may require more effort. In the following we assume that all matrices are of compatible dimensions.

1. Let x be a scalar circular complex Gaussian random variable (as defined in the above Section A.3). Prove that $\bar{x}$ and $\tilde{x}$ must be independent and have the same variance. Prove also that the converse is true, i.e. if $\bar{x}$ and $\tilde{x}$ have the previous properties, then x is a circular Gaussian random variable.

2. Give an example of a complex Gaussian random vector that is *not* circularly symmetric. Also, conclude that the p.d.f. of a (zero-mean) complex Gaussian vector $\boldsymbol{x}$ is *not* determined by its covariance matrix $E\left[\boldsymbol{x}\boldsymbol{x}^H\right]$ unless it is circularly symmetric.

3. Let x be a scalar circularly symmetric Gaussian random variable with zero mean and variance σ^2. Prove that $|x|$ and $|x|^2$ are Rayleigh and exponentially distributed, respectively.

4. Let $n(t)$ be a real-valued stationary stochastic noise process that has a flat spectral density over two narrow bands centered around $\pm f_c$. Suppose that a complex baseband representation of this noise is defined in the same way as in Section A.1. Prove that $\bar{n}_b(t)$ and $\tilde{n}_b(t)$ are uncorrelated and that they have the same variance (hence the noise $n_b(t)$ is *circularly symmetric*).

   Hint: Show first that

$$n(t) = \operatorname{Re}\left\{n_b(t)e^{i2\pi f_c t}\right\} = \bar{n}_b(t)\cos(2\pi f_c t) - \tilde{n}_b(t)\sin(2\pi f_c t) \tag{A.4.1}$$

   and that

$$\begin{aligned} E\left[|n(t)|^2\right] =& \frac{1}{2}\Big(E\left[(\bar{n}_b(t))^2\right](1+\cos(4\pi f_c t)) \\ &+E\left[(\tilde{n}_b(t))^2\right](1-\cos(4\pi f_c t)) \\ &+E\left[\bar{n}_b(t)\tilde{n}_b(t)\right]\sin(4\pi f_c t)\Big) \end{aligned} \tag{A.4.2}$$

   and use the assumption on the stationarity of $n(t)$.

5. Prove Equation (A.3.7).

6. Prove Equation (A.1.10).

7. Prove that $\operatorname{Tr}\{\boldsymbol{X}\boldsymbol{Y}\} = \operatorname{Tr}\{\boldsymbol{Y}\boldsymbol{X}\}$.

8. What is the relation between the singular value decomposition of $\boldsymbol{X}$ and the eigenvalue-decompositions of $\boldsymbol{X}^H\boldsymbol{X}$ and $\boldsymbol{X}\boldsymbol{X}^H$, respectively?

9. Prove that $\|\boldsymbol{X}\|^2 = \operatorname{Tr}\{\boldsymbol{X}^H\boldsymbol{X}\}$.

10. Prove that $\mathrm{Tr}\{\boldsymbol{X}\}$ is equal to the sum of the eigenvalues of $\boldsymbol{X}$.

11. Explain how the singular value decomposition of a Hermitian matrix reduces to the eigenvalue-decomposition of the same matrix.

12. Prove that $\mathrm{vec}(\boldsymbol{ABC}) = (\boldsymbol{C}^T \otimes \boldsymbol{A})\,\mathrm{vec}(\boldsymbol{B})$

13. Prove that $\mathrm{Tr}\{\boldsymbol{A}^H\boldsymbol{B}\} = \mathrm{vec}^H(\boldsymbol{A})\,\mathrm{vec}(\boldsymbol{B})$

14. Prove that $\mathrm{Tr}\{\boldsymbol{A}\otimes\boldsymbol{B}\} = \mathrm{Tr}\{\boldsymbol{A}\}\mathrm{Tr}\{\boldsymbol{B}\}$

15. Prove that $\mathrm{Tr}\{\boldsymbol{ABCD}\} = \mathrm{vec}^T(\boldsymbol{A}^T)(\boldsymbol{D}^T\otimes\boldsymbol{B})\,\mathrm{vec}(\boldsymbol{C})$

16. Prove that $(\boldsymbol{A}\otimes\boldsymbol{C})(\boldsymbol{B}\otimes\boldsymbol{D}) = (\boldsymbol{AB}\otimes\boldsymbol{CD})$

17. Prove that $(\boldsymbol{A}\otimes\boldsymbol{B})^{-1} = (\boldsymbol{A}^{-1}\otimes\boldsymbol{B}^{-1})$

18. Prove that $(\boldsymbol{A}\otimes\boldsymbol{B})^H = (\boldsymbol{A}^H\otimes\boldsymbol{B}^H)$

19. Show that for square $\boldsymbol{A}$ and $\boldsymbol{B}$ of dimensions $m\times m$ and $n\times n$, respectively, it holds that:

$$|\boldsymbol{A}\otimes\boldsymbol{B}| = |\boldsymbol{A}|^n\cdot|\boldsymbol{B}|^m \tag{A.4.3}$$

20. Suppose that $\|\boldsymbol{Ax}\|^2 = \|\boldsymbol{x}\|^2$ for all $\boldsymbol{x}$. Show that $\boldsymbol{A}$ must be a unitary matrix.

21. Assume that $\boldsymbol{X}$ and $\boldsymbol{Y}$ are square matrices of dimension $n\times n$ and that both of them have rank m. What can we say about

(a) the eigenvalues of $\boldsymbol{XX}^H$?

(b) the rank of $\boldsymbol{XY}$?

(c) the determinant $|\boldsymbol{XY}|$?

Can we say anything about $\mathrm{Tr}\{\boldsymbol{XY}\}$?

22. Assume that $\boldsymbol{X}$ is *skew-Hermitian*: $\boldsymbol{X}^H = -\boldsymbol{X}$, and let $\boldsymbol{Y}$ be a Hermitian matrix. What can we say about $\mathrm{Tr}\{\boldsymbol{XY}\}$?

23. Let $\boldsymbol{X}$ be a positive definite matrix of dimension $N\times N$. Prove that

$$|\boldsymbol{X}| \le \prod_{n=1}^{N}\boldsymbol{X}_{n,n} \tag{A.4.4}$$

(This inequality is called the Hadamard inequality.)

24. Let $\boldsymbol{B}$ be a matrix with full column rank and let $\boldsymbol{A}$ be another given matrix. Prove that

$$\|\boldsymbol{A}-\boldsymbol{B}\boldsymbol{X}\|^2 \geq \|(\boldsymbol{I}-\boldsymbol{B}(\boldsymbol{B}^H\boldsymbol{B})^{-1}\boldsymbol{B}^H)\boldsymbol{A}\|^2 \tag{A.4.5}$$

with equality if

$$\boldsymbol{X}=(\boldsymbol{B}^H\boldsymbol{B})^{-1}\boldsymbol{B}^H\boldsymbol{A} \tag{A.4.6}$$

Hint: show first that

$$\begin{aligned}\|\boldsymbol{A}-\boldsymbol{B}\boldsymbol{X}\|^2 &= \left\|\Pi_{\boldsymbol{B}}(\boldsymbol{A}-\boldsymbol{B}\boldsymbol{X})+\Pi_{\boldsymbol{B}}^{\perp}(\boldsymbol{A}-\boldsymbol{B}\boldsymbol{X})\right\|^2 \\ &= \|\Pi_{\boldsymbol{B}}(\boldsymbol{A}-\boldsymbol{B}\boldsymbol{X})\|^2+\left\|\Pi_{\boldsymbol{B}}^{\perp}(\boldsymbol{A}-\boldsymbol{B}\boldsymbol{X})\right\|^2\end{aligned} \tag{A.4.7}$$

25. Prove that $|\boldsymbol{I}+\boldsymbol{A}\boldsymbol{B}|=|\boldsymbol{I}+\boldsymbol{B}\boldsymbol{A}|$.

26. Let $\boldsymbol{X}$ be a positive definite matrix and let $\boldsymbol{Y}$ be a positive semi-definite matrix. Prove that $|\boldsymbol{X}+\boldsymbol{Y}|\geq|\boldsymbol{X}|$.

27. Prove that a right circulant matrix has the eigenvalue-decomposition (A.2.10). Extend the result to left circulant matrices and to *block* circulant matrices.

28. Prove that

$$\left|\mathrm{Tr}\{\boldsymbol{X}\boldsymbol{Y}\}\right|^2 \leq \|\boldsymbol{X}\|^2\cdot\|\boldsymbol{Y}\|^2 \tag{A.4.8}$$

(This inequality is called the matrix Cauchy-Schwarz inequality.) Determine a necessary and sufficient condition for equality to hold in (A.4.8).

29. Prove Equation (A.2.7).

30. Prove that the eigenvalues of a projection matrix $\Pi_{\boldsymbol{X}}$ are either zero or one.

31. Show that $\Pi_{\boldsymbol{X}}^{\perp}\Pi_{\boldsymbol{X}}^{\perp}=\Pi_{\boldsymbol{X}}^{\perp}$, that $\Pi_{\boldsymbol{X}}^{\perp}\Pi_{\boldsymbol{X}}=\boldsymbol{0}$ and that $\Pi_{\boldsymbol{X}}^{\perp}\boldsymbol{X}=\boldsymbol{0}$.

32. Let $\boldsymbol{X}(\theta)$ be a matrix-valued function of a real-valued scalar variable θ and let $\boldsymbol{A}$ and $\boldsymbol{B}$ be given constant matrices. Prove that

$$\begin{aligned}\frac{d}{d\theta}\mathrm{Tr}\{\boldsymbol{A}\boldsymbol{X}(\theta)\} &= \mathrm{Tr}\left\{\boldsymbol{A}\frac{d\boldsymbol{X}(\theta)}{d\theta}\right\} \\ \frac{d}{d\theta}\mathrm{Tr}\{\boldsymbol{X}^H(\theta)\boldsymbol{A}\boldsymbol{X}(\theta)\} &= 2\mathrm{Re}\mathrm{Tr}\left\{\boldsymbol{X}^H(\theta)\boldsymbol{A}\frac{d\boldsymbol{X}(\theta)}{d\theta}\right\} \\ \frac{d}{d\theta}\log|\boldsymbol{X}(\theta)| &= \mathrm{Tr}\left\{\boldsymbol{X}^{-1}(\theta)\frac{d\boldsymbol{X}(\theta)}{d\theta}\right\} \\ \frac{d}{d\theta}\mathrm{Tr}\{\boldsymbol{X}^{-1}(\theta)\} &= -\mathrm{Tr}\left\{\boldsymbol{X}^{-2}(\theta)\frac{d\boldsymbol{X}(\theta)}{d\theta}\right\}\end{aligned} \tag{A.4.9}$$

33. Prove that $\mathrm{Re}\{\boldsymbol{X}\}\mathrm{Re}\{\boldsymbol{Y}\} = \frac{1}{2}\mathrm{Re}\{\boldsymbol{XY} + \boldsymbol{XY}^*\}$.

34. Prove the "matrix inversion lemma":

$$(\boldsymbol{A} - \boldsymbol{C}\boldsymbol{B}^{-1}\boldsymbol{D})^{-1} = \boldsymbol{A}^{-1} + \boldsymbol{A}^{-1}\boldsymbol{C}(\boldsymbol{B} - \boldsymbol{D}\boldsymbol{A}^{-1}\boldsymbol{C})^{-1}\boldsymbol{D}\boldsymbol{A}^{-1} \tag{A.4.10}$$

35. Let $\boldsymbol{A}$ be an arbitrary constant matrix and let $\boldsymbol{X}$ be a matrix whose elements are independent and circularly symmetric Gaussian random variables with mean zero and variance γ^2. Prove that

$$\mathrm{Tr}\{\boldsymbol{AX}\} \sim N_C(0, \gamma^2\mathrm{Tr}\{\boldsymbol{AA}^H\}) \tag{A.4.11}$$

Appendix B

THE THEORY OF AMICABLE ORTHOGONAL DESIGNS

An *amicable orthogonal design* is a set of $n_t \times N$ matrices $\{\boldsymbol{A}_n, \boldsymbol{B}_n\}$ that satisfy the following equations:

$$\begin{aligned} \boldsymbol{A}_n\boldsymbol{A}_n^H &= \boldsymbol{I}, \boldsymbol{B}_n\boldsymbol{B}_n^H = \boldsymbol{I} \\ \boldsymbol{A}_n\boldsymbol{A}_p^H &= -\boldsymbol{A}_p\boldsymbol{A}_n^H, \boldsymbol{B}_n\boldsymbol{B}_p^H = -\boldsymbol{B}_p\boldsymbol{B}_n^H, \quad n \neq p \\ \boldsymbol{A}_n\boldsymbol{B}_p^H &= \boldsymbol{B}_p\boldsymbol{A}_n^H \end{aligned} \tag{B.1}$$

We have seen in previous chapters that the theory of amicable orthogonal designs is a cornerstone for the development of orthogonal STBC. This appendix is devoted to the existence and construction of such designs.

We will start in Section B.1 with considering the design of $\{\boldsymbol{A}_n\}$ alone (neglecting any relation between $\{\boldsymbol{A}_n\}$ and $\{\boldsymbol{B}_n\}$). In a strict sense, this is only relevant if the symbols $\{s_n\}$ that we want to transmit are real-valued; however, note that by mapping a complex symbol onto two real ones, the transmission of a block of n_s complex symbols becomes equivalent to the transmission of a block of $2n_s$ real symbols. Next, in Section B.2 we go on to consider the case of complex symbols and the *joint* design of $\{\boldsymbol{A}_n, \boldsymbol{B}_n\}$ to satisfy the conditions in (B.1).

The algebraic theory that underlies the existence of matrices that satisfy (B.1) is quite intricate. The relevant results related to the construction of such matrices are summarized in Section B.3. A reader interested only in the applications of orthogonal designs may go directly to Section B.3.

B.1 The Case of Real Symbols

In this section we assume that the symbols are real and that only $\{\boldsymbol{A}_n\}$ are of interest. In this case, it would be desirable if we could choose the matrices $\{\boldsymbol{A}_n\}$ to be real-valued, since in that case all quantities under consideration would become real-valued. It turns out that when we consider the case of real symbols we can construct full-rate OSTBC (i.e., with $n_s = N$) using real-valued $\{\boldsymbol{A}_n\}$. Therefore in what follows we will restrict our analysis to the case when $\{\boldsymbol{A}_n\}$ are real-valued matrices. A set of matrices satisfying the corresponding conditions in (B.1) is called an *orthogonal design*. Initially we constrain $\{\boldsymbol{A}_n\}$ to be *square* matrices, that is $N = n_t$, but later we relax this condition and consider the case of non-square $\{\boldsymbol{A}_n\}$ as well. Before we present a more detailed theory concerning the existence of matrices $\{\boldsymbol{A}_n\}$ that satisfy (B.1) we exemplify how such matrices can sometimes be found using heuristic arguments.

Example B.1: A real orthogonal design for $n_t = 2$.
Inspired by the well-known parameterization of real-valued orthogonal 2×2-matrices, we can easily construct an Alamouti-like matrix for real symbols, and hence two matrices $\boldsymbol{A}_1, \boldsymbol{A}_2$ that satisfy (B.1). A real-valued orthogonal matrix of dimension 2×2 takes on the following form:

$$\begin{bmatrix} \cos(\phi) & \sin(\phi) \\ \sin(\phi) & -\cos(\phi) \end{bmatrix} \tag{B.1.1}$$

By multiplying (B.1.1) with a real-valued scalar λ, and setting

$$\begin{aligned} s_1 &= \lambda \cos(\phi) \\ s_2 &= \lambda \sin(\phi) \end{aligned} \tag{B.1.2}$$

(B.1.1) can be written

$$\boldsymbol{X} = \begin{bmatrix} s_1 & s_2 \\ s_2 & -s_1 \end{bmatrix} \tag{B.1.3}$$

which is a matrix with the sought property (i.e., a matrix that satisfies $\boldsymbol{X}\boldsymbol{X}^T = (|s_1|^2 + |s_2|^2) \cdot \boldsymbol{I}$; cf. Theorem 7.1 on page 102). The matrices $\boldsymbol{A}_1$ and $\boldsymbol{A}_2$ associated with (B.1.3) can easily be identified, and it can be verified that these matrices satisfy (B.1); we leave doing so to the reader. ■

Example B.2: A real orthogonal design for $n_t = 4$.
Let $\boldsymbol{X}_1$ and $\boldsymbol{X}_2$ be two matrices having the form (B.1.3). Clearly, $\boldsymbol{X}_1^T = \boldsymbol{X}_1$ and $\boldsymbol{X}_2^T = \boldsymbol{X}_2$. Also, it can easily be verified that $\boldsymbol{X}_1\boldsymbol{X}_2 = \boldsymbol{X}_2\boldsymbol{X}_1$, and hence the matrices with the structure (B.1.3) commute with each other. Inspired by this observation, let us take a matrix of the form (B.1.3) and substitute the scalars s_1

and s_2 by two 2×2 matrices of the form (B.1.3). Doing so, we obtain an orthogonal matrix of dimension 4×4:

$$\begin{bmatrix} \boldsymbol{X}_1 & \boldsymbol{X}_2 \\ \boldsymbol{X}_2 & -\boldsymbol{X}_1 \end{bmatrix} \tag{B.1.4}$$

which is a linear function of four real-valued symbols. Such a matrix is effectively an OSTBC matrix for real symbols and $n_t = 4$ transmit antennas; the corresponding matrices $\{\boldsymbol{A}_k\}$ can easily be found. ■

B.1.1 Square OSTBC Matrices ($n_t = N$)

In this section we impose the constraint that $N = n_t$ and consider the existence of square OSTBC matrices. The main tool for doing so is the theory of Hurwitz-Radon matrices that was formulated almost a century ago.

Let $\{\boldsymbol{A}_n\}$ be a set of n_s matrices of size $N \times N$ which satisfy the following conditions (see (B.1)):

$$\begin{aligned} \boldsymbol{A}_n\boldsymbol{A}_n^H &= \boldsymbol{I}, & n &= 1,\ldots,n_s \\ \boldsymbol{A}_n\boldsymbol{A}_p^H &= -\boldsymbol{A}_p\boldsymbol{A}_n^H, & n &= 1,\ldots,n_s; p = 1,\ldots,n_s; n \neq p \end{aligned} \tag{B.1.5}$$

If $n_s = N$, we can use a set of such matrices for OSTBC encoding and transmit n_s real symbols over N symbol periods; hence we have a full-rate code. If $n_s < N$, the rate of the code will be $R = n_s/N < 1$. If we define $\boldsymbol{V}_n = \boldsymbol{A}_n\boldsymbol{A}_1^H$ for $n = 2,\ldots,n_s$, then the $n_s - 1$ matrices $\{\boldsymbol{V}_n\}$, of size $N \times N$, satisfy the following conditions:

$$\begin{aligned} \boldsymbol{V}_n^H\boldsymbol{V}_n &= \boldsymbol{I}, & n &= 2,\ldots,n_s \\ \boldsymbol{V}_n^H &= -\boldsymbol{V}_n, & n &= 2,\ldots,n_s \\ \boldsymbol{V}_n^H\boldsymbol{V}_p &= -\boldsymbol{V}_p^H\boldsymbol{V}_n, & n &= 2,\ldots,n_s; p = 2,\ldots,n_s; n \neq p \end{aligned} \tag{B.1.6}$$

If $\boldsymbol{A}_n$ (and hence $\boldsymbol{V}_n$) are *real* then the set of $n_s - 1$ matrices $\boldsymbol{V}_n$ constitutes a so-called *Hurwitz-Radon* (H-R) family of matrices of order N. The order N reflects their size; from here onward we say *order* instead of *size* to conform with the terminology in the literature on orthogonal designs (see, e.g., [Geramita and Seberry, 1979]). Clearly, given a set of matrices $\{\boldsymbol{A}_n\}$ we can construct the H-R family by using the above definition of $\{\boldsymbol{V}_n\}$; conversely, if we have a H-R family $\{\boldsymbol{V}_n\}$, $n = 2,\ldots,n_s$ we can augment it with an identity matrix $\boldsymbol{V}_1 = \boldsymbol{I}$ and obtain a set of matrices $\{\boldsymbol{A}_n\}$ that satisfy (B.1.5). Therefore there is a simple mapping between a set of n_s real matrices that satisfy (B.1.5) and the corresponding H-R family.

Next we consider the existence of a H-R family of matrices. The following theorem, which is due to Radon, gives a precise description of the conditions under which a H-R family of order N exists.

Theorem B.1: Existence of Hurwitz-Radon families.
Let N be a given integer and let the integers a, b, c and d be defined implicitly via:

$$\begin{aligned} N &= 2^a b \\ a &= 4c + d, \quad 0 \le d < 4, \quad c \ge 0 \end{aligned} \tag{B.1.7}$$

where b is an odd number. Also, let $\rho(N)$ be defined through:

$$\rho(N) = 8c + 2^d \tag{B.1.8}$$

Then

1. *Any H-R family of order N has fewer than $\rho(N)$ members.*

2. *For any N there is a H-R family of order N having exactly $\rho(N)-1$ members.*

Proof: The proof is given in [RADON, 1922], [GERAMITA AND SEBERRY, 1979] and is not repeated here. However, later in this section we will give a constructive proof of a variant of this theorem. ■

From Theorem B.1 we obtain the following result.

Theorem B.2: Existence of Hurwitz-Radon families, cont'd.
A H-R family of $N-1$ matrices of size $N \times N$ exists only if $N = 2, 4$ or 8.

Proof: Note that $N = 2^a b = 2^{4c+d} b = 2^{4c} 2^d b$ and that $\rho(N) = 8c + 2^d$. If $\rho(N) - 1 = N - 1$, we must have

$$8c + 2^d = 2^{4c} 2^d b \tag{B.1.9}$$

Since b, c and d are non-negative integers and $0 \le d < 4$, the above equation holds if and only if $c = 0$, $b = 1$ and $d = 0, 1, 2$ or 3. The corresponding values of N (and of $\rho(N)$, since we have $\rho(N) = N$) are 1, 2, 4 and 8. Excluding the degenerate case $N = 1$, the theorem follows. ■

If we have a H-R family of $N-1$ matrices of order N, then by including the identity matrix of order N, i.e., $\boldsymbol{A}_1 = \boldsymbol{I}$, we obtain a family of N matrices of order N which satisfy (B.1.5). In the context of OSTBC, this corresponds to a full-rate code for real symbols. Therefore, the implication of Theorem B.2 is that if we constrain $\{\boldsymbol{A}_n\}$ to be square matrices ($n_t = N$), then a square full-rate OSTBC for real symbols exists only when the number of transmit antennas is $n_t = 2, 4$ or

8. However, *non-square* full-rate OSTBC matrices exist for any other values of n_t (see below).

As an aside, note that it is necessary that $N \geq n_t$ (otherwise, $\boldsymbol{A}_n$ cannot satisfy $\boldsymbol{A}_n \boldsymbol{A}_n^H = \boldsymbol{I}$). When $N = n_t$ we say that the corresponding transmission scheme is a *minimal delay* design; the reason for this terminology is that in this case we have the minimum possible decoding delay. Hence, in words, Theorem B.2 states that minimal delay OSTBC schemes for real-valued symbols exist only for $n_t = 2, 4$ or 8.

The code matrices discussed so far are real-valued, but it turns out that we can even constrain their elements to be integers. An *integer matrix* is a matrix whose elements are integers, and a H-R family whose members are all integer matrices is called an *integer* H-R family. Since the members of an integer H-R family are unitary, they can have elements only from the set $\{1, 0, -1\}$. Such integer H-R matrices are especially attractive, since the associated OSTBC matrices $\boldsymbol{X}$ can be formed by the transmitter without any multiplication. The existence of such integer H-R matrices is considered in the following theorem.

Theorem B.3: Existence of integer Hurwitz-Radon families.
For any N, there exists an integer H-R family of order N that has $\rho(N) - 1$ members.

Proof: See [GERAMITA AND SEBERRY, 1979]. In Section B.4 we present a compact and *constructive* proof. ■

We illustrate the result of Theorem B.3 for the case of $n_t = 4$ in the following example.

Example B.3: Real orthogonal design for $n_t = 4$.
For the case of $n_t = 4$ transmit antennas a set of 4 matrices which meet the conditions in (B.1.5) is given by:

$$\boldsymbol{A}_1 = \begin{bmatrix} 1 & 0 & 0 & 0 \\ 0 & 1 & 0 & 0 \\ 0 & 0 & 1 & 0 \\ 0 & 0 & 0 & 1 \end{bmatrix} \qquad \boldsymbol{A}_3 = \begin{bmatrix} 0 & 0 & -1 & 0 \\ 0 & 0 & 0 & -1 \\ 1 & 0 & 0 & 0 \\ 0 & 1 & 0 & 0 \end{bmatrix}$$

$$\boldsymbol{A}_2 = \begin{bmatrix} 0 & 1 & 0 & 0 \\ -1 & 0 & 0 & 0 \\ 0 & 0 & 0 & -1 \\ 0 & 0 & 1 & 0 \end{bmatrix} \qquad \boldsymbol{A}_4 = \begin{bmatrix} 0 & 0 & 0 & -1 \\ 0 & 0 & 1 & 0 \\ 0 & -1 & 0 & 0 \\ 1 & 0 & 0 & 0 \end{bmatrix} \tag{B.1.10}$$

If $\{s_1, s_2, s_3, s_4\}$ is a set of four symbols to be transmitted, then the OSTBC matrix is given by:

$$\boldsymbol{X} = \begin{bmatrix} s_1 & s_2 & -s_3 & -s_4 \\ -s_2 & s_1 & s_4 & -s_3 \\ s_3 & -s_4 & s_1 & -s_2 \\ s_4 & s_3 & s_2 & s_1 \end{bmatrix} \tag{B.1.11}$$

which is slightly different from the "ad-hoc" construction in Example B.2. ■

B.1.2 Non-Square OSTBC Matrices ($n_t < N$)

Next we relax the constraint that $N = n_t$; hence we assume that $n_t < N$. In order to obtain a full-rate OSTBC design, we must find N matrices of size $n_t \times N$ which satisfy the conditions in (B.1.5). From Theorem B.2 we know that full-rate square designs exist only for the case of $n_t = 2, 4$ or 8 transmit antennas. For the case of $n_t = 3$, 5, 6, and 7 we can easily construct "wide" matrices $\{\boldsymbol{A}_n\}$ simply by taking any n_t rows of an orthogonal design for $n_t = 4$ or $n_t = 8$.

Specifically, if $\{\boldsymbol{A}_n^{(4)}\}$ is a set of four 4×4-matrices that satisfy (B.1.5) and we let, for instance,

$$\boldsymbol{\Phi}_3 = \begin{bmatrix} 1 & 0 & 0 & 0 \\ 0 & 1 & 0 & 0 \\ 0 & 0 & 1 & 0 \end{bmatrix} \tag{B.1.12}$$

then

$$\begin{aligned} (\boldsymbol{\Phi}_3\boldsymbol{A}_n^{(4)})(\boldsymbol{\Phi}_3\boldsymbol{A}_n^{(4)})^H &= \boldsymbol{\Phi}_3\boldsymbol{A}_n^{(4)}\boldsymbol{A}_n^{(4)H}\boldsymbol{\Phi}_3^H = \boldsymbol{\Phi}_3\boldsymbol{\Phi}_3^H = \boldsymbol{I} \\ (\boldsymbol{\Phi}_3\boldsymbol{A}_n^{(4)})(\boldsymbol{\Phi}_3\boldsymbol{A}_k^{(4)})^H &= \boldsymbol{\Phi}_3\boldsymbol{A}_n^{(4)}\boldsymbol{A}_k^{(4)H}\boldsymbol{\Phi}_3^H \\ &= -\boldsymbol{\Phi}_3\boldsymbol{A}_k^{(4)}\boldsymbol{A}_n^{(4)H}\boldsymbol{\Phi}_3^H = -(\boldsymbol{\Phi}_3\boldsymbol{A}_k^{(4)})(\boldsymbol{\Phi}_3\boldsymbol{A}_n^{(4)})^H \end{aligned} \tag{B.1.13}$$

Hence, the set of matrices $\{\boldsymbol{\Phi}_3\boldsymbol{A}_n^{(4)}\}$ satisfy (B.1.5) and is therefore an orthogonal design for $n_t = 3$. For $n_t = 5$, 6 and 7, we can obtain a similar construct by taking 5, 6 and 7 rows of the $\{\boldsymbol{A}_n\}$ matrices for $n_t = 8$.

Clearly, the non-square OSTBC designs for $n_t = 3, 5, 6$ and 7 obtained as outlined above achieve full data rate (since $n_s = N$). However, it is a valid question whether they are the best possible constructs in the sense that there is no other construct with a smaller number of columns, that satisfies (B.1.5) as well as achieves the same rate. Such a construct, if it exists, would have a smaller decoding delay at the same data rate. We will not elaborate on this issue in the present text; the reader is referred to [GANESAN AND STOICA, 2001B] for a more detailed discussion on these aspects. The cited paper also presents a method for constructing non-square (wide) OSTBC matrices for *any* n_t. Note that the corresponding set of wide matrices $\{A_n\}$ is called a *generalized orthogonal design*.

We conclude this section by presenting the $\{A_n\}$ construct obtained via the above method for $n_t = 3$.

Example B.4: Non-square real orthogonal design for $n_t = 3$.
Let us consider a system with $n_t = 3$ transmit antennas. Following the construction method outlined above, we get from (B.1.10) and (B.1.12) the following set of 3×4 matrices:

$$A_1 = \begin{bmatrix} 1 & 0 & 0 & 0 \\ 0 & 1 & 0 & 0 \\ 0 & 0 & 1 & 0 \end{bmatrix} \quad A_3 = \begin{bmatrix} 0 & 0 & -1 & 0 \\ 0 & 0 & 0 & -1 \\ 1 & 0 & 0 & 0 \end{bmatrix}$$
$$A_2 = \begin{bmatrix} 0 & 1 & 0 & 0 \\ -1 & 0 & 0 & 0 \\ 0 & 0 & 0 & -1 \end{bmatrix} \quad A_4 = \begin{bmatrix} 0 & 0 & 0 & -1 \\ 0 & 0 & 1 & 0 \\ 0 & -1 & 0 & 0 \end{bmatrix} \tag{B.1.14}$$

■

B.2 The Case of Complex Symbols

In the previous Section B.1 we explained that for any number of transmit antennas n_t, we can always find a set of N matrices $\{A_n\}$ of size $n_t \times N$ (with $N \geq n_t$) that satisfy (B.1), and hence can be used to build a full-rate OSTBC for real symbols. In this section we go on to discuss the joint design of $\{A_n, B_n\}$, which provides us with a means for transmitting complex symbols. We start by observing that it is straightforward to use the full-rate codes obtained for real symbols to get a rate-1/2 code for complex symbols. After doing so, we proceed to discuss the existence of $\{A_n, B_n\}$ and how the design of such matrices relates to the theory of amicable orthogonal designs.

B.2.1 Rate-1/2 Complex OSTBC Designs

A rate-1/2 design can easily be obtained by taking a block of $n_s = N/2$ complex symbols $\{s_1, \ldots, s_{N/2}\}$ and transmitting their real and imaginary parts separately. The resulting code matrix is:

$$X = \begin{bmatrix} \sum_{n=1}^{n_s} \bar{s}_n A_n & \sum_{n=1}^{n_s} \tilde{s}_n A_n \end{bmatrix} \tag{B.2.1}$$

where $\{\boldsymbol{A}_n\}$ are $n_t \times N/2$ matrices designed for the transmission of real symbols (see Section B.1; as explained there N must be chosen as a function of n_t). Hence, the block of $N/2$ complex symbols has been mapped onto a block of N real symbols, which are transmitted over N time intervals; the so-obtained code has a rate of $R = 1/2$. The reader can verify that $\boldsymbol{X}\boldsymbol{X}^H = \|\boldsymbol{s}\|^2 \cdot \boldsymbol{I}$ and hence that (B.1) is satisfied (cf. Theorem 7.1 on page 102).

Example B.5: Rate-1/2 OSTBC for $n_t = 4$.
For the case of $n_t = 4$ transmit antennas, the code matrix in (B.2.1) takes on the following form (cf. (B.1.11)):

$$\begin{bmatrix} \bar{s}_1 & \bar{s}_2 & -\bar{s}_3 & -\bar{s}_4 & \tilde{s}_1 & \tilde{s}_2 & -\tilde{s}_3 & -\tilde{s}_4 \\ -\bar{s}_2 & \bar{s}_1 & \bar{s}_4 & -\bar{s}_3 & -\tilde{s}_2 & \tilde{s}_1 & \tilde{s}_4 & -\tilde{s}_3 \\ \bar{s}_3 & -\bar{s}_4 & \bar{s}_1 & -\bar{s}_2 & \tilde{s}_3 & -\tilde{s}_4 & \tilde{s}_1 & -\tilde{s}_2 \\ \bar{s}_4 & \bar{s}_3 & \bar{s}_2 & \bar{s}_1 & \tilde{s}_4 & \tilde{s}_3 & \tilde{s}_2 & \tilde{s}_1 \end{bmatrix} \tag{B.2.2}$$

■

B.2.2 Full-Rate Designs

The simple design in (B.2.1) makes only use of the matrices $\{\boldsymbol{A}_n\}$. Of course, the important question is whether the rate can be improved if we design both $\{\boldsymbol{A}_n\}$ and $\{\boldsymbol{B}_n\}$ jointly, so that they satisfy the conditions in (B.1), instead of designing $\{\boldsymbol{A}_n\}$ alone. As in the previous section we first consider the case of square matrices; that is we set $N = n_t$.

Let $\{\boldsymbol{A}_n\}_{n=1}^{s}$ and $\{\boldsymbol{B}_n\}_{n=1}^{t}$ be a set of $N \times N$ matrices which satisfy the conditions in (B.1). If we have such a set of matrices then we can transmit $s+t$ *real* symbols by re-mapping these $s+t$ real symbols onto $\min\{s,t\}$ complex symbols and $(\max\{s,t\} - \min\{s,t\})$ real symbols in an appropriate way. Since we have a scheme for transmitting N real symbols during N symbol periods, in order to benefit from the joint design of $\{\boldsymbol{A}_n, \boldsymbol{B}_n\}$ we need that $s+t \geq N$.

Suppose that we have a set of $s+t$ *real* matrices $\{\boldsymbol{A}_n\}_{n=1}^{s}$, $\{\boldsymbol{B}_n\}_{n=1}^{t}$ that satisfy (B.1). If we define $\boldsymbol{V}_n = \boldsymbol{A}_n\boldsymbol{A}_1^H, n = 2, \ldots, s$ and $\boldsymbol{W}_p = \boldsymbol{B}_p\boldsymbol{A}_1^H, p = 1, \ldots, t$, then the $N \times N$ matrices $\{\boldsymbol{V}_n\}$ and $\{\boldsymbol{W}_p\}$ satisfy the following conditions:

$$\begin{aligned}
\boldsymbol{V}_n &= -\boldsymbol{V}_n^H, \quad n = 2, \ldots, s \\
\boldsymbol{W}_n &= \boldsymbol{W}_n^H, \quad n = 1, \ldots, t \\
\boldsymbol{V}_n \boldsymbol{V}_n^H &= 2\boldsymbol{I}, \quad n = 2, \ldots, s \\
\boldsymbol{W}_n \boldsymbol{W}_n^H &= \boldsymbol{I}, \quad n = 1, \ldots, t \\
\boldsymbol{V}_n \boldsymbol{V}_p^H + \boldsymbol{V}_p \boldsymbol{V}_n^H &= \boldsymbol{0}, \quad n = 2, \ldots, s;\ p = 2, \ldots, s;\ n \neq p \\
\boldsymbol{W}_n \boldsymbol{W}_p^H + \boldsymbol{W}_p \boldsymbol{W}_n^H &= \boldsymbol{0}, \quad n = 1, \ldots, t;\ p = 1, \ldots, t;\ n \neq p \\
\boldsymbol{V}_n \boldsymbol{W}_p^H - \boldsymbol{W}_p \boldsymbol{V}_n^H &= \boldsymbol{0}, \quad n = 2, \ldots, s;\ p = 1, \ldots, t
\end{aligned} \tag{B.2.3}$$

Such a family of real matrices is called a H-R family of type $(s-1, t)$ of order N [GERAMITA AND SEBERRY, 1979], or simply an H-R$(s-1,t)$ family. If we have an H-R$(s-1,t)$ family, then by including the $N \times N$ identity matrix, i.e., letting $\boldsymbol{V}_1 = \boldsymbol{I}$, we obtain a family of $s+t$ matrices which satisfy (B.1). Hence there is a one-to-one mapping between $s+t$ real matrices that satisfy (B.1) and a H-R$(s-1,t)$ family.

The existence of H-R families that satisfy (B.2.3) is linked to the representation theory of Clifford algebras and was addressed in [KAWADA AND IWAHORI, 1950]. The associated mathematics is far from trivial. The following theorem summarizes one of the key results in a manner suitable to our problem.

Theorem B.4: Existence of $H-R(s-1,t)$ families.
Let N be an arbitrary integer. Define c and d, where $0 \le d < 4$, implicitly through the relation $N = 2^{4c+d}$, and let $\rho_t(N) - 1 = 8c - t + \delta$, where, for a given t and d, the values of δ are as shown in table B.1. Then for any $t \le \rho_t(N)$, there exists an H-R$(\rho_t(N)-1,t)$ family of integer matrices of order N.

Proof: See [GERAMITA AND SEBERRY, 1979]. ■

If we have an H-R$(s-1,t)$ family of order $N \times N$ and with $s = t = N$, we have effectively a full-rate OSTBC for complex symbols. Such a design, if it exists, is called a full-rate design. Since we have $N = n_t$, these full-rate designs will also be delay optimal. Clearly, the existence of $\{\boldsymbol{V}_j\}$ and $\{\boldsymbol{W}_j\}$ with $s = t = N$ implies the existence of two distinct "regular" H-R families of order N (as defined in (B.1.6)), each with $N-1$ members. Since an H-R family of order N with $N-1$ members exists only for $N = 2, 4$ or 8 (see Theorem B.2), it follows immediately that it is impossible to construct a full-rate square OSTBC design for complex

	d			
$t \pmod 4$	0	1	2	3
0	0	1	3	7
1	1	2	3	5
2	-1	3	4	5
3	-1	1	5	6

Table B.1. δ as function of d and t (see Theorem B.4).

symbols for any other values of N. In fact, it turns out that the *only* value of N for which a full-rate square design exists is $N = 2$. We state this result in the following theorem.

Theorem B.5: Nonexistence of full-rate OSTBC for $n_t > 2$.
A full-rate square OSTBC design for complex symbols exists only for $n_t = N = 2$.

Proof: Following the argument in the paragraph preceding this theorem, we need to consider only the cases of $N = 2, 4, 8$. For a full-rate design we need $s = t = N$ in (B.2.3). If we fix $t = N$, then the maximum value of s is given by $\rho_N(N)$. For a full data rate design we require that $\rho_N(N) = N$ (see Theorem B.4). From the definition of $\rho_t(N)$ we note that

$$\rho_2(2) = 2 \tag{B.2.4}$$

$$\rho_4(4) = 0 \tag{B.2.5}$$

$$\rho_8(8) = 0 \tag{B.2.6}$$

which completes the proof. ■

An example of a full-rate design when $n_t = N = 2$ is the Alamouti code discussed in Section 6.3.1. In Example 7.2 (see page 103) we gave the matrices $\boldsymbol{A}_n$ and $\boldsymbol{B}_n$ associated with this code. Theorem B.5 asserts that the construction in [ALAMOUTI, 1998] cannot be extended to more than two transmit antennas.

B.2.3 Achieving Rates Higher than 1/2 for $n_t \geq 2$

We know that a complex design of rate 1/2 can always be obtained (see (B.2.1)). In the previous section, we proved that a full-rate square design for complex symbols exists only for $n_t = 2$ transmit antennas. We next go on to see if we can get rates higher than 1/2 by simultaneously designing $\{\boldsymbol{A}_n, \boldsymbol{B}_n\}$ when $n_t > 2$. The first step towards this end is the following existence theorem. Note that we still restrict our discussion to the case of $n_t = N$.

Theorem B.6: Wolfe's slide lemma.
Assume that there exist n_s pairs of matrices $\{\boldsymbol{A}_n\}_{n=1}^{n_s}$, $\{\boldsymbol{B}_n\}_{n=1}^{n_s}$ of size $n_t \times n_t$ which satisfy the conditions in (B.1). Then there exist also n_s+1 pairs of matrices of size $2n_t \times 2n_t$ which satisfy the conditions in (B.1).

Proof: We give a constructive proof that is based on [Geramita and Seberry, 1979].

Let $\{\boldsymbol{A}_n\}_{n=1}^{n_s}$ and $\{\boldsymbol{B}_n\}_{n=1}^{n_s}$ be a set of $n_t \times n_t$ matrices which satisfy the conditions in (B.1). Define

$$\begin{aligned} \boldsymbol{V}_n &= \boldsymbol{A}_n \boldsymbol{A}_1^H & n &= 2, \ldots, n_s \\ \boldsymbol{W}_k &= \boldsymbol{B}_k \boldsymbol{A}_1^H & k &= 1, \ldots, n_s \end{aligned}$$

and let

$$\boldsymbol{M} = \begin{bmatrix} 0 & 1 \\ -1 & 0 \end{bmatrix}, \qquad \boldsymbol{P} = \begin{bmatrix} 0 & 1 \\ 1 & 0 \end{bmatrix}, \qquad \boldsymbol{R} = \begin{bmatrix} 1 & 0 \\ 0 & -1 \end{bmatrix}$$

Consider the $n_s + 1$ pairs of matrices of size $2n_t \times 2n_t$ given by:

$$\begin{aligned} \breve{\boldsymbol{A}}_1 &= \boldsymbol{I}_{2n_t} \\ \breve{\boldsymbol{A}}_n &= \boldsymbol{P} \otimes \boldsymbol{V}_n \qquad 2 \le n \le n_s \\ \breve{\boldsymbol{A}}_{n_s+1} &= \boldsymbol{M} \otimes \boldsymbol{I}_{n_t} \end{aligned} \qquad \begin{aligned} \breve{\boldsymbol{B}}_k &= \boldsymbol{P} \otimes \boldsymbol{W}_k \qquad 1 \le k \le n_s \\ \breve{\boldsymbol{B}}_{n_s+1} &= \boldsymbol{R} \otimes \boldsymbol{I}_{n_t} \end{aligned} \tag{B.2.7}$$

Then it can be verified that the matrices $\{\breve{\boldsymbol{A}}_n, \breve{\boldsymbol{B}}_n\}$ in (B.2.7) satisfy the conditions in (B.1). ■

Theorem B.6 is used in the next examples to obtain some useful OSTBC designs.

Example B.6: Rate-3/4 OSTBC for $n_t = 4$.
From (7.4.7) we have two pairs of 2×2 matrices which satisfy the conditions in (B.1). Using the construction method given in the proof of Theorem B.6 we can construct, starting from (7.4.7), three pairs of 4×4 matrices which satisfy the

conditions in (B.1). These three pairs of matrices are given by:

$$A_1 = \begin{bmatrix} 1 & 0 & 0 & 0 \\ 0 & 1 & 0 & 0 \\ 0 & 0 & 1 & 0 \\ 0 & 0 & 0 & 1 \end{bmatrix} \qquad B_1 = \begin{bmatrix} 1 & 0 & 0 & 0 \\ 0 & 1 & 0 & 0 \\ 0 & 0 & -1 & 0 \\ 0 & 0 & 0 & -1 \end{bmatrix}$$

$$A_2 = \begin{bmatrix} 0 & 0 & 1 & 0 \\ 0 & 0 & 0 & 1 \\ -1 & 0 & 0 & 0 \\ 0 & -1 & 0 & 0 \end{bmatrix} \qquad B_2 = \begin{bmatrix} 0 & 0 & 1 & 0 \\ 0 & 0 & 0 & -1 \\ 1 & 0 & 0 & 0 \\ 0 & -1 & 0 & 0 \end{bmatrix} \tag{B.2.8}$$

$$A_3 = \begin{bmatrix} 0 & 0 & 0 & -1 \\ 0 & 0 & 1 & 0 \\ 0 & -1 & 0 & 0 \\ 1 & 0 & 0 & 0 \end{bmatrix} \qquad B_3 = \begin{bmatrix} 0 & 0 & 0 & -1 \\ 0 & 0 & -1 & 0 \\ 0 & -1 & 0 & 0 \\ -1 & 0 & 0 & 0 \end{bmatrix}$$

The associated code matrix $\boldsymbol{X}$ is given by:

$$\boldsymbol{X} = \begin{bmatrix} s_1 & 0 & s_2 & -s_3 \\ 0 & s_1 & s_3^* & s_2^* \\ -s_2^* & -s_3 & s_1^* & 0 \\ s_3^* & -s_2 & 0 & s_1^* \end{bmatrix} \tag{B.2.9}$$

which is a rate-3/4 OSTBC for complex symbols for $n_t = 4$ transmit antennas. ■

Example B.7: Rate-1/2 OSTBC for $n_t = 8$.
Proceeding in the same way we can construct four pairs of 8×8 matrices using the three pairs of 4×4 matrices in (B.2.8). The result is best summarized in terms of the associated OSTBC matrix, which is given by:

$$\boldsymbol{X} = \begin{bmatrix} s_1 & 0 & 0 & 0 & s_4 & 0 & s_2 & -s_3 \\ 0 & s_1 & 0 & 0 & 0 & s_4 & s_3^* & s_2^* \\ 0 & 0 & s_1 & 0 & -s_2^* & -s_3 & s_4^* & 0 \\ 0 & 0 & 0 & s_1 & s_3^* & -s_2 & 0 & s_4^* \\ -s_4^* & 0 & s_2 & -s_3 & s_1^* & 0 & 0 & 0 \\ 0 & -s_4^* & s_3^* & s_2^* & 0 & s_1^* & 0 & 0 \\ -s_2^* & -s_3 & -s_4 & 0 & 0 & 0 & s_1^* & 0 \\ s_3^* & -s_2 & 0 & -s_4 & 0 & 0 & 0 & s_1^* \end{bmatrix} \tag{B.2.10}$$

Note that the decoding delay associated with (B.2.10) is equal to 8, whereas an OSTBC matrix for $n_t = 8$ derived in a manner similar to (B.2.1) would have a decoding delay of 16. ■

With the code in Example B.7, we transmit four complex symbols over eight symbol periods, and hence the rate is equal to $R = 4/8 = 1/2$. It turns out that there is no point in using this construction method for more than 8 transmit antennas. In particular, if we extend the construction to more than 8 transmit antennas then the rate becomes less than 1/2, and hence it is better to use the construction in (B.2.1). For example, using the four pairs of 8×8 matrices designed for $n_t = 8$, we can apply the construction method in the proof of Theorem B.6 to design five pairs of 16×16 matrices. However, in that case we will transmit only 5 symbols over 16 symbol periods; consequently the rate becomes 5/16 which is less than 1/2, and hence it is better to use the code in (B.2.1).

We started with a full-rate (square) design for $n_t = 2$ transmit antennas using *real* matrices $\{\boldsymbol{A}_n, \boldsymbol{B}_n\}$ and constructed a square rate-3/4 design for $n_t = 4$ transmit antennas and a square rate-1/2 design for $n_t = 8$ transmit antennas. The main tool for doing that was the algorithm described in the proof of Theorem B.6, but this method considered only the case when $\boldsymbol{A}_n$ and $\boldsymbol{B}_n$ are real matrices. Can we gain anything if we let $\{\boldsymbol{A}_n, \boldsymbol{B}_n\}$ be complex? Unfortunately, the answer to this question is no, as explained below.

Consider a set of s matrices $\{\boldsymbol{A}_n\}$ and t matrices $\{\boldsymbol{B}_n\}$ of size $n_t \times n_t$ which satisfy the conditions in (B.1). We let $\{\boldsymbol{A}_n, \boldsymbol{B}_n\}$ be *complex* $n_t \times n_t$ matrices. Define

$$\begin{aligned} \boldsymbol{P}_k &= \boldsymbol{A}_k & k &= 1, \ldots, s \\ \boldsymbol{P}_{s+k} &= i\, \boldsymbol{B}_k & k &= 1, \ldots, t \end{aligned}$$

The $s + t$ complex matrices $\{\boldsymbol{P}_k\}_{k=1}^{s+t}$ satisfy the conditions

$$\begin{aligned} \boldsymbol{P}_k \boldsymbol{P}_k^H &= \boldsymbol{I} & & k = 1, \ldots, s+t \\ \boldsymbol{P}_k \boldsymbol{P}_n^H &= -\boldsymbol{P}_n \boldsymbol{P}_k^H & & k = 1, \ldots, s+t;\ n = 1, \ldots, s+t;\ n \neq k \end{aligned} \tag{B.2.11}$$

We have the following theorem.

Theorem B.7: Herstein theorem.
Let $n_t = 2^a b$ where b is odd. Let $s + t$ be the number of complex matrices $\{\boldsymbol{P}_k\}$ of size $n_t \times n_t$ which satisfy the conditions in (B.2.11). Then $s + t \leq 2a + 2$.

Proof: See [HERSTEIN, 1968]. ■

For $n_t = 2$ transmit antennas we have $a = 1$, and we already know a design with $s = t = 2$. Now by Theorem B.7, $s + t \leq 2a + 2 = 4$. For the design we have, $s + t = 4$. Thus we have achieved the bound specified by Theorem B.7 for the case of $n_t = 2$ antennas. If we set $s = t$, then the bound in Theorem B.7 becomes, $2t \leq 2a + 2$ or $t \leq a + 1$. Suppose we have a construction with $t = a + 1$ for some $n_t = 2^a$. Then by using the construction in the proof of Theorem B.6 we can obtain a design with $t = a + 2$ for $n_t = 2^{a+1}$. Since we have a real design with $s = t = a + 1 = 2$ for $a = 1$, we can obtain a real design with $s = t = a + 1$ for any $n_t = 2^a$. Thus the bound in Theorem B.7 can be achieved for any $n_t = 2^a$ using *real integer matrices* $\{\boldsymbol{A}_n, \boldsymbol{B}_n\}$. It follows that we do not gain anything by letting $\{\boldsymbol{A}_n, \boldsymbol{B}_n\}$ be complex valued; moreover for the case of $n_t = 2^a$, there exists no other method to construct (square) $\{\boldsymbol{A}_n, \boldsymbol{B}_n\}$ that is better than the one described in the proof of Theorem B.6.

What happens when $n_t \neq 2^a$? If we restrict the matrices to be square, then the achievable rate may be much smaller than 1/2. For example if we have $n_t = 5$ transmit antennas, then $a = 0, b = 5$ and we get $s + t \leq 2a + 2 = 2$. The code would transmit $n_s = s = t = 1$ symbol over $n_t = 5$ symbol periods giving a rate of 1/5. To remedy this situation we have to consider non-square matrices, which we do next.

B.2.4 Generalized Designs for Complex Symbols

So far we have restricted the matrices $\{\boldsymbol{A}_n\}$ and $\{\boldsymbol{B}_n\}$ to be square. In this case we obtained a full-rate code for $n_t = 2$ transmit antennas, a rate 3/4 code for $n_t = 4$ transmit antennas and a rate 1/2 code for $n_t = 8$ transmit antennas. When $n_t \neq 2^a$ the rates could be much less than 1/2 which would lead to code matrices of no use since we already have a scheme for rate-1/2 codes (see (B.2.1)).

If we remove the restriction that $\{\boldsymbol{A}_n, \boldsymbol{B}_n\}$ be square we can get rates which are greater than or equal to 1/2. For instance, for $n_t = 3$ transmit antennas, we can easily obtain a rate-3/4 code by taking the first three rows of the design for four transmit antennas in (B.2.9). In the case of $n_t = 5, 6, 7$ transmit antennas, we can get a rate-1/2 code by taking the first 5,6 and 7 rows, respectively, of the code matrix for eight transmit antennas in (B.2.10).

B.2.5 Discussion

In the previous subsections, we have presented a number of existence theorems and construction methods for complex OSTBC. Only in the case of $n_t = 2$ were we able to find a full-rate code; for $n_t = 3, 4$ we found a rate-3/4 code and for a larger number of transmit antennas we used a construction that gives rate $R = 1/2$. For real symbols, we could always find a full-rate code. Is it possible to get a full-rate code for complex symbols when $n_t > 2$? The answer to this question appears to be negative. For example, it was proved in [SU AND XIA, 2003] that no amicable orthogonal design for $n_t > 2$ exists with rate $> 3/4$; hence the rate-3/4 code for complex symbols for $n_t = 3$ and 4 transmit antennas reaches the maximum rate. Further results on achievable rates for large number of transmit antennas can be found in [TIRKKONEN AND HOTTINEN, 2002].

B.3 Summary

We summarize in Table B.2 the best known OSTBC designs and their properties for real and complex symbols, for different number of transmit antennas. The table also shows whether a design has the minimal delay property, and it also gives a reference to the equation or theorem that can be used to construct the corresponding code matrices.

B.4 Proofs

Proof of Theorem B.3

Observe first that we can write any number N as $N = 2^a b$, where a and b are integers and b is odd. Then from the definition of $\rho(N)$ it can be seen that $\rho(N) = \rho(2^a)$ since b does not enter in the definition of $\rho(N)$ (in other words, $\rho(N)$ depends solely on a).

Suppose we have a Hurwitz-Radon family with $\rho(N) - 1$ members for $N = 2^a$. The matrices in this family are of size $N \times N$. Now suppose we have another number $M = 2^a b = Nb$ where b is odd. Then we know that $\rho(M) = \rho(2^a b) = \rho(2^a) = \rho(N)$. If we take the Kronecker product of $\boldsymbol{I}_b$ with each matrix in the Hurwitz-Radon family of order N, then we obtain another Hurwitz-Radon family of $Nb \times Nb$ matrices. This new family will have $\rho(M) = \rho(N)$ members and will be of order $M = Nb$. It follows that we need to prove the proposition only for $N = 2^a$.

Let $N_1 = 2^{4s+3}$, $N_2 = 2^{4(s+1)}$, $N_3 = 2^{4(s+1)+1}$, $N_4 = 2^{4(s+1)+2}$ and $N_5 = 2^{4(s+1)+3}$, where $s \geq 0$ is an integer. Then:

$$\begin{aligned}
\rho(N_2) &= \rho(N_1) + 1 \\
\rho(N_3) &= \rho(N_1) + 2 \\
\rho(N_4) &= \rho(N_1) + 4 \\
\rho(N_5) &= \rho(N_1) + 8
\end{aligned}$$

For any matrices $\boldsymbol{V}$ and $\boldsymbol{W}$, let $\boldsymbol{V} \otimes \boldsymbol{W}$ denote the Kronecker product of $\boldsymbol{V}$

Symbols	n_t	N	Rate	MD	Construction rule
Real	2	2	1	yes	Eq. (B.1.3)
	3	4	1	no	Eq. (B.1.14)
	4	4	1	yes	Eq. (B.1.10)
	5	8	1	no	Sec. B.1.2
	6	8	1	no	Sec. B.1.2
	7	8	1	no	Sec. B.1.2
	8	8	1	yes	Th. B.3
	>8	N chosen such that $n_t \leq \rho(N)$.	1	no	[GANESAN AND STOICA, 2001B]
Complex	2	2	1	yes	Eq. (7.4.7)
	3	4	3/4	no	Eq. (7.4.8)
	4	4	3/4	yes	Eq. (7.4.10)
	5	8	1/2	no	Sec. B.2.4
	6	8	1/2	no	Sec. B.2.4
	7	8	1/2	no	Sec. B.2.4
	8	8	1/2	yes	Eq. (B.2.10)
	>8	N chosen such that $n_t \leq \rho(N)$.	1/2	no	Eq. (B.2.1)

Table B.2. Summary of the best known OSTBC for real and complex symbols, respectively, and for different values of the number of transmit antennas n_t. In the table, MD stands for "minimal delay."

and $\boldsymbol{W}$. Let

$$\boldsymbol{C} = \begin{bmatrix} 0 & 1 \\ -1 & 0 \end{bmatrix} \qquad \boldsymbol{P} = \begin{bmatrix} 0 & 1 \\ 1 & 0 \end{bmatrix} \qquad \boldsymbol{R} = \begin{bmatrix} 1 & 0 \\ 0 & -1 \end{bmatrix} \tag{B.4.1}$$

The following results can be verified by a tedious check.

(i) $\{\boldsymbol{C}\}$ is an H-R family of $\rho(2) - 1$ integer matrices of order 2.

(ii) $\{\boldsymbol{C}\otimes\boldsymbol{I}_2, \boldsymbol{P}\otimes\boldsymbol{C}, \boldsymbol{R}\otimes\boldsymbol{C}\}$ is an H-R family of $\rho(4)-1$ integer matrices of order 4.

(iii) $\{\boldsymbol{I}_2 \otimes \boldsymbol{C} \otimes \boldsymbol{I}_2, \boldsymbol{I}_2 \otimes \boldsymbol{P} \otimes \boldsymbol{C}, \boldsymbol{R} \otimes \boldsymbol{R} \otimes \boldsymbol{C}, \boldsymbol{P} \otimes \boldsymbol{R} \otimes \boldsymbol{C}, \boldsymbol{C} \otimes \boldsymbol{P} \otimes \boldsymbol{R}, \boldsymbol{C} \otimes \boldsymbol{P} \otimes \boldsymbol{P}, \boldsymbol{C} \otimes \boldsymbol{R} \otimes \boldsymbol{I}_2\}$ is an H-R family of $\rho(8) - 1$ integer matrices of order 8.

The results above handle the cases $N = 2t, 4t, 8t$, t odd. It can be easily verified that if $\{\boldsymbol{M}_1, \ldots, \boldsymbol{M}_s\}$ is an H-R family of order N, then

(1) $\{\boldsymbol{C} \otimes \boldsymbol{I}_N\} \bigcup \{\boldsymbol{R} \otimes \boldsymbol{M}_l | l = 1, \ldots, s\}$ is an H-R family of $s+1$ integer matrices of order $2N$.

If, in addition, $\{\boldsymbol{L}_1, \ldots, \boldsymbol{L}_m\}$ is an H-R family of integer matrices of order k, then

(2) $\{\boldsymbol{P} \otimes \boldsymbol{I}_k \otimes \boldsymbol{M}_l | 1 \leq l \leq s\} \bigcup \{\boldsymbol{R} \otimes \boldsymbol{L}_p \otimes \boldsymbol{I}_N | 1 \leq p \leq m\} \bigcup \{\boldsymbol{C} \otimes \boldsymbol{I}_{Nk}\}$ is an H-R family of $s + m + 1$ integer matrices of order $2Nk$.

Using the definition of $N_1, \ldots, N_5$, we may proceed by induction: Starting with the fact (iii) above gives us the case $N_1 = 2^3$.

Now from the definition of N_1 and N_2 and also from the definition of $\rho(N_1)$ and $\rho(N_2)$ we can see that if we have $\rho(N_1) - 1$ matrices of order N, then by using (1) above, we can construct $\rho(N_2) - 1$ matrices of order N_2. Similarly by using (2) with $k = N_1$, $N = 2$, we can construct $\rho(N_3) - 1$ matrices of order N_3. Using (2) with $k = N_1$ and $N = 4$ gives us $\rho(N_4) - 1$ matrices of order N_4. Finally by using (2) with $k = N_1$ and $N = 8$ we can construct $\rho(N_5) - 1$ matrices of order N_5.

Thus starting with a construction for $N = N_1 = 2^3$ we got matrices for orders $N_2 = 2^4$, $N_3 = 2^5$, $N_4 = 2^6$, and $N_5 = 2^7$. Note that the construction for getting matrices of orders $N_2, \ldots, N_5$ will work as long as we have a set of matrices with $\rho(N_1) - 1$ members for $N_1 = 2^{4s+3}$ for some non-negative s. We started with a set of matrices for $N_1 = 2^3$ (and hence $s = 0$) and ended up with a set of matrices for $N_5 = 2^7$. Now $2^7 = 2^{4s+3}$ with $s = 1$. Therefore we can repeat the construction and proceed similarly to get matrices for higher powers of 2.

REFERENCES

[AGRAWAL ET AL., 1998] Agrawal, D., Tarokh, V., Naguib, A., and Seshadri, N. (1998). Space-time coded OFDM for high data rate wireless communication over wideband channels. In *Proc. of IEEE Vehicular Technology Conference*, volume 3, pages 2232–2236, Ottawa, Canada.

[AGRELL ET AL., 2002] Agrell, E., Eriksson, T., Vardy, A., and Zeger, K. (2002). Closest point search in lattices. *IEEE Transactions on Information Theory*, 48(7):2201–2214.

[AL-DHAHIR AND SAYED, 2000] Al-Dhahir, N. and Sayed, A. H. (2000). The finite-length multi-input multi-output MMSE-DFE. *IEEE Transactions on Signal Processing*, 48(10):2921–2936.

[ALAMOUTI, 1998] Alamouti, S. M. (1998). A simple transmit diversity technique for wireless communications. *IEEE Journal on Selected Areas in Communications*, 16(8):1451–1458.

[ANDERSON, 1971] Anderson, T. W. (1971). *The Statistical Analysis of Time Series.* Wiley, New York.

[BESSON AND STOICA, 2002] Besson, O. and Stoica, P. (2002). Channel and frequency offsets estimation in MIMO time-selective channels. In *Proc. of the IEEE Sensor Array and Multichannel Signal Processing Workshop*, Rosslyn, VA.

[BHASHYAM ET AL., 2002] Bhashyam, S., Sabharwal, A., and Aazhang, B. (2002). Feedback gain in multiple antenna systems. *IEEE Transactions on Communications*, 50(5):785–798.

[BIGLIERI ET AL., 1991] Biglieri, E., Divsalar, D., McLane, P., and Simon, M. (1991). *Introduction to Trellis-Coded Modulation with Applications.* Macmillan, New York, NY.

[BIGLIERI ET AL., 2002] Biglieri, E., Taricco, G., and Tulino, A. (2002). Performance of space-time codes for a large number of antennas. *IEEE Transactions on Information Theory*, 48(7):1794–1803.

[BLISS ET AL., 2002] Bliss, D. W., Forsythe, K. W., Hero, III, A. O., and Yegulalp, A. F. (2002). Environmental issues for MIMO capacity. *IEEE Transactions on Signal Processing*, 50(9):2128–2142.

[BLUM, 2002] Blum, R. S. (2002). Analytical tools for the design of space-time convolutional codes. *IEEE Transactions on Communications*, 50(10):1593–1599.

[BLUM ET AL., 2001] Blum, R., Geoffrey, L., Winters, J., and Yan, Q. (2001). Improved space-time coding for MIMO-OFDM wireless communications. *IEEE Transactions on Communications*, 49(11):1873–1878.

[BÖLCSKEI AND PAULRAJ, 2000A] Bölcskei, H. and Paulraj, A. (2000a). Performance analysis of space-time codes in correlated Rayleigh fading environments. In *Proc. of Asilomar Conference on Signals, Systems and Computers*, volume 1, pages 687–693, Pacific Grove, CA.

[BÖLCSKEI AND PAULRAJ, 2000B] Bölcskei, H. and Paulraj, A. (2000b). Space-frequency coded broadband OFDM systems. In *Proc. of IEEE Wireless Communications and Networking Conference*, volume 1, pages 1–6, Chicago, IL.

[BÖLCSKEI ET AL., 2002] Bölcskei, H., Gesbert, D., and Paulraj, A. J. (2002). On the capacity of OFDM-based spatial multiplexing systems. *IEEE Transactions on Communications*, 50(2):225–234.

[BOUTROS AND VITERBO, 1998] Boutros, J. and Viterbo, E. (1998). Signal space diversity: A power- and bandwidth-efficient diversity technique for the Rayleigh fading channel. *IEEE Transactions on Information Theory*, 44(4):1453–1467.

[BOUTROS ET AL., 1996] Boutros, J., Viterbo, E., Rastello, C., and Belfiore, J.-C. (1996). Good lattice constellations for both Rayleigh fading and Gaussian channels. *IEEE Transactions on Information Theory*, 42(2):502–518.

[BREHLER AND VARANASI, 2001] Brehler, M. and Varanasi, M. K. (2001). Asymptotic error probability analysis of quadratic receivers in Rayleigh-fading channels with applications to a unified analysis of coherent and noncoherent space-time receivers. *IEEE Transactions on Information Theory*, 47(6):2383–2399.

[CAIRE ET AL., 1998] Caire, G., Taricco, G., and Biglieri, E. (1998). Bit-interleaved coded modulation. *IEEE Transactions on Information Theory*, 44(3):927–946.

[CAIRE ET AL., 1999] Caire, G., Taricco, G., and Biglieri, E. (1999). Optimum power control over fading channels. *IEEE Transactions on Information Theory*, 45(5):1468–1489.

[CHUGG AND POLYDOROS, 1996] Chugg, K. M. and Polydoros, A. (1996). MLSE for an unknown channel, Part I: Optimality considerations. *IEEE Transactions on Communications*, 44(7):836–846.

[CORREIA, 2001] Correia, L., editor (2001). *Wireless Flexible Personalised Communications, COST 259: European Co-operation in Mobile Radio Research.* John Wiley and Sons, West Sussex, England.

[COUPECHOUX AND BRAUN, 2000] Coupechoux, M. and Braun, V. (2000). Space-time coding for the EDGE mobile radio system. In *Proc. of IEEE International Conference on Personal Wireless Communications*, pages 28–32, Hyderabad, India.

[COVER AND THOMAS, 1991] Cover, T. M. and Thomas, J. A. (1991). *Elements of Information Theory.* John Wiley and Sons.

[DAM ET AL., 1999] Dam, H., Berg, M., Andersson, S., Bormann, R., Fredrich, M., Ahrens, F., and Henss, T. (1999). Performance evaluation of adaptive antenna base stations in a commercial GSM network. In *Proc. of IEEE Vehicular Technology Conference.*

[DAMEN ET AL., 2000] Damen, O., Chkeif, A., and Belfiore, J.-C. (2000). Lattice code decoder for space-time codes. *IEEE Communications Letters*, 4(5):161–163.

[DERRYBERRY ET AL., 2002] Derryberry, R. T., Gray, S. D., Ionescu, D. M., Mandyam, G., and Raghothaman, B. (2002). Transmit diversity in 3G CDMA systems. *IEEE Communications Magazine*, 40(4):68–75.

[DURGIN, 2002] Durgin, G. D. (2002). *Space-time wireless channels.* Prentice-Hall.

[ENG ET AL., 1996] Eng, T., Kong, N., and Milstein, L. (1996). Comparison of diversity combining techniques for Rayleigh fading channels. *IEEE Transactions on Communications*, 44(9):1117–1129.

[Ertel et al., 1998] Ertel, R. B., Cardieri, P., Sowerby, K. W., Rappaport, T. S., and Reed, J. H. (1998). Overview of spatial channel models for antenna array communication systems. *IEEE Personal Communications*, 5(1):10–22.

[ETSI, 2002] ETSI (2002). Digital cellular telecommunications system: Radio transmission and reception (GSM 05.05). ETSI standardization document.

[Fincke and Pohst, 1985] Fincke, U. and Pohst, M. (1985). Improved methods for calculating vectors of short length in a lattice, including a complexity analysis. *Mathematics of Computation*, 44(170):463–471.

[Flore and Lindskog, 2000] Flore, D. and Lindskog, E. (2000). Time-reversal space-time block coding vs. transmit delay diversity – a comparison based on a GSM-like system. In *Proc. of the Digital Signal Processing Workshop*, Hunt, Texas.

[Forney, 1991] Forney, G. D. (1991). Geometrically uniform codes. *IEEE Transactions on Information Theory*, 37(5):1241 –1260.

[Foschini, Jr., 1996] Foschini, Jr., G. J. (1996). Layered space-time architecture for wireless communication in a fading environment when using multi-element antennas. *Bell Labs Tech. Journal*, 1:41–59.

[Foschini, Jr. and Gans, 1998] Foschini, Jr., G. and Gans, M. J. (1998). On the limits of wireless communications in a fading environment when using multiple antennas. *Wireless Personal Communications*, 6(3):311–355.

[Gamal and Damen, 2002] Gamal, H. E. and Damen, M. O. (2002). An algebraic number theoretic framework for space-time coding. In *IEEE International Symposium on Information Theory*, pages 132–132, Lausanne, Switzerland.

[Gamal and Hammons, Jr., 2002] Gamal, H. E. and Hammons, Jr., A. R. (2002). On the design and performance of algebraic space-time codes for BPSK and QPSK modulation. *IEEE Transactions on Communications*, 50(6):907–913.

[Ganesan and Stoica, 2001a] Ganesan, G. and Stoica, P. (2001a). Space-time block codes: a maximum SNR approach. *IEEE Transactions on Information Theory*, 47(4):1650–1656.

[Ganesan and Stoica, 2001b] Ganesan, G. and Stoica, P. (2001b). Space-time diversity. In *Signal Processing Advances in Wireless and Mobile Communications, vol. 2*, NJ. Prentice-Hall. Eds.: G. Giannakis, Y. Hua, P. Stoica and L. Tong.

[Ganesan and Stoica, 2001c] Ganesan, G. and Stoica, P. (2001c). Space-time diversity using orthogonal and amicable orthogonal designs. *Wireless Personal Communications*, 18(2):165–178.

[Ganesan and Stoica, 2002a] Ganesan, G. and Stoica, P. (2002a). Differential detection based on space-time block codes. *Wireless Personal Communications*, 21(2):163–180.

[Ganesan and Stoica, 2002b] Ganesan, G. and Stoica, P. (2002b). Differential modulation using space-time block codes. *IEEE Signal Processing Letters*, 9(2):57–60.

[Ganesan et al., 2002] Ganesan, G., Stoica, P., and Larsson, E. G. (2002). Diagonally weighted orthogonal space-time block codes. In *Proc. of Asilomar Conference on Signals, Systems and Computers*, Pacific Grove, CA.

[Geramita and Geramita, 1978] Geramita, A. V. and Geramita, J. M. (1978). Complex orthogonal designs. *Journal of Combinatorial Theory, Series A*, 25:211–225.

[Geramita and Seberry, 1979] Geramita, A. V. and Seberry, J. (1979). *Orthogonal Designs, Quadratic Forms and Hadamard Matrices*, volume 43 of *Lecture notes in pure and applied mathematics*. Marcel Dekker, New York and Basel.

[Gesbert et al., 2002] Gesbert, D., Bölcskei, H., Gore, D., and Paulraj, A. (2002). Outdoor MIMO wireless channels: Models and performance prediction. *IEEE Transactions on Communications*, 50(12):1926–1934.

[Ginis and Cioffi, 2001] Ginis, G. and Cioffi, J. (2001). On the relation between V-BLAST and the GDFE. *IEEE Communications Letters*, 5(9):364–366.

[Golub and van Loan, 1989] Golub, G. H. and van Loan, C. F. (1989). *Matrix Computations*. The Johns Hopkins University Press, Maryland, USA.

[Gore and Paulraj, 2002] Gore, D. A. and Paulraj, A. J. (2002). MIMO antenna subset selection with space-time coding. *IEEE Transactions on Signal Processing*, 50(10):2580–2588.

[Gore et al., 2001] Gore, D., Sandhu, S., and Paulraj, A. (2001). Delay diversity code for frequency selective channels. *Electronics Letters*, 37(20):1230–1231.

[Gorokhov, 2002] Gorokhov, A. (2002). Antenna selection algorithms for MEA transmission systems. In *Proc. of International Conference on Acoustics, Speech and Signal Processing (ICASSP)*, pages 2857–2860, Orlando, FL.

[GUEY ET AL., 1999] Guey, J. C., Fitz, M. P., Bell, M. R., and Kuo, W. Y. (1999). Signal design for transmitter diversity wireless communication systems over Rayleigh fading channels. *IEEE Transactions on Communications*, 47(4):527–537.

[HASSIBI AND HOCHWALD, 2000] Hassibi, B. and Hochwald, B. M. (2000). Optimal training in space-time systems. In *Proc. of Asilomar Conference on Signals, Systems and Computers*, volume 1, pages 743–747, Pacific Grove, CA. Also submitted to IEEE Transactions on Information Theory.

[HASSIBI AND HOCHWALD, 2002A] Hassibi, B. and Hochwald, B. M. (2002a). Cayley differential unitary space-time codes. *IEEE Transactions on Information Theory*, 48(6):1485–1503.

[HASSIBI AND HOCHWALD, 2002B] Hassibi, B. and Hochwald, B. M. (2002b). High-rate codes that are linear in space and time. *IEEE Transactions on Information Theory*, 48(7):1804–1824.

[HASSIBI AND VIKALO, 2002] Hassibi, B. and Vikalo, H. (2002). On the expected complexity of integer least-squares problems. In *Proc. of International Conference on Acoustics, Speech and Signal Processing (ICASSP)*, pages 1497–1500, Orlando, FL.

[HEATH AND PAULRAJ, 2002] Heath, R. W. and Paulraj, A. J. (2002). Linear dispersion codes for MIMO systems based on frame theory. *IEEE Transactions on Signal Processing*, 50(10):2429–2441.

[HEATH ET AL., 2001A] Heath, R. W., Bölcskei, H., and Paulraj, A. J. (2001a). Space-time signaling and frame theory. In *Proc. of International Conference on Acoustics, Speech and Signal Processing (ICASSP)*, volume 4, pages 2445–2448, Salt Lake City, UT.

[HEATH ET AL., 2001B] Heath, R. W., Sandhu, S., and Paulraj, A. (2001b). Antenna selection for spatial multiplexing systems with linear receivers. *IEEE Communications Letters*, 5(4):142–144.

[HERSTEIN, 1968] Herstein, I. N. (1968). *Non Commutative Rings*. Number 15 in Carus Mathematical Monograph. Mathematical Association of America, Washington, D. C.

[HOCHWALD AND MARZETTA, 2000] Hochwald, B. M. and Marzetta, T. L. (2000). Unitary space-time modulation for multiple-antenna communications in Rayleigh flat fading. *IEEE Transactions on Information Theory*, 46(2):543–564.

[HOCHWALD AND SWELDENS, 2000] Hochwald, B. M. and Sweldens, W. (2000). Differential unitary space-time modulation. *IEEE Transactions on Communications*, 48(12):2041–2052.

[HOCHWALD AND TEN BRINK, 2001] Hochwald, B. M. and ten Brink, S. (2001). Achieving near-capacity on a multiple-antenna channel. In *Proc. of Allerton Conference*, Champaign-Urbana, IL. Also submitted to IEEE Transactions on Communications.

[HOCHWALD ET AL., 2000] Hochwald, B. M., Marzetta, T. L., Richardson, T. J., Sweldens, W., and Urbanke, R. (2000). Systematic design of unitary space-time constellations. *IEEE Transactions on Information Theory*, 46(6):1962–1973.

[HORN AND JOHNSON, 1985] Horn, R. A. and Johnson, C. R. (1985). *Matrix Analysis*. Cambridge University Press, Cambridge, UK.

[HUGHES, 2000] Hughes, B. L. (2000). Differential space-time modulation. *IEEE Transactions on Information Theory*, 46(7):2567–2578.

[JAFAR ET AL., 2001] Jafar, S., Vishwanath, S., and Goldsmith, A. (2001). Channel capacity and beamforming for multiple transmit and multiple receive antennas with covariance feedback. In *Proc. of IEEE International Conference on Communications*, volume 7, pages 2266–2270, Helsinki, Finland.

[JAFARKHANI AND TAROKH, 2001] Jafarkhani, H. and Tarokh, V. (2001). Multiple transmit antenna differential detection from generalized orthogonal designs. *IEEE Transactions on Information Theory*, 47(6):2626–2631.

[JÖNGREN ET AL., 2002] Jöngren, G., Skoglund, M., and Ottersten, B. (2002). Combining beamforming with orthogonal space-time block coding. *IEEE Transactions on Information Theory*, 48(3):611–627.

[KARLSSON AND HEINEGÅRD, 1996] Karlsson, J. and Heinegård, J. (1996). Interference rejection combining for GSM. In *Proc. of IEEE Conference on Universal Personal Communications*, pages 433–437.

[KAWADA AND IWAHORI, 1950] Kawada, Y. and Iwahori, N. (1950). On the Structure and Representation of Clifford Algebras. *Journal of Mathematical Society of Japan*, 2:34–43.

[KAY, 1993] Kay, S. M. (1993). *Fundamentals of Statistical Signal Processing: Estimation Theory*. Prentice Hall, Englewood Cliffs.

[KAY, 1998] Kay, S. M. (1998). *Fundamentals of Statistical Signal Processing: Detection Theory.* Prentice Hall, Englewood Cliffs.

[KERMOAL ET AL., 2002] Kermoal, J. P., Schumacher, L., Pedersen, K. I., Mogensen, P. E., and Fredriksen, F. (2002). A Stochastic MIMO Radio Channel Model With Experimental Validation. *IEEE Journal on Selected Areas in Communications*, 20(6):1211–1226.

[KYRITSI AND COX, 2001] Kyritsi, P. and Cox, D. C. (2001). Correlation properties of MIMO radio channels for indoor scenarios. In *Proc. of Asilomar Conference on Signals, Systems and Computers*, pages 994–998, Pacific Grove, CA.

[KYRITSI ET AL., 2002] Kyritsi, P., Cox, D. C., Valenzuela, R. A., and Wolniansky, P. W. (2002). Effect of antenna polarization on the capacity of a multiple element system in an indoor environment. *IEEE Journal on Selected Areas in Communications*, 20(6):1227–1239.

[LARSSON AND LI, 2001] Larsson, E. G. and Li, J. (2001). Preamble design considerations for multiple-antenna OFDM based WLANs with null subcarriers. *IEEE Signal Processing Letters*, 8(11):285–288.

[LARSSON AND STOICA, 2003] Larsson, E. G. and Stoica, P. (2003). Mean square error optimality of orthogonal space-time block codes. *IEEE Signal Processing Letters.* In press.

[LARSSON ET AL., 2002A] Larsson, E. G., Ganesan, G., Stoica, P., and Wong, W.-H. (2002a). On the performance of orthogonal space-time block coding with quantized feedback. *IEEE Communications Letters*, 6(11):487–489.

[LARSSON ET AL., 2002B] Larsson, E. G., Stoica, P., and Li, J. (2002b). On maximum-likelihood detection and decoding for space-time coding systems. *IEEE Transactions on Signal Processing*, 50(4):937–944.

[LARSSON ET AL., 2002C] Larsson, E. G., Stoica, P., Lindskog, E., and Li, J. (2002c). Space-time block coding for frequency-selective channels. In *Proc. of International Conference on Acoustics, Speech and Signal Processing (ICASSP)*, pages 2405–2408, Orlando, FL.

[LARSSON ET AL., 2003] Larsson, E. G., Stoica, P., and Li, J. (2003). Space-time block codes: ML detection for unknown channels and unstructured interference. *IEEE Transactions on Signal Processing*, 51(2):362–372.

[Li et al., 1999] Li, Y. G., Seshadri, N., and Ariyavisitakul, S. (1999). Channel estimation in OFDM systems with transmitter diversity in mobile wireless channels. *IEEE Journal on Selected Areas in Communications*, 17(3):461–471.

[Liang and Xia, 2002] Liang, X.-B. and Xia, X.-G. (2002). Unitary signal constellations for differential space-time modulation with two transmit antennas: parametric codes, optimal designs, and bounds. *IEEE Transactions on Information Theory*, 48(7):2291–2322.

[Liew and Hanzo, 2002] Liew, T. H. and Hanzo, L. (2002). Space–time codes and concatenated channel codes for wireless communications. *Proceedings of the IEEE*, 90(2):187–219.

[Lin and Costello, 1983] Lin, S. and Costello, Jr., D. J. (1983). *Error Control Coding: Fundamentals and Applications.* Prentice-Hall, Englewood Cliffs, NJ.

[Lindskog and Paulraj, 2000] Lindskog, E. and Paulraj, A. (2000). A transmit diversity scheme for channels with intersymbol interference. In *Proc. of IEEE International Conference on Communications*, pages 307–311, New Orleans, LA.

[Liu et al., 2001a] Liu, Z., Giannakis, G. B., and Hughes, B. L. (2001a). Double Differential Space-Time Block Coding For Time-Selective Fading Channels. *IEEE Transactions on Communications*, 49(9):1529–1539.

[Liu et al., 2001b] Liu, Z., Giannakis, G. B., Zhou, S., and Muquet, B. (2001b). Space-time coding for broadband wireless communications. *Wireless Communications and Mobile Computing*, 1(1):35–53.

[Marzetta and Hochwald, 1999] Marzetta, T. L. and Hochwald, B. M. (1999). Capacity of a mobile multiple-antenna communication link in Rayleigh flat-fading. *IEEE Transactions on Information Theory*, 45(1):139–157.

[May and Rohling, 2001] May, T. and Rohling, H. (2001). Orthogonal frequency division multiplexing. In Molisch, A. F., editor, *Wideband Wireless Digital Communications.* Prentice-Hall, Upper Saddle River, New Jersey.

[Meyr et al., 1998] Meyr, H., Moeneclaey, M., and Fechtel, S. A. (1998). *Digital Communication Receivers: Synchronization, Channel Estimation and Signal Processing.* John Wiley & Sons, Inc., New York, NY.

[Moguez and Castedo, 2002] Moguez, J. and Castedo, L. (2002). Semiblind space-time decoding in wireless communications: A maximum likelihood approach. *Signal Processing*, 82(1):1–18.

[Molisch et al., 2001] Molisch, A. F., Win, M. Z., and Winters, J. H. (2001). Capacity of MIMO systems with antenna selection. In *Proc. of IEEE International Conference on Communications*, pages 570–574, Helsinki, Finland.

[Molisch et al., 2002] Molisch, A. F., Steinbauer, M., Toeltsch, M., Bonek, E., and Thoma, R. S. (2002). Capacity of MIMO systems based on measured wireless channels. *IEEE Journal on Selected Areas in Communications*, 20(3):561–569.

[Moon and Stirling, 2000] Moon, T. K. and Stirling, W. C. (2000). *Mathematical Methods and Algorithms for Signal Processing.* Prentice-Hall, Upper Saddle River, New Jersey, USA.

[Mouly and Pautet, 1992] Mouly, M. and Pautet, M.-B. (1992). *The GSM System for Mobile Communications.* Telecom Publishing, Olympia, WA.

[Moustakas and Simon, 2002] Moustakas, A. L. and Simon, S. H. (2002). Optimizing multi-transmitter-single-receiver (MISO) antenna systems with partial channel knowledge. *Bell Laboratories Technical Memorandum.*

[Mukkavilli et al., 2001] Mukkavilli, K. K., Sabharwal, A., and Aazhang, B. (2001). Design of multiple antenna coding schemes with channel feedback. In *Proc. of Asilomar Conference on Signals, Systems and Computers*, pages 1009–1013, Pacific Grove, CA.

[Nabar et al., 2002] Nabar, R. U., Bölcskei, H., Erceg, V., Gesbert, D., and Paulraj, A. J. (2002). Performance of multiantenna signaling techniques in the presence of polarization diversity. *IEEE Transactions on Signal Processing*, 50(10):2553–2562.

[Naguib et al., 2000] Naguib, A. F., Seshadri, N., and Calderbank, A. R. (2000). Increasing data rate over wireless channels: Space-time coding and signal processing for high data rate wireless communications. *IEEE Signal Processing Magazine*, 17(3):76–92.

[Narula et al., 1998] Narula, A., Lopez, M. J., Trott, M. D., and Wornell, G. W. (1998). Efficient use of side information in multiple-antenna data transmission over fading channels. *IEEE Journal on Selected Areas in Communications*, 16(8):1423–1436.

[Narula et al., 1999] Narula, A., Trott, M. D., and Wornell, G. W. (1999). Performance limits of coded diversity methods for transmitter antenna arrays. *IEEE Transactions on Information Theory*, 45(7):2418–2433.

[NEGI AND CIOFFI, 1998] Negi, R. and Cioffi, J. (1998). Pilot tone selection for channel estimation in a mobile OFDM system. *IEEE Transactions on Consumer Electronics*, 44(3):1122–1128.

[OTTERSTEN, 1996] Ottersten, B. (1996). Array processing for wireless communications. In *Proc. of IEEE Signal Processing Workshop on Statistical Signal and Array Processing*, pages 466–473.

[PAPOULIS, 2002] Papoulis, A. (2002). *Probability, Random Variables, and Stochastic Processes.* McGraw-Hill, New York, fourth edition.

[PAULRAJ AND KAILATH, 1993] Paulraj, A. and Kailath, T. (1993). Increasing capacity in wireless broadcast systems using distributed transmission/directional reception. U.S. Patent no. 5,345,599.

[PAULRAJ ET AL., 2003] Paulraj, A., Nabar, R., and Gore, D. (2003). *Introduction to space-time communications.* Cambridge University Press, Cambridge, UK.

[PAULRAJ AND PAPADIAS, 1997] Paulraj, A. J. and Papadias, C. B. (1997). Space-time processing for wireless communications. *IEEE Signal Processing Magazine*, 14(6):49–83.

[PEEL AND SWINDLEHURST, 2001] Peel, C. B. and Swindlehurst, A. L. (2001). Performance of unitary space-time modulation in a continuously changing channel. In *Proc. of International Conference on Acoustics, Speech and Signal Processing (ICASSP)*, volume 4, pages 2433–2436, Salt Lake City, UT.

[PROAKIS, 2001] Proakis, J. G. (2001). *Digital Communications.* McGraw Hill, Inc., New York, NY, fourth edition.

[RADON, 1922] Radon, J. (1922). Lineare scharen orthogonaler Matrizen. *Abhandlungen aus dem Mathematischen Seminar der Hamburgishen Universität*, 1:1–14.

[RALEIGH AND CIOFFI, 1998] Raleigh, G. G. and Cioffi, J. M. (1998). Spatial-temporal coding for wireless communication. *IEEE Transactions on Communications*, 46(3):357–366.

[RAPPAPORT, 2002] Rappaport, T. (2002). *Wireless Communications: Principles and Practice.* Prentice-Hall, Upper Saddle River, NJ, second edition.

[RUDIN, 1976] Rudin, W. (1976). *Principles of Mathematical Analysis.* Mc-Graw Hill.

[SAMPATH, 2001] Sampath, H. (2001). *Linear Precoding and Decoding for Multiple-Input Multiple-Ouput (MIMO) channels.* PhD thesis, Stanford Univeristy, Stanford, CA.

[SAMPATH AND PAULRAJ, 2002] Sampath, H. and Paulraj, A. (2002). Linear precoding for space-time coded systems with known fading correlations. *IEEE Communications Letters*, 6(6):239–241.

[SANDHU, 2002] Sandhu, S. (2002). *Signal design for multiple-input multiple-output wireless: a unified perspective.* PhD thesis, Stanford University.

[SANDHU AND PAULRAJ, 2000] Sandhu, S. and Paulraj, A. (2000). Space-time block codes: a capacity perspective. *IEEE Communications Letters*, 4(12):384–386.

[SANDHU ET AL., 2000] Sandhu, S., Nabar, R. U., Gore, D. A., and Paulraj, A. (2000). Near-optimal selection of transmit antennas for a MIMO channel based on Shannon capacity. In *Proc. of Asilomar Conference on Signals, Systems and Computers*, volume 1, pages 567–571, Pacific Grove, CA.

[SANDHU ET AL., 2002] Sandhu, S., Paulraj, A., and Pandit, K. (2002). On non-linear space-time block codes. In *Proc. of International Conference on Acoustics, Speech and Signal Processing (ICASSP)*, volume 3, pages 2417–2420, Orlando, FL.

[SCAGLIONE ET AL., 2002] Scaglione, A., Stoica, P., Barbarossa, S., Giannakis, G. B., and Sampath, H. (2002). Optimal designs for space-time linear precoders and decoders. *IEEE Transactions on Signal Processing*, 50(5):1051–1064.

[SIMON AND ALOUINI, 2000] Simon, M. K. and Alouini, M.-S. (2000). *Digital communications over fading channels: a unified approach to performance analysis.* John Wiley and Sons, New York, NY.

[SIMON AND MOUSTAKAS, 2002] Simon, S. H. and Moustakas, A. L. (2002). Optimizing MIMO antenna systems with channel covariance feedback. *Bell Laboratories Technical Memorandum.*

[SKLAR, 1997] Sklar, B. (1997). A primer on turbo code concepts. *IEEE Communications Magazine*, 35(12):94–102.

[SÖDERSTRÖM AND STOICA, 1989] Söderström, T. and Stoica, P. (1989). *System Identification.* Prentice Hall International, Hemel Hempstead, UK.

[SRINIVASAN ET AL., 2001] Srinivasan, R., Heikkilä, M. J., and Pirhonen, R. (2001). Performance evaluation of space-time coding for EDGE. In *Prof. of International Conference of Communications*, volume 10, pages 3056–3060, Helsinki, Finland.

[STAMOULIS ET AL., 2001] Stamoulis, A., Al-Dhahir, N., and Calderbank, A. R. (2001). Further results on interference cancellation and space-time block codes. In *Proc. of Asilomar Conference on Signals, Systems and Computcrs*, pages 257–261, Pacific Grove, CA.

[STOICA AND GANESAN, 2002A] Stoica, P. and Ganesan, G. (2002a). Maximum-SNR spatial-temporal formatting designs for MIMO channels. *IEEE Transactions on Signal Processing*, 50(12):3036–3042.

[STOICA AND GANESAN, 2002B] Stoica, P. and Ganesan, G. (2002b). Trained space-time block decoding for flat fading channels with frequency offsets. In *Proc. of the European Signal Processing Conference (EUSIPCO)*, Toulouse, France.

[STOICA AND GANESAN, 2003] Stoica, P. and Ganesan, G. (2003). Space-time block codes: Trained, blind and semi-blind detection. *Digital Signal Processing*, 13(1):93–105.

[STOICA AND LINDSKOG, 2002] Stoica, P. and Lindskog, E. (2002). Space-time block coding for channels with intersymbol interference. *Digital Signal Processing*, 12(4):616–627.

[STOICA AND MOSES, 1997] Stoica, P. and Moses, R. (1997). *Introduction to Spectral Analysis.* Prentice Hall, Upper Saddle River, NJ.

[STRIDH AND OTTERSTEN, 2000] Stridh, R. and Ottersten, B. (2000). Spatial characterization of indoor radio channel measurements at 5 GHz. In *Proc. of IEEE Sensor Array and Multichannel Signal Processing Workshop*, pages 58–62, Cambridge, MA.

[STÜBER, 2001] Stüber, G. (2001). *Principles of Mobile Communications.* Kluwer Academic Publishers.

[SU AND XIA, 2003] Su, W. and Xia, X.-G. (2003). On Space-Time Block Codes from Complex Orthogonal Designs. *Wireless Personal Communications.* To appear.

[SWINDLEHURST AND LEUS, 2002] Swindlehurst, A. L. and Leus, G. (2002). Blind and semi-blind equalization for generalized space-time block codes. *IEEE Transactions on Signal Processing*, 50(10):2489–2498.

[SWINDLEHURST ET AL., 2001] Swindlehurst, A. L., German, G., Wallace, J., and Jensen, M. (2001). Experimental measurements of capacity for MIMO indoor wireless channels. In *Proc. of IEEE Signal Processing Workshop on Signal Processing Advances in Wireless Communications*, pages 30–33, Taoyuan, Taiwan.

[TAROKH AND JAFARKHANI, 2000] Tarokh, V. and Jafarkhani, H. (2000). A differential detection scheme for transmit diversity. *IEEE Journal on Selected Areas in Communications*, 18(7):1169–1174.

[TAROKH ET AL., 1998] Tarokh, V., Seshadri, N., and Calderbank, A. R. (1998). Space-time codes for high data rate wireless communications: Performance criterion and code construction. *IEEE Transactions on Information Theory*, 44(2):744–765.

[TAROKH ET AL., 1999A] Tarokh, V., Jafarkhani, H., and Calderbank, A. R. (1999a). Space-time block codes from orthogonal designs. *IEEE Transactions on Information Theory*, 45(5):1456–1467.

[TAROKH ET AL., 1999B] Tarokh, V., Naguib, A., Seshadri, N., and Calderbank, A. R. (1999b). Space-time codes for high data rate wireless communication: Performance criteria in the presence of channel estimation errors, mobility, and multiple paths. *IEEE Transactions on Communications*, 47(2):199–207.

[TELATAR, 1999] Telatar, I. E. (1999). Capacity of multi-antenna Gaussian channels. *European Transaction on Telecommunications*, 10(6):585–595.

[TIRKKONEN AND HOTTINEN, 2002] Tirkkonen, O. and Hottinen, A. (2002). Square-matrix embeddable space-time block codes for complex signal constellations. *IEEE Transactions on Information Theory*, 48(2):384–395.

[TONG ET AL., 1994] Tong, L., Xu, G., and Kailath, T. (1994). Blind identification and equalization based on second-order statistics: A time domain approach. *IEEE Transactions on Information Theory*, 40(2):340–349.

[TUNG ET AL., 2001] Tung, T.-L., Yao, K., and Hudson, R. E. (2001). Channel estimation and adaptive power allocation for performance and capacity improvement of multiple-antenna OFDM systems. In *Proc. of IEEE Signal Processing Workshop on Signal Processing Advances in Wireless Communications*, pages 82–85, Taoyuan, Taiwan.

[van Nee et al., 1999] van Nee, R., Awater, G., Morikura, M., Takanashi, H., Webster, M., and Halford, K. (1999). New high-rate wireless LAN standards. *IEEE Communications Magazine*, 37(12):82–88.

[Van Trees, 2002] Van Trees, H. L. (2002). *Optimum Array Processing (Detection, Estimation, and Modulation Theory, Part IV)*. John Wiley and Sons, Inc., New York, NY.

[Verdú, 1998] Verdú, S. (1998). *Multiuser detection*. Cambridge University Press, Cambridge, UK.

[Visotsky and Madhow, 2001] Visotsky, E. and Madhow, U. (2001). Space-time transmit precoding with imperfect feedback. *IEEE Transactions on Information Theory*, 47(6):2632–2639.

[Viterbo and Boutros, 1999] Viterbo, E. and Boutros, J. (1999). A universal lattice code decoder for fading channels. *IEEE Transactions on Information Theory*, 45(5):1639–1642.

[Wang and Xia, 2002] Wang, H. and Xia, X.-G. (2002). Upper bounds on rates of complex orthogonal space-time block codes. In *Proc. of International Symposium on Information Theory*, page 303, Lausanne, Switzerland. Also submitted to IEEE Transactions on Information Theory.

[Winters et al., 1994] Winters, J., Salz, J., and Gitlin, R. (1994). The impact of antenna diversity on the capacity of wireless communications systems. *IEEE Transactions on Communications*, 42(2):1740–1751.

[Wittneben, 1991] Wittneben, A. (1991). Base station modulation diversity for digital SIMULCAST. In *Proc. of IEEE Vehicular Technology Conference*, pages 848–853.

[Xia, 2001] Xia, X.-G. (2001). Precoded and vector OFDM systems robust to channel spectral nulls and with reduced cyclic prefix length. *IEEE Transactions on Communications*, 49(8):1363–1374.

[Xin et al., 2001] Xin, Y., Wang, Z., and Giannakis, G. B. (2001). Space-time constellation-rotating codes maximizing diversity and coding gains. In *Proc. of IEEE Global Telecommunication Conference*, pages 455–459, San Antonio, TX.

[Yang and Roy, 1993] Yang, J. and Roy, S. (1993). On joint transmitter and receiver optimization for multiple input multiple output (MIMO) transmission systems. *IEEE Transactions on Communications*, 42(12):3221–3231.

[Yu and Ottersten, 2002] Yu, K. and Ottersten, B. (2002). Models for MIMO propagation channels: a review. *Wireless communications and mobile computing*, 2(7):653–666.

[Yu et al., 2001] Yu, K., Bengtsson, M., Ottersten, B., McNamara, D., Karlsson, P., and Beach, M. (2001). Second order statistics of NLOS indoor MIMO channels based on 5.2 GHz measurements. In *Proc. of IEEE Global Telecommunications Conference*, volume 1, pages 156–160, San Antonio, TX.

[Zangwill, 1967] Zangwill, W. I. (1967). *Nonlinear Programming: A Unified Approach.* Prentice-Hall, Inc., Englewood Cliffs, N.J.

[Zetterberg, 1996] Zetterberg, P. (1996). *Mobile cellular communications with base station antenna arrays: spectrum efficiency, algorithms and propagation models.* PhD thesis, Royal Institute of Technology, Stockholm, Sweden.

[Zhou and Giannakis, 2001] Zhou, S. and Giannakis, G. B. (2001). Space-time coding with maximum diversity gains over frequency-selective fading channels. *IEEE Signal Processing Letters*, 8(10):269–272.

INDEX